# The Chemistry and Metabolism of 4′-Deoxypyridoxine

Author
Stephen P. Coburn, Ph.D.
Director
Biochemistry Department
Fort Wayne State Hospital and Training Center
Fort Wayne, Indiana

CRC Press
Taylor & Francis Group
Boca Raton London New York

CRC Press is an imprint of the
Taylor & Francis Group, an **Informa** business

First published 1981 by CRC Press
Taylor & Francis Group
6000 Broken Sound Parkway NW, Suite 300
Boca Raton, FL 33487-2742

Reissued 2018 by CRC Press



**Library of Congress Cataloging in Publication Data**

Coburn, Stephen P.

The chemistry and metabolism of 4' deoxypyridoxine.

Bibliography: p.
Includes indexes.
1. Deoxypyridoxine. 2. Deoxypyridoxine—
Metabolism. 3. Vitamin B6 deficiency.
I. Title. [DNLM: 1. Pyridoxine—Anatagonists
and inhibitors. QU 195 C658c]
QP772.P9C6 612'.399 80-36793
ISBN 0-8493-5783-7 AACR1

A Library of Congress record exists under LC control number: 80036793



ISBN 13: 978-1-315-89145-3 (hbk)
ISBN 13: 978-1-351-07055-3 (ebk)

Visit the Taylor & Francis Web site at http://www.taylorandfrancis.com and the
CRC Press Web site at http://www.crcpress.com

# PREFACE

Because it serves as a cofactor for a wide variety of enzymes, vitamin $B_6$ is one of the most pharmacologically active vitamins and has been extensively studied both in basic research and in relation to clinical situations such as epilepsy, cancer, toxemia of pregnancy, oral contraceptive use, and activity of the immune system. Recently, it has been proposed as a factor in arteriosclerosis. Since it is difficult to produce $B_6$ deficiency symptoms in adult animals and humans solely by use of a $B_6$-deficient diet, it has been quite common to supplement the diet with $B_6$ antagonists, particularly 4′-deoxypyridoxine. In fact in the 40 years since Moeller et al.,[330-333] first reported the use of deoxypyridoxine, over 500 articles dealing with various aspects of the chemistry and biological activity of vitamin $B_6$ antagonists have appeared in the literature.

In vitro studies showed that 4′-deoxypyridoxine could be phosphorylated by pyridoxine kinase[522] and that deoxypyridoxine phosphate[306] could inhibit some enzymes requiring pyridoxal phosphate. These results led to general acceptance of the hypothesis that deoxypyridoxine exerted its effects in vivo as a competitive inhibitor of pyridoxal phosphate enzymes. In some cases acceptance of this hypothesis has been so complete that authors made no distinction between the effects of a dietary $B_6$ deficiency and symptoms produced by deoxypyridoxine. For example, in their recent report of the effects of vitamin $B_6$ deficiency on thymic epithelial cells and T-lymphocyte differentiation, Willis-Carr and St. Pierre[552] consistently speak of a dietary $B_6$ deficiency. The fact that the drinking water contained 0.1 mg 4′-deoxypyridoxine per milliliter was noted only in the methods section with the comment that the deoxypyridoxine was used "to suppress the uptake and utilization of vitamin $B_6$ being made available from other, uncontrollable sources." Thus the assumption that deoxypyridoxine simply reinforces the effects of a dietary deficiency continues right up to the present.

However, other workers have noted rather disturbing discrepancies. As early as 1949 Umbreit and Waddell[522] stated " . . . it hardly seemed likely that a simple concept of substrate competition would suffice to describe the action of desoxypyridoxine." Beaton et al.,[38] commented, "These observations might be interpreted to mean that although deoxypyridoxine accentuates external manifestations of deficiency, it is not a true antivitamin in accentuating biochemical abnormalities." In his review of vitamin $B_6$ antagonists, Umbreit[521] went so far as to state, "One might almost say that while vitamin $B_6$ is an antagonist for desoxypyridoxine, it looks as though desoxypyridoxine is not an antagonist for vitamin $B_6$." Similarly, in discussing competitive inhibition Rosen et al.[421] stated that "While this concept might hold in certain in vitro systems, it does not seem sufficient to explain some of the effects of deoxypyridoxine in animals." Despite these repeated warnings the frequent use of deoxypyridoxine in studies of vitamin $B_6$ has continued with little attempt to examine the mode of action of deoxypyridoxine in vivo. In addition, the biochemical mechanisms involved in simple $B_6$ deficiency alone are not well understood as indicated by this quote from Milholland et al.*: "Functional changes in the central nervous system, atrophy of lymphoid tissue, inhibition of tumor growth, anemia, and impairment of the antibody responses are some of the effects produced in various mammalian species fed diets deficient in vitamin $B_6$ and in some cases receiving $B_6$ antimetabolites. As yet, none of these effects is understood in terms of the inhibition of specific pyridoxine dependent enzymes."

Our interest in this problem developed from our work with vitamin $B_6$ metabolism in Down's syndrome. In connection with this work, we determined the pyridoxal phosphate concentration in lymphocytes and polymorphonuclear leukocytes. The only previous data on this subject came from a study by Johnston et al.[229] involving deoxypyr-

* From Milholland, R., Rosen, F., and Nichol, C., *Ann. N.Y. Acad. Sci.*, 166, 126, 1969. With permission.

idoxine. Included with their leukocyte data were data showing that deoxypyridoxine did not reduce the concentrations of pyridoxal phosphate in the liver even though it did increase clinical signs of a $B_6$ deficiency. If the accepted mechanism for deoxypyridoxine action were incorrect, the results obtained with deoxypyridoxine in Down's syndrome might have to be reinterpreted. Therefore, we became more deeply involved in examining just how deoxypyridoxine acts in vivo. As we developed our own work and expanded our knowledge of the literature on deoxypyridoxine, it became clear that although there were a number of intriguing observations about deoxypyridoxine action in the literature, most workers had little interest in deoxypyridoxine itself, and therefore did not pursue these observations. It also became exident that because the use of deoxypyridoxine in many experiments was considered to be equivalent to a $B_6$ deficiency, deoxypyridoxine often was not used as an indexing term nor was its use routinely mentioned in the abstracts. A number of the articles included in this review such as the report by Willis-Carr and St. Pierre[552] which was discussed above were discovered solely by chance. Others were located through the bibliographies of other articles.

One result of the information explosion is that often even the literature which is readily accessible is ignored if it is more than ten years old. This is unfortunate because thorough knowledge of the historical development of a topic often provides increased inderstanding and sometimes can prevent repetition of unsuccessful experiments which may be mentioned in the original papers but are usually omitted from the abstracts. Also, in the case of deoxypyridoxine the early literature contains some human studies which probably could not be repeated under the more stringent controls now in effect. Therefore, we hope that one benefit of this review will be to increase awareness of the older literature.

At first glance much of the literature on 4′-deoxypyridoxine seems very contradictory. However, the effects of deoxypyridoxine are highly dependent on factors such as the absolute dosage, the ratio between deoxypyridoxine and vitamin $B_6$, and the timing of the dosage. When these factors are carefully compared, much of the conflict can be resolved. As a consequence of the need to compare such factors, we have included more detail about experimental procedures than is usually found in this type of review.

As we compiled this material we discovered some rather ambiguous uses of the term, inhibitor. For example, since pyridoxine and deoxypyridoxine can serve as substrates for pyridoxal kinase with $K_m$ values comparable to those of pyridoxal,[509] should they be considered as inhibitors? In discussing competitive inhibitors Mahler and Cordes* state:

> The inhibitor may resemble one of the substrates sufficiently so as to be bound in its stead at the active site. It forms an EI (enzyme inhibitor complex), instead of an ES (enzyme-substrate complex), and thus competes with the substrate (S) for the free enzyme. This EI complex is either incapable of being converted to products, in which case the inhibitor is referred to as a dead-end inhibitor, or its rate of conversion is much smaller than that of the ES complex.

Similarly, concerning competitive inhibition Lehninger[271] states, "The inhibitor molecule is not chemically changed by the enzyme," and from Dixon[119] " . . . by definition of a competitive inhibitor EI does not break down . . . " Thus, all of these texts assume that the inhibitor is not efficiently transformed by the enzyme. Distinguishing between true inhibition as defined by these authors and competing substrates may be only a minor technicality for in vitro studies of individual enzymes. However, one of the common purposes of in vitro studies is to understand the in vivo situation. Casual

* From Mahler, H. R. and Cordes, E. H., *Biological Chemistry*, Harper & Row, New York, 1966, 250. With permission.

application of the term, inhibitor, to compounds which are actually good substrates for an enzyme may lead to incorrect interpretation of results in complex systems. For example, let us assume that under optimum conditions a given kinase system will produce 100 nmol pyridoxal phosphate per minute. We now add sufficent 4′-deoxypyridoxine to reduce the formation of pyridoxal phosphate to 50 nmol/min. If we accept the usual definition of competitive inhibitor, at the end of one min the reaction mixture would contain 50 nmol pyridoxal phosphate plus the original amount of unchanged deoxypyridoxine. However, since deoxypyridoxine is a substrate, in reality we would have 50 nmol pyridoxal phosphate and approximately an equal amount of 4′-deoxypyridoxine phosphate. Because deoxypyridoxine phosphate might interfere with a number of subsequent reactions of pyridoxal phosphate and/or might be further metabolized, it is important to recognize its existence in this system. Frequent, unqualified use of the term, inhibitor, in a case such as this tends to obscure the fact that the substance being described is actually a good substrate for the enzyme and that the metabolites of the inhibitor as well as the form initially added to the system may be influencing the results. Therefore, in this presentation we have attempted to distinguish between those cases in which the apparent reduction in activity meets the usual criteria for an inhibitor and those in which it is due to competing substrates.

While 4′-deoxypyridoxine is the most widely used $B_6$ antagonist and is our major interest, there has been some work with the 3- and 5′-deoxy derivatives. Since comparisons between these derivatives may provide some additional information about their mechanisms of action, we have incorporated all of these derivatives in this review. Since the great majority of the following discussion deals with 4′-deoxypyridoxine, the unmodified term, deoxypyridoxine, will be assumed to refer to 4′-deoxypyridoxine.

One of the most challenging aspects of studying metabolism in vivo is evaluating the many complex interrelationships. Any discussion of these relationships is hampered by the fact that the writer can present only one topic at a time while a living organism is influenced by dozens of factors simultaneously. Therefore, the discussion must be divided rather arbitrarily into individual topics. Recognizing that this choice is arbitrary, we have provided a detailed subject index for both the text and bibliography to help insure that the reader can locate information of interest.

In view of the complexity of trying to coordinate the results of over 500 articles, it is almost inevitable that some pertinent comments or data have been lost. We apologize both to the original authors and to the reader for any such errors.

Considering the number of articles which we discovered more or less by chance, I would hesitate to call the bibliography exhaustive. It does contain all of the articles which we have encountered. In over 90% of the cases we were able to locate the original article. For the remainder we relied on abstracts.

Our primary goal in preparing this review was to organize the literature in order to provide additional insights for developing our own work. In addition, we felt that providing a thorough analysis of previous work with deoxypyridoxine would help others to use deoxypyridoxine more effectively as a tool for examining various metabolic pathways and to explore further its potential clinical use in cancer therapy, immunosuppression, or other areas. Finally, we hope that this review will also stimulate some new insights into the action of vitamin $B_6$ itself.

**Stephen P. Coburn**

# THE AUTHOR

**Stephen P. Coburn,** Ph.D., is the Director of the Biochemistry Department at the Fort Wayne State Hospital and Training Center, a residential institution for the mentally retarded in Fort Wayne, Indiana, and a visiting lecturer in chemistry at Indiana University—Purdue University, Fort Wayne. Dr. Coburn received his B.S. from Rutgers University, New Brunswick, New Jersey, and his M.S. and Ph.D. in biochemistry from Purdue University. Dr. Coburn is a member of numerous professional societies including the American Institute of Nutrition, American Chemical Society, American Association on Mental Deficiency, American Association for the Advancement of Science, Biochemical Society, and the American Association for Clinical Chemistry. Dr. Coburn's current research interests include continuing studies of the effects of various antagonists, drugs, and diseases on vitamin $B_6$ metabolism in animals and man.

# ACKNOWLEDGMENTS

I am very appreciative of the help provided by Michael Sheets, our librarian; Dr. Kenneth Modesitt and Debbie Ross, who provided the format and programs used to generate the subject index; Dr. Chi Chung Lin, who translated some of the Japanese articles; Dr. Avanelle Kirksey, who reviewed the manuscript; Sharon Hudson and Elaine Leykauf, who did most of the typing and clerical work; and others who assisted in various ways.

TABLE OF CONTENTS

Chapter 1

# CHEMISTRY

## I. 4′-DEOXYPYRIDOXINE AND DERIVATIVES

### A. Syntheses

#### *1. Unlabeled*

Moeller[330-333] first reported the use of 4′-deoxypyridoxine, which he obtained from E. Merck & Co. However, the earliest published synthesis appears to be that of Harris[184] who prepared 4′-deoxypyridoxine by hydrogenating pyridoxine or 4′-ethoxypyridoxine in alcohol. The fact that the 4′-position is reduced while leaving the 5′-hydroxy group untouched is an example of the increased reactivity of the 4′-position. However, Heyl et al.[195] observed that hydrogenation of pyridoxine in aqueous dilute hydrochloric acid produced a 42% yield of 4-deoxypyridoxine and a 24% yield of 5′-deoxypyridoxine hydrochloride, suggesting that the reduction is not always completely specific for the 4′-position. Hydrogenation of pyridoxal-5′-phosphate produced 5′-deoxypyridoxine rather than 4′-deoxypyridoxine. Direct reduction of pyridoxine probably accounts for most of the commercially available 4′-deoxypyridoxine. However, there is always the possibility that the product prepared by this method will be contaminated with pyridoxine. Such contamination has occasionally been verified. Price et al.[399] noted that two different lots of deoxypyridoxine had a 20-fold difference in antipyridoxine activity. Gailani et al.[154] reported that some of their deoxypyridoxine contained 0.05% pyridoxine. Smejkal and Smejkalova[462] reported detecting pyridoxal in their material. It is possible that some of the conflicting results reported later could be explained by varying amounts of pyridoxine contamination.

Several other syntheses based on reduction of pyridoxine or its derivatives have been reported. Matsumoto[302] used electrolytic reduction of pyridoxine or pyridoxal in acidic solution. Morisawa et al.[342] used catalytic reduction of 3,4-diacetylpyridoxine hydrochloride. Prostenik and Butula[400] examined the reduction of a variety of $B_6$-related compounds and have developed conditions for controlling the selectivity of the zinc catalyzed reduction. Taborsky[492] found that pyridoxine hydrochloride or its 4′-methoxymethyl derivative would react with hydrazine to give 4′-deoxypyridoxine in high yield rather than the desired 4′-hydrazinmethyl derivative. When deuterated hydrazine was used, it was discovered that the hydrogens in the 2- and 4-methyl groups and at the 6-position were exchanged presumably through quinone methide intermediates.[245] Under the same conditions, 4′-deoxypyridoxine itself was similarly deuterated indicating that the hydrogenolysis step was not essential to the exchange process.

A number of alternative syntheses have been proposed. Some have the advantage of avoiding any possibility of contamination with pyridoxine but most are complex and give low yields. Zima and Jung[314,569] reduced 2,4-dimethyl-3-nitro-5-cyano-6-chloropyridine to the diamine then diazotized and hydrolyzed to obtain 4′-deoxypyridoxine. Van Wagtendonk and Wibaut[525] independently developed a similar approach. Probably because their paper is more detailed and more readily available, it seems to be cited more frequently than the Zima and Jung work. While the procedure is fairly straightforward the overall yield is only about 5%.

Wibaut et al.[549] improved the procedure of van Wagtendonk and Wibaut to give an overall yield of 15%. Cyanoacetamide (21 g, 0.25 mole) was dissolved in 150 m*l* of 96% ethanol and 3 m*l* of piperidine was added. The solution was brought to gentle reflux and 25 g (0.25 mole) 2,4-pentanedione was slowly added through the condenser. Refluxing was continued for 30 min. After cooling in a refrigerator, 2,4-demethyl-5-

cyano-6-hydroxpyridine was collected by suction and washed with 10 mℓ ice-cold ethanol (yield 36 g [97%]; M.P. 294°C).

A key factor in improving the yield was temperature control during the nitration reaction. Excessively high temperatures caused losses due to oxidation and at low temperatures the reaction was incomplete. The product from above was finely powdered and 74 g (0.5 mole) was suspended in 300 mℓ of acetic anhydride. The nitration mixture was prepared by adding 44 mℓ (1.03 moles) concentrated nitric acid (specific gravity 1.52) to 40 mℓ acetic anhydride with cooling. About 20% (15 to 20 mℓ) of the nitrating mixture was added rapidly with stirring to the suspension of the pyridine derivative to raise the temperature. Ice cooling was started when the temperature reached 35°C. The remainder of the nitrating mixture was added over 8 to 12 min while maintaining the temperature at 45 to 50°C. Stirring and occasional cooling were continued for 10 min while the temperature fell to 39 to 40°C.

The reaction mixture was poured onto 1 kg of crushed ice and stirred for 3 to 4 hr. The solid product was filtered off under suction and washed thoroughly with water. Then it was rapidly washed with 75 to 100 mℓ of an ethanol ether mixture (1:2), followed by four washings with 50 mℓ portions of anhydrous ether. After being suction dried for a short time, the 2,4-dimethyl-3-nitro-5-cyano-6-hydroxypyridine was dried at 60°C in the dark for 2 to 3 hr. Additional product was recovered by adding water to the washings (yield: 65.5 to 67.5 g [68 to 70%]; M.P. 265 to 268°C [decomposition]).

The yield of the chlorination step was also markedly improved. The dry nitro-derivative (50 g, 0.26 mole) was thoroughly mixed with 65 g (0.31 mole) phosphorus pentachloride. The mixture was moistened with 30 mℓ of phosphorus oxychloride and placed in an oil bath at 130°C. Gentle reflux was maintained for about 45 min until the evolution of hydrogen chloride had nearly ceased. Most of the oxychloride was distilled off *in vacuo*. The reaction product was carefully poured into water with cooling in order to decompose any remaining halophosphorus compounds. The solid was washed with water and crystallized from a small amount of ethanol. Cooling to −15°C may be necessary to facilitate crystallization of the 2,4-dimethyl-3-nitro-5-cyano-6-chloropyridine (yield: 39 g [70%]; M.P. 112 to 113°C).

For reduction, 7.1 g (0.034 moles) of the chloro-derivative was dissolved in 200 mℓ of methanol plus 17 mℓ 2 *N* hydrochloric acid. Five grams 4% palladium-on-charcoal was added and the mixture was shaken in an atmosphere of hydrogen at room temperature. Within 30 min, four equivalents of hydrogen were absorbed. After another hour, two more equivalents were absorbed. The catalyst was removed by filtration and washed with 50 mℓ of methanol. The filtrates were concentrated until crystallization of 2,4-dimethyl-3-amino-5-aminomethyl-pyridine-dihydrochloride-monohydrate occurred. Additional product was obtained by adding ether (yield: 5.6 g [70 to 75%]; M.P. 310°C [decomposition]).

Another improvement cited by the authors was the use of barium nitrite in place of sodium nitrite in the diazotization reaction. The amino derivative (3.5 g, 0.0144 moles) was dissolved in 20 mℓ distilled water and 35 mℓ 1.4 *N* sulfuric acid. A solution of 8 g (0.035 moles) of barium nitrite in an unspecified volume of water was added by drops while maintaining the temperature slightly below 0°C. After keeping the mixture at 0°C for an hour, it was maintained at 15°C for another hour. Urea was then added to remove free nitrous acid.

The solution was heated at 90°C for six hours and finally brought to a boil for a few minutes. Barium sulfate was removed by filtration and a few drops of dilute sulfuric acid were added to the filtrate in order to precipitate any remaining barium ions. An excess of sulfuric acid should be avoided. The clear filtrate was treated with charcoal and evaporated to near dryness. The 4′-deoxypyridoxine hydrochloride was recrystallized from ethanol (yield: 45%; M.P. 264°C). It seems likely that the charcoal treatment may have contributed to the low yield reported for this step.

Kolodynska and Wieniawski[242] further modified this synthesis by nitrating with a 65:35 mixture of nitric acid and acetic anhydride at 40 to 45°C in order to reduce the explosion hazard. They also noted that the yield of the chlorination step could be improved by lengthening the reaction time and increasing the amount of phosphorous oxychloride, but the details were not specified.

Balyakina et al.,[28] used a similar route. They cited the original publication of van Wagtendonk and Wibaut[525] but apparently were unaware of the later improvements published by Wibaut et al.,[549] which were discussed above. Since the optimum conditions found by Balyakina et al. produced only a 56% yield for the nitration and 62% for the chlorination compared with 70% for each of the steps in the procedure of Wibaut et al.,[549] the procedure of Balyakina et al. offers no advantage in these steps. However, Balyakina et al. did not isolate the amino derivative between the hydrogenation and diazotization steps. They obtained a 42% yield of 4′-deoxypyridoxine based on the nitro-derivative compared with about 34% for Wibaut et al. Since Balyakina et al. used solium nitrite, the barium nitrite modification suggested by Wibaut et al. appears unnecessary. Balyakina et al. used the following procedure: hydrogen was passed for 20 min through a solution of 0.1 g palladium chloride in 3 mℓ concentrated hydrochloric acid and 25 mℓ of ice-cold water containing 0.4 g activated carbon. Ice-cold water (25 mℓ) containing 2.4 g 2,4-dimethyl-3-nitro-5-cyano-6-chloropyridine was added, and hydrogenation continued until 1350 mℓ of hydrogen had been absorbed (theoretical amount is 1320 mℓ). The catalyst was filtered off and washed with 2 mℓ of water. The washings were combined with the colorless solution, and 2.4 mℓ of concentrated hydrochloric acid was added. The solution was heated to 80 to 85°C and 1.6 g sodium nitrite in 5 mℓ of water was added over 1.5 hr. The heating was continued for an additional 30 min (test with starch-iodide paper). The solution was evaporated *in vacuo* (10 torr). The residue was extracted with 36 mℓ anhydrous ethanol. The extracts were treated with activated carbon, concentrated until crystallization initiated, and left at 0 to 4°C for 8 hr. The precipitate was separated, washed with 1 mℓ alcohol at 0°C, and dried (yield: 0.7 g [42%]; M.P. 256 to 257°C).

Mori et al.,[337] modified the final steps by using nitrogen trioxide ($N_2O_3$) instead of nitrite salt in the diazotization step. Coover and Bowman[101] reacted cyanoacetamide with 3-acetyl-2,4-pentanedione to yield 2,4-dimethyl-3-acetyl-5-cyano-6-hydroxypyridine. This was converted to the 6-chloro derivative with phosphorus pentachloride, followed by hydrogenation, diazotization, and hydrolysis to yield 4′-deoxypyridoxine.

Now available commercially is 2,4-dimethyl-5-cyano-6-hydroxypyridine, thus eliminating the first step of the synthesis. This appears to be the best method for totally eliminating the possibility of pyridoxine contamination. Starting with 50 g (0.338 moles) 3-cyano-4, 6-dimethyl-2-hydroxypyridine and using the procedure of Wibaut et al.,[549] we have obtained yields of 55.5% for the nitration and 66.8% for the chlorination. We then used the reduction and diazotization conditions suggested by Balyakina et al.,[28] since their yield was comparable to the yield reported by Wibaut et al. and the procedure of Balyakina et al. avoids isolation of the amino compound. The catalyst (800 mg 10% palladium on charcoal) was suspended in 25 mℓ water followed by addition of 5 g 2,4-dimethyl-3-nitro-5-cyano-6-chloropyridine. An additional 75 mℓ water was added slowly with stirring to insure a uniform suspension. The bottle was placed in a Parr hydrogenator and shaken at room temperature under hydrogen (initial pressure 3 atm) until uptake stopped. The catalyst was filtered off and washed with water. Although the filtrate showed multiple spots on thin layer chromatography, continuation with the procedure of Balyakina et al.[28] yielded a good product. To the pale yellow filtrate was added 5.0 mℓ concentrated hydrochloric acid. The mixture was heated to

80 to 85°C. This temperature was maintained, while 3.5 g sodium nitrite in 10 mℓ water was added gradually with stirring over 90 min. Stirring was continued at the same temperature for 30 more min followed by gradual cooling. The solution was evaporated to dryness, extracted with ethanol, and brought to pH 7 with sodium hydroxide. The white precipitate was extracted several times with hot ethanol. The extracts were evaporated to a dark green oil, dissolved in 20 mℓ water, passed through a Dowex® 2-X8 (50 to 100 mesh) column (chloride form, 1.4 × 23 cm), and eluted with water. The eluates were evaporated to dryness, and 4′-deoxypyridoxine was recrystallized as the hydrochloride from acidic ethanol with about 36% yield based on the chloro-derivative.

In the course of identifying a sulfur containing derivative of vitamin $B_6$ produced in heat sterilized milk, Wendt and Bernhart[546] demonstrated that 4′-deoxypyridoxine was produced by hydrogenolysis of bis-4′-pyridoxyldisulfide. Another example of the unusual reactivity of the 4′-position of pyridoxine-type compounds is the observation of McCasland et al.[284] that heating 2-methyl-3-amino-4,5-di(hydroxymethyl) pyridine hydrochloride with hydroiodic acid produced 2,4- dimethyl-3- amino-5- hydroxymethylpyridine hydroiodide instead of the desired 2-methyl-3-amino-4,5-di(iodomethyl)pyridine hydroiodide. The product was converted to deoxypyridoxine by diazotization followed by hydrolysis.

Singh and Korytnyk[457] used the alpha[4], 3-*O*-isopropylidene derivative to permit selective esterification of the 5′-position of pyridoxine. The isopropylidene group was then removed and the 4′-position was chlorinated with thionyl chloride followed by catalytic hydrogenation and hydrolysis of the ester to yield 4′-deoxypyridoxine. The authors stated that bioassay with *Saccharomyces carlsbergensis* (ATCC 9080) indicated that deoxypyridoxine prepared by this route was more pure than the product of other routes. The increased number of steps compared with the direct reduction of pyridoxine probably reduces the amount of unchanged pyridoxine which might appear in the final product. However, it would seem preferable to avoid involving pyridoxine completely.

While the Diels Alder condensation between oxazole derivatives and fumaric or maleic acid esters[139] has received general acceptance as one of the best routes to the total synthesis of pyridoxine, there has been only one published attempt to use this procedure for the synthesis of deoxypyridoxine.[564] Reacting 4-methyloxazole with ethyl crotonate in a sealed tube at 90°C for 20 hr produced 4′-deoxypyridoxine in only 10% yield. The possibility of obtaining 5′-deoxypyridoxine in addition to or instead of 4′-deoxypyridoxine may be one factor in the low yield. Ethyl crotonate is also a less active dienophile than ethyl fumarate. In our laboratory we have attempted an analogous synthesis by reacting 4-methyl-5-ethoxyoxazole with crotonyl chloride, crotonaldehyde, or ethylcrotonate. While ultraviolet spectroscopy suggests that the oxazole reacts, we have been unable to isolate the desired product. Padyukova et al.[380] have used the Diels Alder condensation to obtain a 24% yield of some $B_6$ analogs after reacting for 10 days at room temperature. The details of the synthesis of a 5′-deoxypyridoxine derivative are presented later and may provide some ideas for improving the effectiveness of the Diels Alder condensation. In one case the yield was increased to 48% and the time reduced to 5 hr at room temperature when the pressure was increased to 10,000 atm.

Balyakina et al.[28] cited Buesing[74] as reporting preparation of 4′-deoxypyridoxine but we did not find such information in the article. Considering the limited success achieved with the Diels Alder condensation, we have concluded that combining the best steps from the procedures of Wibaut et al.[549] and Balyakina[28] is currently the most efficient synthesis for 4′-deoxypyridoxine which completely avoids the possibility of contamination with pyridoxine.

*2. Labeled*

The problem of pyridoxine contamination of deoxypyridoxine has already been mentioned. The reverse situation also occurs. 4′- and 5′-deoxypyridoxines have been identified as impurities in pyridoxine labeled with tritium[15] and carbon-14.[97] Argoudelis and Kummerow[15] felt the specific activity of the impurities was probably much higher than that of the pyridoxine itself. Failure to recognize such impurities could lead to errors in the interpretation of data obtained with these pyridoxine preparations.

Conversely, deliberate use of tritiated deoxypyridoxine was a key factor in our discovery of the deoxypyridoxine sulfate derivative in rat urine.[92] The deoxypyridoxine was tritiated by Amersham/Searle (Arlington Heights, Illinois) using acid catalyzed exchange in an aqueous medium. Even though this technique usually produces fewer side reactions than the Wilzbach technique, only 20% of the activity in the crude preparation ran with deoxypyridoxine on silica gel plates. The crude material was chromatographed repeatedly on 2 mm preparative plates of silica gel H until a single labeled spot was obtained. We later discovered that Korytnyk and Ahrens[245] had observed an exhange reaction promoted by hydrazine. This appears to be very promising as an efficient way of producing tritiated 4′-deoxypyridoxine. It is particularly attractive because the 5′-position is not labeled. Therefore, there should be no loss of label even in those species which oxidize the 5′-position during metabolism. While deuterated hydrazine is readily available, the tritiated compound is not listed in the catalogs of the major suppliers. Preliminary studies in our laboratory suggest that the exchange will also occur in less concentrated solutions of hydrazine than were used by Korytnyk and Ahrens.[245] Therefore, it might be possible to use solutions of unlabeled hydrazine in tritiated water as a source of labeled 4′-deoxypyridoxine.

## B. Physical, Chemical, and Spectral Characteristics

1. Structure:

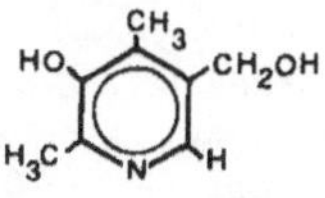


2. Empirical formula: $C_8H_{11}NO_2$
3. Molecular weight:
   a. Free base: 153.18
   b. Hydrochloride: 189.64
4. Percentage composition:

| | Free base | Hydrochloride |
|---|---|---|
| Carbon | 62.73 | 50.66 |
| Hydrogen | 7.24 | 6.38 |
| Hitrogen | 9.14 | 7.39 |
| Oxygen | 20.89 | 16.87 |
| Chloride | — | 18.70 |

5. Systematic name: 5-hydroxy-4,6-dimethyl-3-pyridinemethanol
6. Melting point of the hydrochloride:

| | |
|---|---|
| 235°C | Singh and Korytnyk[457] |
| 254°C | Moeller et al.[333] |
| 257°C | Balyakina et al.,[28] Mori et al.,[337] and van Wagtendonk and Wibaut[525] |
| 264°C | Wibaut et al.,[549] Wendt and Bernhart[546] |
| 267—268°C | Harris[184] |
| 268—270°C | Kolodynska and Wieniawski[242] |
| 273—275°C | Taborsky[492] |

## Table 1
## ULTRAVIOLET SPECTRAL DATA OF 4′-DEOXYPYRIDOXINE AND RELATED COMPOUNDS

### 4′-Deoxypyridoxine

| Solvent | $\lambda_{max}$ (nm) | E | $\lambda_{max}$ | E | Ref. |
|---|---|---|---|---|---|
| 0.1 *N* HCl | 224 | 2600 | 282 | 8300 | 282 |
| | — | — | 283 | 8200 | 387 |
| pH 4.97 | 282 | 6500 | 313 | 2170 | 282 |
| pH 5.50 | 282 | 4340 | 313 | 4930 | 282 |
| 0.1 *M* $PO_4$, pH 7 | 255 | 3900 | 315 | 7800 | 387 |
| pH 7.13 | 255 | 3860 | 313 | 8060 | 282 |
| pH 8.85 | 254 | 3840 | 313 | 8000 | 282 |
| pH 9.40 | 251 | 4190 | 310 | 7230 | 282 |
| pH 10.13 | 246 | 5240 | 305 | 6660 | 282 |
| pH 13.0 | 244 | 6870 | 301 | 7100 | 282 |
| 0.1 *N* NaOH | 245 | 6900 | 300 | 7000 | 387 |
| **4′-Deoxypyridoxine-5′-Phosphate Monohydrate** | | | | | |
| 0.1 *N* HCl | — | — | 283 | 8600 | 387 |
| 0.1 *M* $PO_4$, pH 7 | 253 | 3800 | 315 | 8400 | 387 |
| 0.1 *N* HCl | 245 | 6900 | 300 | 7500 | 387 |

Considering that previous reports contained much higher values and that no comment on the discrepancy was included, it seems likely that the value of 235°C reported by Singh and Korytnyk was a misprint, possibly a transposition of 253°C.

7. Fluorescence in absolute ethanol (Sanders et al.[434]):

| | |
|---|---|
| Excitation maximum: | 290 nm |
| Emission maximum: | 442 nm |
| Limit of detection: | 0.076 μg/mℓ |

8. Ionization constants (pK):

| | Lunn and Morton[282] | Metzler et al.[315] |
|---|---|---|
| Cation | 5.4 | 5.35 |
| Zwitterion | 9.98 | 9.73 |

Metzler and Snell[317] commented that the $pK_2$ of 9.98 reported by Lunn and Morton[282] was high compared with data on related compounds.

9. Spectra:
   Ultraviolet: See Table 1
   A detailed discussion of the application of band shape analysis of evaluating tautomeric equilibria in 3-hydroxypyridines including a variety of pyridoxine analogs was presented by Metzler et al.[315]
   Infrared: See Figure 1
   Proton magnetic resonance: See Table 2
   Mass fragmentation: See Scheme 1

The fragmentation pattern of 4′-deoxypyridoxine resembled that of *m*-hydroxybenzyl alcohol (DeJongh et al.[112]).

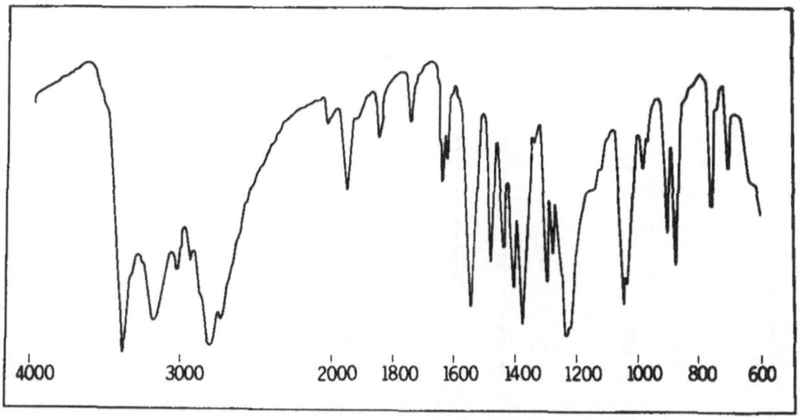


FIGURE 1. Infrared spectrum of 4′-deoxypyridoxine hydrochloride.

## C. Analytical Procedures

A wide variety of systems for identifying deoxypyridoxine through paper and thin layer chromatography have been reported (Tables 3 and 4). However, there has been little interest in developing good quantitative techniques. Korytnyk et al.[243,247] found that diacetyl deoxypyridoxine and the trimethylsilyl derivative of 4′-deoxypyridoxine-5′-phosphate were suitable for gas chromatography on a 4′ × ¼″ glass column packed with 1.2% silicone gum rubber SE-30 on 80- to 100-mesh Chromosorb G®. The acetylation was typically conducted by adding 50 to 60 mg of each $B_6$ compound to 10 mℓ of a 1:1 mixture of pyridine and acetic anhydride. The suspension was shaken until solution was achieved (about 30 min), and then allowed to stand for four hours. A 1- to 2- μℓ aliquot of the reaction mixture was chromatographed without further purification. The retention times of several acetylated derivatives at a helium flow rate of 75 mℓ/min and column temperature of 125°C are listed in Table 5. Pyridoxal acetate and 4-pyridoxic acid lactone gave a linear response over the range of 0.8 to 15.0 μg. The phosphate derivatives were separated at 175°C (Table 6).

Patzer and Hilker[384] used *N*-methyl-bis-trifluoroacetamide to produce volatile derivatives of $B_6$ compounds including deoxypyridoxine. Samples dried in microvials were treated with 40 μℓ absolute ethanol, sealed, and heated at 125°C for 15 min to convert the pyridoxal to the hemiacetal. The ethanol was then evaporated at 70° under nitrogen. After adding 30 μℓ *N*-methyl-bis-trifluoroacetamide, the vials were sealed and heated for 20 min at 125°C. A 6 ft × 2 mm glass column packed with 5% silicone oil DC-550 on Chromosorb P AW DMCS® (80 to 100 mesh) was used at 150°C with the injector at 220°C. At a flow rate of 40 mℓ/min, deoxypyridoxine had a retention time of about 4.8 min compared with 6, 12, and 14 min for pyridoxine, pyridoxal, and pyridoxamine, respectively. Samples of at least 250 ng were required for detection by the flame ionization detector used.

Tadera et al.[493] included deoxypyridoxine in their automated ion-exchange procedure. Diazotized 5-chloroaniline-2,4-disulfonylchloride was used as the color reagent.

Nakamura and Inoue[360] have reported the separation of 4′-deoxypyridoxine and several B vitamins by high-performance liquid chromatography on Zipax® SCX, a cation exchange column made by connecting two 50 cm pieces of 2.1 mm diameter tubing (Table 7). The 5 μℓ injected sample contained 200 μg of each compound. A 254 nm detector was used. The retention times for pyridoxal, pyridoxine, and pyridoxamine decreased rapidly from over 30 min at a buffer strength (pH 3.3) of $10^{-2}$ *M* to less than 10 min at $10^{-1}$ *M*. The retention time of pyridoxamine still remained over 20 min. Using 0.01 *M* buffer at a flow rate of 1 mℓ/min, the retention times of pyridoxal and

## Table 2
## PROTON MAGNETIC RESONANCE SPECTRA

| Compound | 2 $CH_3$ | | | 4 $CH_3$ | | | 5 $CH_2$ | | | 6H | | | Ref. |
|---|---|---|---|---|---|---|---|---|---|---|---|---|---|
| | Acid | Neutral | Alk. | Acid | Neutral | Alk. | Acid | Neutral | Alk. | Acid | Neutral | Alk. | |
| 4′-Deoxypyridoxine | −156 | −138 | −137 | −142 | −142 | −148 | −285 | −274 | −271 | −488 | −452 | −444 | 256 |
| | −156 | — | — | −141 | — | — | −285 | — | — | −487 | — | — | 394 |
| | — | 7.63 | — | — | 7.83 | — | — | 5.52 | — | 2.12 | — | — | 131 |
| 4′-Deoxypyridoxine phosphate | −155 | −143 | −138 | −142 | −131 | −130 | −296 | −285 | −283 | −488 | −457 | −454 | 256 |
| | | | | 4$CH_2$ | | | | | | | | | |
| 3-Deoxypyridoxine | −165 | −145 | −145 | −283 | —[a] | —[a] | −285 | −276 | −276 | −511 | −494 | −494 | 256 |
| | | | | | | | | | | 3H | | | |
| | | | | | | | | | | −477 | −438 | −438 | 256 |
| | | | | 5$CH_3$ | | | 4$CH_2$ | | | 6H | | | |
| 5′-Deoxypyridoxine | | 7.67 | | | 7.82 | | | 5.27 | | | 2.18 | | 131 |
| | | −138[b] | | | −130[b] | | | −281 | | | −464 | | 504 |
| 4,3-0-Isopropylidene 5′-Deoxypyridoxine | | −142 | | | −125 | | | −283 | | | −469 | | 253 |
| | | | | | | | 4CH | | | | | | |
| 5′-Deoxypyridoxal | | −164 | −147 | −164 | — | −136 | −638 | — | −613 | −492 | — | −431 | 219 |

[a] Peak obscured by HOD between −279 and −284 c.p.s.

[b] Authors not certain which peak is 2-ME or 5-ME.

SCHEME 1 Fragmentation of 4′ - deoxypyridoxine[a]

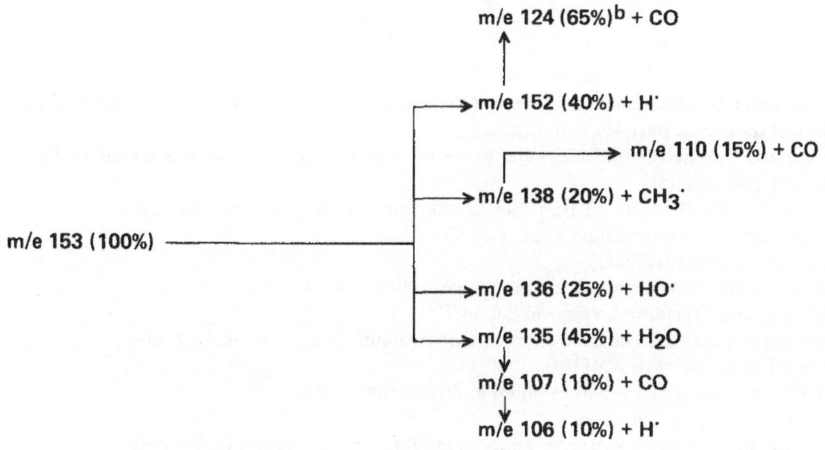


[a] Adapted from DeJongh et al[112]

[b] Values in parentheses are the approximate relative intensities of the peaks

## TABLE 3
## $R_f$ VALUES OF DEOXYPYRIDOXINE AND RELATED COMPOUNDS

| | | Compound | | | | |
|---|---|---|---|---|---|---|
| Support | Solvent | 4-DOP[a] | DOPP[b] | DOPS[c] | DOPH[d] | Ref. |
| Whatman # 1 | A | 0.74—0.80 | 0.54—0.60 | — | — | 387 |
| | B | 0.82—0.90 | 0.04—0.10 | — | — | 387 |
| | C | 0.27 | — | — | — | 409 |
| | D | 0.32 | — | — | — | 409 |
| | E | 0.01 | — | — | — | 409 |
| | F | 0.02 | — | — | — | 409 |
| | G | 0.0 | — | — | — | 409 |
| | H | 0.0 | — | — | — | 409 |
| | O | 0.68 | — | — | — | 15 |
| | P | 0.71 | — | — | — | 15 |
| | Q | 0.49 | — | — | — | 15 |
| S & S 598 | A | 0.84—0.91 | 0.63—0.72 | — | — | 387 |
| | B | 0.77—0.85 | 0.08—0.15 | — | — | 387 |
| Silica gel H | I | 0.40 | 0.95 | 0.95 | 0.95 | 92 |
| | J | 0.45 | 0.05 | 0.05 | 0.05 | 92 |
| | K | 0.65 | 0.05 | 0.53 | 0.18 | 92 |
| | L | 0.70 | 0.39 | 0.63 | 0.62 | 92 |
| Paper (type | M | 0.54 | — | — | — | 97 |
| unspecified | N | 0.55 | | | | 97 |

*Note:*

(A) t-butanol:water:89% formic acid (70:15:15);
(B) 2,6-lutidine:water:ethanol:diethylamine (55:25:20:1);
(C) 2-butanone:acetone:formic acid:water (40:2:1:6);
(D) 2-butanone:dietylamine:water (921:2:77);
(E) 4-methyl-2-pentanone:formic acid:water (10 parts ketone saturated with 1 part 4% formic acid);

## TABLE 3 (continued)
## $R_f$ VALUES OF DEOXYPYRIDOXINE AND RELATED COMPOUNDS

(F) chloroform:methanol:formic acid:water (10 parts chloroform saturated with a mixture of 1 part methanol and 1 part 4% formic acid);
(G) benzene:2-butanone:formic acid:water (9 parts benzene and 1 part butanone saturated with 1 part 2% formic acid);
(H) benzene:formic acid:water (10 parts benzene saturated with 1 part 2% formic acid);
(I) water:concentrated ammonium hydroxide (200:1);
(J) chloroform:methanol (75:25);
(K) 2-butanol:water:concentrated ammonium hydroxide (300:118:14);
(L) 1-butanol:water:pyridine:acetic acid (30:24:20:6);
(M) t-amyl alcohol:acetone:water:concentrated ammonium hydroxide (40:35:20:5);
(N) 1-butanol saturated with 1 *N*HCl;
(O) ethanol:water:concentrated ammonium hydroxide (80:16:4);
(P) 1-butanol saturated with water;
(Q) t-amyl alcohol:acetone:water: concentrated ammonium hydroxide (40:35:20:5).

[a] 4′-Deoxypyridoxine.
[b] 4′-Deoxypyridoxine-5′-phosphate.
[c] 4′-Deoxypyridoxine-3-sulfate.
[d] 4′-Deoxy-5-pyridoxic acid.

## Table 4
## DETECTION REAGENTS FOR CHROMATOGRAPHY

| Reagent | Color | Ref. |
|---|---|---|
| 0.3% Diazotized sulfanilic acid in dioxane:water (1:2) | Orange | 409 |
| 0.6% Diazotized *o*-dianisidine in dioxane:water (1:2) | Dark violet | 409 |
| 0.4% *p*-Nitrobenzenediazonium fluoroborate in dioxane:water (1:2) | Scarlet lake turning to madder carmine | 409 |
| 0.5% 2,6-Dibromoquinone-4-chloroimide in dioxaneetone (4:1) | Prussian blue | 409 |
| 2% Aqueous ferric chloride | Burnt sienna | 409 |
| 2% Aqueous phosphomolybdic acid | Silver grey | 409 |
| Diazotized *p*-nitroaniline (4 m*l* 0.7% *p*-nitroaniline in 1.08*N* HCl added dropwise to 5 m*l* ice-cold 1% $NaNO_2$ and diluted to 100 m*l* with cold water.) | Orange | 537 |
| 1% Dichloroquinonechlorimide in toluene followed by dilute ammonium hydroxide | | 418 |

pyridoxine were reduced from over 30 min at pH 3.9 and 4.5 respectively, to about 10 min at pH 4.5 (pyridoxal) and 5.4 (pyridoxine). Deoxypyridoxine dropped from over 30 min at pH 5.1 to about 14 min at pH 5.8.

**Table 5**
**RETENTION TIMES OF ACETYLATED PYRIDOXINE DERIVATIVES[a]**

| Parent compound | Retention time (min) |
|---|---|
| 4-Pyridoxic acid lactone | 2.2 |
| 5-Pyridoxic acid lactone | 2.8 |
| 4′-Deoxypyridoxine | 3.0 |
| Pyridoxal | 3.1 |
| Pyridoxine | 8.5 |
| Pyridoxamine | 19.6 |

[a] Adapted from Korytnyk et al.[247]

**Table 6**
**RETENTION TIMES OF SILYLATED PHOSPHATE DERIVATES[a]**

| Parent compound | Retention time (min) |
|---|---|
| 4′-Deoxypyridoxine-5′-phosphate | 3.4 |
| Pyridoxine-5′-phosphate | 5.0 |
| Pyridoxal-5′-phosphate | 6.9 |

[a] Adapted from Korytnyk et al.[247]

**Table 7**
**RETENTION TIMES ON ZIPAX® SCX WITH 0.05 *M* PHOSPHATE BUFFER (pH 3.3) AT A FLOW RATE OF 1 m*l*/min[a]**

| Compound | Retention time (min) |
|---|---|
| Pyridoxine | 7 |
| Pyridoxal | 11 |
| 4′-Deoxypyridoxine | 15 |

[a] Adapted from Nakamura and Inoue.[360]

Yoshida et al.[563] included deoxypyridoxine as an internal standard in their liquid chromatographic technique for $B_6$ compounds in blood. An 8 × 500 mm column of Toyo® Soda Gel LS-160 was used with a pH 5.5 buffer (0.0045 *M* sodium perchlorate, 0.04 *M* potassium dihydrogenphosphate and 2 *M* sodium chloride) as an eluent. The column separated pyridoxine, pyridoxamine, pyridoxal, pyridoxic acid, and deoxypyridoxine but did not separate the phosphates. After deproteinization with trichloroacetic acid, the excess acid was extracted with ether and hexane. A 100 μ*l* aliquot was injected and monitored by fluorescence at 325 nm excitation and 385 nm emission. Detection limits were 0.1 nmol/m*l* blood for pyridoxal, 0.2 nmol/m*l* for pyridoxamine, and pyridoxine and 0.3 nmol/m*l* for pyridoxic acid.

We have found that concentrations of 4′-deoxypyridoxine exceeding 0.1 mg/m*l* in urine can be readily measured spectrophotometrically by subtracting the reading at 315 nm at pH 1 from the reading at pH 7. Argoudelis and Kummerow[15] reported that at pH 5.1 and 120 V for 8 hr on paper, deoxypyridoxine migrated −6 cm compared with −4 cm for pyridoxine.

## D. Derivatives

In conjunction with our attempts to identify the metabolites of deoxypyridoxine, we employed the following reactions for selectively blocking either the 3- or 5′-position.[92] Ethyl chloroformate will react only with the phenol group of 4′-deoxypyridoxine. In a typical example, 1 g (5.25 mmol) of deoxypyridoxine hydrochloride was dissolved in 700 m*l* of tetrahydrofuran (redistilled over ferrous sulfate to remove peroxides) containing 1.5 m*l* (10.8 mmol) of triethylamine. Absorbance at 315 nm was used to determine when solution was complete. Ethylchloroformate 0.55 m*l* (5.8 mmol) was added with stirring. Stirring at room temperature was continued for 15 to 30 min after the yellow color faded. After filtering off the precepitated triethylammonium chloride, the solvent was removed *in vacuo* leaving a yellowish oil. The lack of absorbance at 315 nm indicated that the reaction was quantitative. If desired, the hydrochloride could be isolated by dissolving the oil in ethanol containing a stoichoimetric amount of hydrochloric acid and crystallizing with ether (M.P. 180°C decomposition). If further syntheses were planned, it was more convenient to leave it as the free base. For larger scale preparations of this derivative, it was more convenient to use absolute alcohol as the solvent and sodium ethoxide as the base.

The ethylchloroformate reaction is not specific for the phenol group in pyridoxine or other compounds with a hydroxymethyl group in the 4-position because the acyl group will migrate from the 3- to the 4′-position.[252] Since the migration is complete, using a 1:1 ratio between the ethylchloroformate and the free base form of pyridoxine will esterify the 4′-position and leave the phenol free. Addition of a second mole of ethylchloroformate will esterify the phenol group.

If the 5′-position needs to be blocked, the 4′-deoxypyridoxine-3-(ethyl carbonate) prepared above can be acetylated in the 5′-position. Treatment of the diester with *N*-methylpiperazine at room temperature will remove the carbonate ester from the phenol and leave the acetate at the 5′-position. In a typical case, to 1.35 g (6 mmol) of deoxypyridoxine-3-(ethylcarbonate) dissolved in 5 m*l* of dry pyridine were added 2 m*l* (21 mmol) of acetic anhydride. The solution was allowed to stand overnight at room temperature. After evaporating to dryness *in vacuo*, the residue was dissolved in 10 m*l* of ether and washed twice with ice-cold saturated, aqueous sodium chloride. Base was added as necessary to keep the wash slightly alkaline in order to prevent formation of a water soluble salt. The ether was dried with magnesium sulfate and evaporated to dryness *in vacuo* yielding 1.47 g (93%) of oil. To a solution of the oil (5.5 mmol) in ethyl acetate were added 1.2 m*l* (10.8 mmol) of *N*-methylpiperazine. The moisture was allowed to stand overnight at room temperature and then evaporated to dryness. The impurities were extracted with ether and the residue was dried *in vacuo* yielding 768 mg (72%) of 4′-deoxypyridoxine-5′-acetate (M.P. 158 to 160°C).

Probably the most important derivative of deoxypyridoxine is the 5′-phosphate since this is the form which often shows inhibitory activity against enzymes requiring pyridoxal phosphate. We have successfully used the following adaptation of the procedure of Peterson and Sober.[387] Suspend 16.0 g phosphorus pentoxide in 12.3 m*l* 85% phosphoric acid. Add 1.5 g deoxypyridoxine hydrochloride to 10 m*l* of the solution and heat at 60°C for 2 hr. Dissolve the viscous product in 100 m*l* hydrochloric acid and immerse in a boiling water bath 30 min. Cool and dilute to 250 m*l* with water. Add 30 g activated charcoal and mix to absorb the product. Add another 15 g charcoal until absorbance of solution stabilizes. Suction filter, wash the charcoal cake with about 700 m*l* water until the pH of the filtrate is slightly acid (pH 5). Elute the cake with 2% ammonia until the absorbance of the emerging solution is less than 0.4. Concentrate the solution *in vacuo* and place on an Amberlite® CG-50 column (3.9 × 98 cm) and elute with water. Evaporate the main peak to a very small volume, precipitate with alcohol and ether, and wash crystals with ether (yield: 447 mg [20%]). This was a much lower yield than the 75% reported by Peterson and Sober.[387] However, we have only tried the procedure once. The yield might improve with experience.

4′-Deoxypyridoxine is considerably more resistant than pyridoxal-5′-phosphate to hydrolysis. In 6 *N* sulfuric acid at 100 °C, pyridoxal-5′-phosphate at a concentration between 1.2 to 1.9 mg/m*l* was 50% hydrolyzed in 4 hr while 4′-deoxypyridoxine-5′-phosphate required 21 hr to reach the same point.[387] Both compounds were quite resistant to alkaline hydrolysis. Only about 5% of the phosphate was released after 15 hr in 1 *N* sodium hydroxide at 100°C.

The ability of $B_6$ compounds to chelate divalent metal ions has been a major factor in studies of the mechanism of $B_6$ action. Colleter et al.[96] studied the copper complexes of pyridoxine, pyridoxal, pyridoxamine, and deoxypyridoxine. The authors noted that even though its structure prohibits the same type of chelation found in other $B_6$ compounds, 4′-deoxypyridoxine does form a relatively stable complex. Table 8 lists the derivatives of 4′-deoxypyridoxine which we have found in the literature.

## II. 5′-DEOXYPYRIDOXINE

### A. Syntheses

#### *1. Unlabeled*

The first reported synthesis of 5′-deoxypyridoxine appears to be that of Heyl et al.,[195] The 4-position was protected by a methoxy group followed by replacement of the 5′-hydroxy group with chlorine using thionyl chloride. The chloro-derivative was then reduced to 2,5-dimethyl-3-hydroxy-4-methoxymethylpyridine by hydrogenation in the presence of palladium on charcoal. The authors noted that the solutions for reduction must be dilute. For example, 11.35 g of 2-methyl-3-hydroxy-4-methoxymethyl-5-chloromethyl pyridine hydrochloride was reduced in 600 m*ℓ* of methanol with 5 g of 5% palladium on charcoal. The total yield of two such batches was 19.0 g (94%).

The methoxy group was removed by heating 1.47 g 2,5-dimethyl-3-hydroxy-4-methoxymethyl pyridine hydrochloride in 50 m*ℓ* 4 *N* hydrochloric acid in a sealed tube at 180 to 190°C for three hours (70% yield).

5′-Deoxypyridoxamine was prepared by heating 3.1 g 5′-deoxypyridoxine, 50 m*ℓ* methanol, and 50 m*ℓ* liquid ammonia in a bomb at 130°C for 18 hr (yield 60%, M.P. 160 to 161°C). 5′-Deoxypyridoxal was synthesized by oxidizing 5.7 g 5′-deoxypyridoxine hydrochloride with 2.8 g manganese dioxide in 1.5 m*ℓ* sulfuric acid and 75 m*ℓ* water at 60 to 70°C for 2 hr (yield 29%; M.P. 108 to 109°C). For 5′-deoxypyridoxal hydrochloride M.P. was 191 to 193°C.

Hydrogenation of pyridoxal-5′-phosphate produced only 5-deoxypyridoxine, whereas hydrogenation of pyridoxal or pyridoxine produced a mixture of 4′- and 5′-deoxypyridoxine. Therefore, the authors concluded that the reduction of the phosphate involved direct hydrogenolysis rather than hydrolysis followed by reduction. Cunningham and Thanassi[108] have recently confirmed that catalytic hydrogenation of either pyridoxal-5′-phosphate or pyridoxamine-5′-phosphate will yield 5′-deoxypyridoxine or 5′-deoxypyridoxamine, respectively. This is probably the most convenient approach for small-scale syntheses. However, for larger scale preparations, the high cost of the phosphorylated compounds needed here might make the routes discussed earlier more attractive.

Muhlradt and Snell[348] and Korytnyk et al.[253] both used alpha$^4$-3-O-isopropylidene pyridoxine to protect the 3 and 4′-positions. The 5′-position was then chlorinated with thionyl chloride followed by hydrogenation and hydrolysis. The procedure was also presented in detail by Korytnyk and Ikawa.[251] We have used the latter procedure successfully with the exception that we found it necessary to reflux the thionyl chloride reaction for 2 hr in order to get a satisfactory yield of the chloride.

In a typical run, 600 m*ℓ* of acetone was distilled directly into a 1000- m*ℓ* round bottom flask containing 45 g pyridoxine hydrochloride. After cooling to 0°C in brine, dry hydrogen chloride was bubbled through a gas diffusion tube with stirring until a 15 to 17% solution was obtained. Stirring was continued for 30 min in the ice bath and for one hour at room temperature. The reaction can be monitored by chromatog-

## Table 8
## DERIVATIVES OF 4′- DEOXYPYRIDOXINE

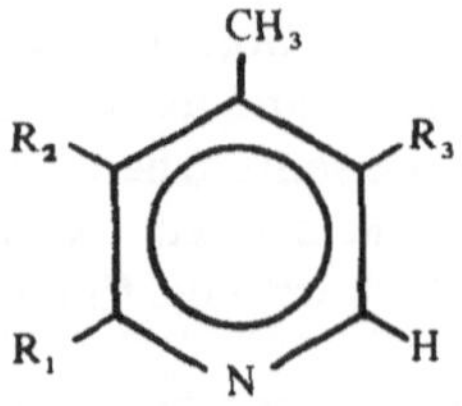

| Modifications in $R_1$ | Ref. |
|---|---|
| $R_2$ = OH  $R_3$ = $CH_2OH$ | |
| $R_1$ | |
| $CH_3CH_2-$ | 196, 339 |
| $CH_3-CH(CH_3)-CH_2-$ | 196 |
| $CH_3CH_2CH_2CH_2CH_2-$ | 196 |
| $HC(=O)-$ | 254 |
| $CH_2{=}CH-$ | 254 |
| $HOCH_2$ | 254 |
| $H_2NC(=S)NHNCH-$ | 254 |

| | Modifications in $R_2$ | Ref. |
|---|---|---|
| | $R_1$ = $CH_3$  $R_3$ = $CH_2OH$ | |
| | $R_2$ | |
| 4′-Deoxypyridoxine-3-sulfate | $-OSO_3H$ | 92 |
| 4′-Deoxy-3-0-(2-sulfoethyl)-pyridoxine | $-OCH_2CH_2SO_3H$ | 92 |
| 4′-Deoxy-3-(ethyl carbonate) | $-OC(=O)OCH_2CH_3$ | 92 |
| 4′-Deoxypyridoxine-3-acetate | $-OC(=O)CH_3$ | 340 |

## Table 8 (continued)
## DERIVATIVES OF 4′-DEOXYPYRIDOXINE

| | Modifications in $R_2$ | Ref. |
|---|---|---|
| 4′-Deoxypyridoxine-3-pivalate | $-O\overset{O}{\overset{\|}{C}}C(CH_3)_3$ | 340 |
| 4′-Deoxypyridoxine-3-benzoate | 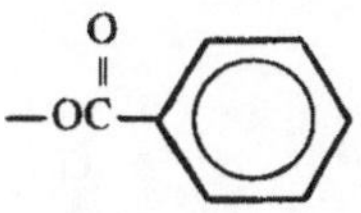 | 340 |
| | $-NH_2$ | 284 |
| 4′-Deoxypyridoxine-3-phosphate | $-OPO_3H_2$ | 279 |
| | **Modifications in $R_3$** | |
| | $R_1 = CH_3$ $R_2 = -OH$ | |
| | **$R_3$** | |
| 4′-Deoxypyridoxine-5′-phosphate | $-CH_2OPO_3H$ | 41, 387, 522 |
| 4′-Deoxypyridoxine-5′-sulfate | $-CH_2OSO_3H$ | 262 |
| 4′-Deoxy-5′-pyridoxic acid | $-\overset{O}{\overset{\|}{C}}OH$ | 92 |
| 4′-Deoxypyridoxine-5′-acetate | $-CH_2O\overset{O}{\overset{\|}{C}}CH_3$ | 279, 258, 429, 340, 92 |
| 4′-Deoxypyridoxine-5′-benzoate | 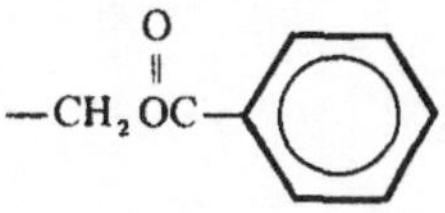 | 457, 340 |
| 4′-Deoxypyridoxine-5′-(4 -nitrobenzoate) | 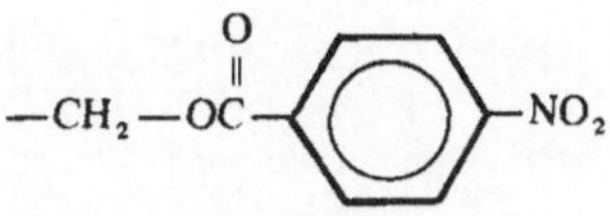 | 457, 340 |
| | 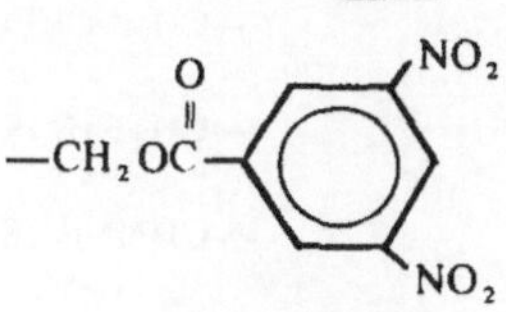 | 340 |
| | 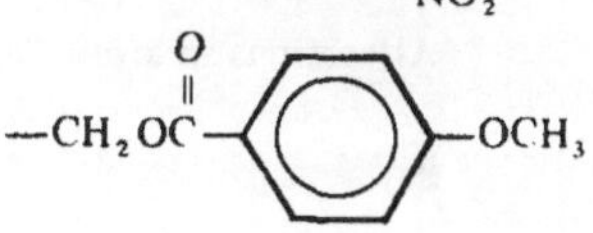 | 340 |
| | 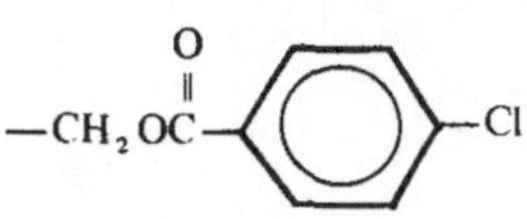 | 340 |

**Table 8 (continued)**
**DERIVATIVES OF 4′-DEOXYPYRIDOXINE**

| Modifications in $R_3$ | Ref. |
|---|---|
| 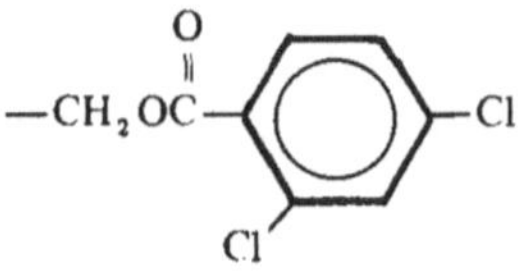 | 340 |
| 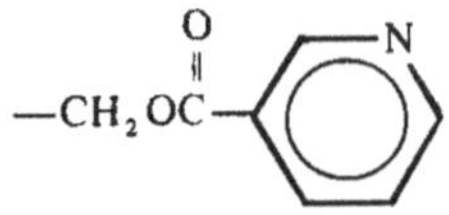 | 340 |
| 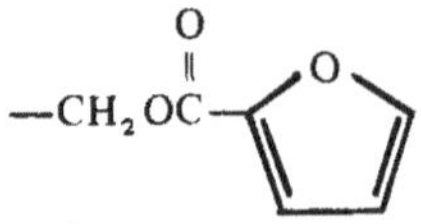 | 340 |
| $-\overset{O}{\overset{\Vert}{C}}H$ | 483, 341 |
| $-\overset{O}{\overset{\Vert}{C}}CH_3$ | 341 |
| $-\overset{OH}{\overset{\vert}{C}}HCH_3$ | 341 |
| $-CH_2CH_2OH$ | 245 |
| $-CH_2CH_2CH_2OH$ | 245 |
| 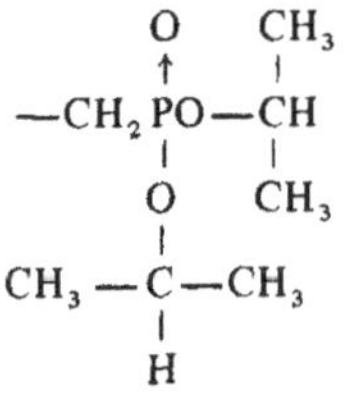 | 45 |
| $-CH{=}CHPO_3H_2$ | 483 |
| $-CH_2CH_2PO_3H$ | 483 |
| $-CHNNHCSNH_2$ | 159 |

**Alterations In More Than One Location**

| $R_1$ | $R_2$ | $R_3$ | |
|---|---|---|---|
| $-CH_3$ | $-O\overset{O}{\overset{\Vert}{C}}CH_3$ | $-CH_2O\overset{O}{\overset{\Vert}{C}}CH_3$ | 247, 340 |
| $-CH_3$ | $-O\overset{O}{\overset{\Vert}{C}}CH_2CH_3$ | $-CH_2O\overset{O}{\overset{\Vert}{C}}CH_2CH_3$ | 340 |

## Table 8 (continued)
## DERIVATIVES OF 4′-DEOXYPYRIDOXINE

**Alterations In More Than One Location**

| $R_1$ | $R_2$ | $R_3$ | Ref. |
|---|---|---|---|
| $-CH_3$ | 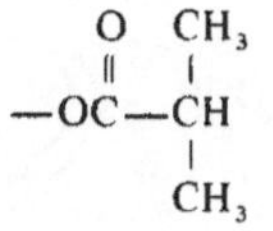 | 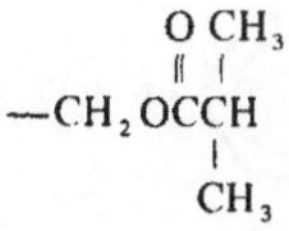 | 340 |
| $-CH_3$ | $-O\overset{O}{\overset{\|}{C}}CH_2CH_2CH_2CH_2CH_3$ | $-CH_2O\overset{O}{\overset{\|}{C}}CH_2CH_2CH_2CH_2CH_3$ | 340 |
| $-CH_3$ | 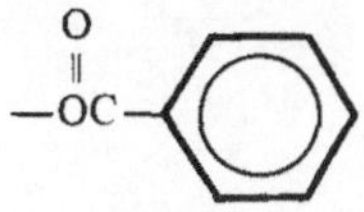 | 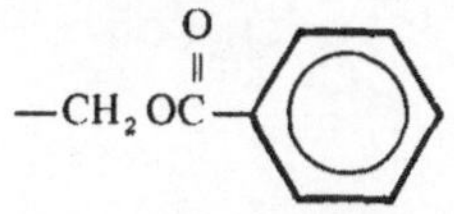 | 340 |
| $-CH_3$ | $-O\overset{O}{\overset{\|}{C}}CH_3$ | 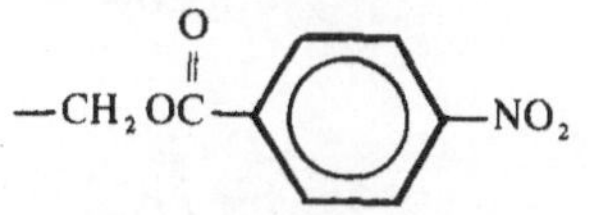 | 340 |
| $-CH_3$ | $-O\overset{O}{\overset{\|}{C}}CH_3$ | 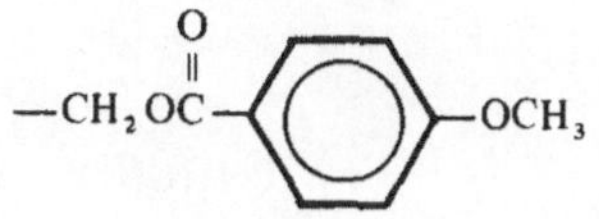 | 340 |
| $-CH_3$ | $-O\overset{O}{\overset{\|}{C}}CH_3$ | 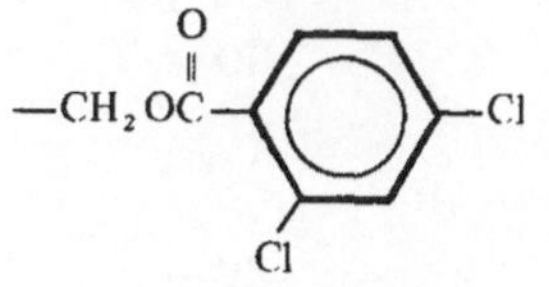 | 340 |
| $CH_3-$ | 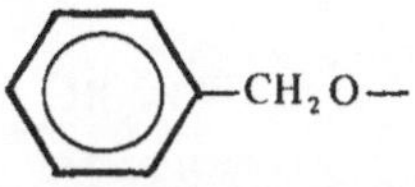 | 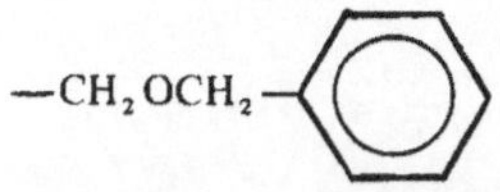 | 254, 339 |
| $CH_3\overset{O}{\overset{\|}{C}}OCH_2-$ | 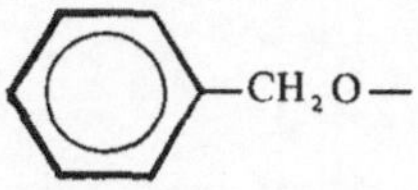 | 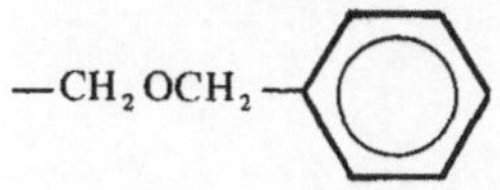 | 339 |
| $HOCH_2$ | 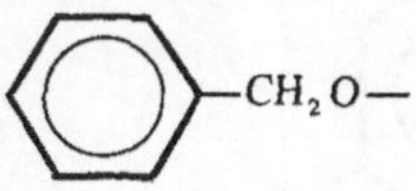 | 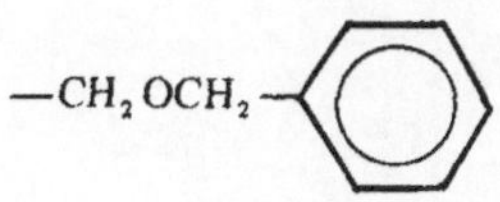 | 254, 339 |
| $H\overset{O}{\overset{\|}{C}}-$ | 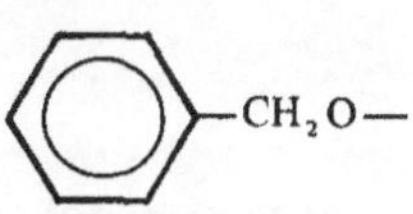 | 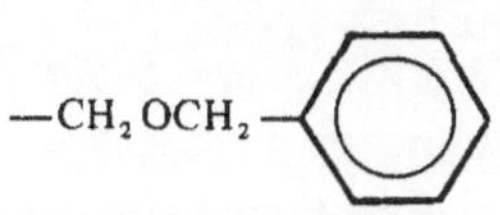 | 254, 339 |
| $CH_3\underset{OH}{\underset{\|}{C}}H-$ | 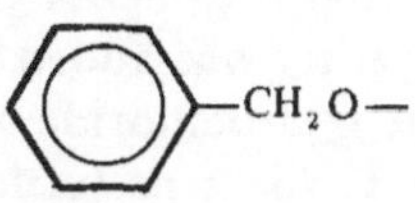 | 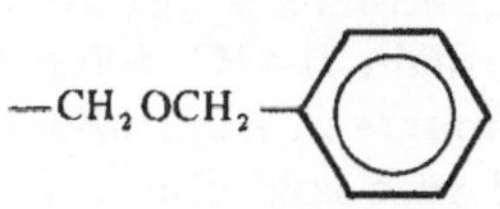 | 339 |
| $CH_3\underset{Cl}{\underset{\|}{C}}H-$ | 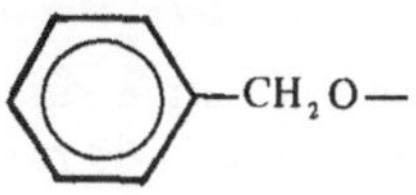 | 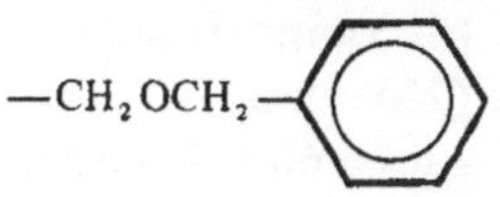 | 339 |

## Table 8 (continued)
## DERIVATIVES OF 4′-DEOXYPYRIDOXINE

**Alterations In More Than One Location**

| $R_1$ | $R_2$ | $R_3$ | Ref. |
|---|---|---|---|
| $CH_2{=}CH-$ | 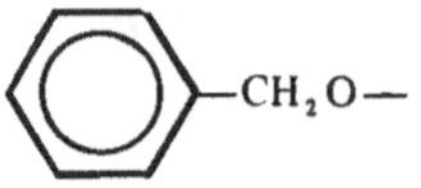 | 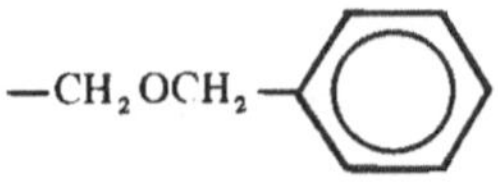 | 254 |
| $CH_2{=}CH-$ | HO— | $-CH_2OPO_3H_2$ | 254 |
| $CH_3-$ | $CH_3\overset{\overset{O}{\|}}{C}O-$ | $-\overset{\overset{O}{\|}}{C}H$ | 483 |
| $CH_3-$ | $CH_3\overset{\overset{O}{\|}}{C}O-$ | $-CH{=}CHPO_3(CH_2CH_3)_2$ | 483 |
| $CH_3-$ | $CH_3\overset{\overset{O}{\|}}{C}O-$ | $-CH_2CH_2PO_3(CH_2CH_3)_2$ | 483 |
| $CH_3-$ | $CH_3(CH_2)_{14}\overset{\overset{O}{\|}}{C}O-$ | $-CH_2O\overset{\overset{O}{\|}}{C}(CH_2)_{14}CH_3$ | 431 |
| $HOCH_2-$ | HO— | $-CH_2OPO_3H_2$ | 394 |
| $H\overset{\overset{O}{\|}}{C}-$ | HO— | $-CH_2OPO_3H_2$ | 394 |
| $HOCH_2-$ | HO— | $-CH_2OH$ | 394 |

**Miscellaneous**

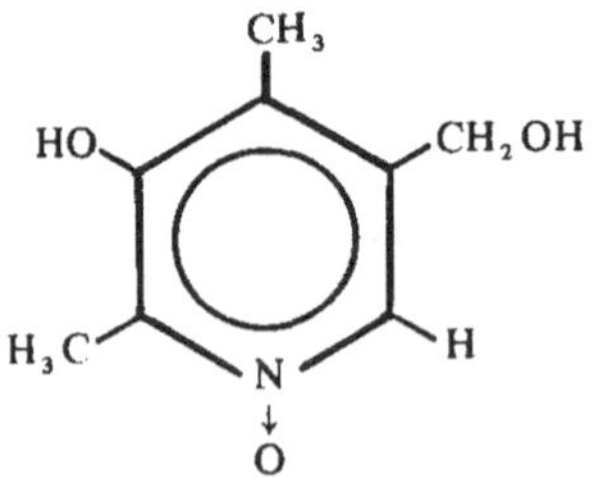


248

raphy and/or ultraviolet spectroscopy to verify that the phenol group is blocked. After standing overnight at 4°C, 400 m*l* of cold, dry ether was added to the cold suspension and the precipitate of isopropylidenepyridoxine hydrochloride was collected by filtration, washed with cold ether, and dried at 65°C for 2 hr (yield 50.4 g [94%]; M.P. 210 to 212°C [decomposition]).

To 50.4 g (205 mmol) of the isopropylidene derivative suspended in 500 m*l* dry benzene was added 60 m*l* (830 mmol) thionyl chloride in 60 m*l* benzene and refluxed for 2 hr. The mixture was cooled to 0°C, filtered, washed with cold ether, and dried at 65°C for 2 hr (yield 53.1 g [98%]; M.P. 192 to 194°C [dec]). (Decomposition = dec.)

The alpha[4]-3-*O*-isopropylidene-5′-deoxy-5′-chloropyridoxine was converted to the free base by dissolving 25 g in cold aqueous sodium bicarbonate, and extracting with ethyl acetate. The organic layer was dried with magnesium sulfate and evaporated *in vacuo* to yield a dark oil. This was dissolved in 150 mℓ ethanol and hydrogenated with 1 g 10% Pd-C catalyst for 7.5 hr. The catalyst was removed by filtration and the filtrate evaporated *in vacuo*. The solid was washed with ethanol and then with ether yielding 20.6 g (95%). Recrystallization from ethanol and ether mixtures may be needed for purification (M.P. 215 to 216°C [dec]).

The isopropylidene group was removed by heating 12.5 g in 60 mℓ 0.25 *N* hydrochloric acid on a steam bath for 1 hr. The solution was filtered if necessary and the filtrate evaporated to dryness *in vacuo*. Recrystallization from ethanol-ether yielded 9.3 g (89%); M.P. 140 to 144°C. If the hydrochloric acid solution is neutralized to pH 6.5 before crystallization, the free base is obtained.[218] Kuroda[263] synthesized several 5′-deoxy compounds including 5′-deoxypyridoxine by catalytic reduction of pyridoxamine to 5′-deoxypyridoxamine followed by removal of the amino group to give 5′-deoxypyridoxine.

*2. Labeled*

Mulligan and Snell[349] obtained tritiated 5′-derivatives by reducing alpha[4]-3-*O*-isopropylidene-5′-deoxy-5′-chloropyridoxine with tritium gas. The chloro-compound (260 mg) was dissolved in 4.5 mℓ absolute methanol with 82 mg anhydrous sodium acetate and 26 mg 10% palladium on charcoal. This mixture was treated with 10 Ci tritium gas (3 Ci/nmol) for one hour at 1 atm. The catalyst was filtered off. The filtrate was concentrated *in vacuo* and allowed to stand overnight at room temperature in 1 *N* aqueous hydrochloric acid. After filtering off any precipitate, the solution was heated 15 min at 80°C and evaporated almost to dryness *in vacuo*. The product was obtained in 80% yield and specific activity of 1.98 Ci/nmol upon crystallization from ethanol and ether.

## B. Physical, Chemical, and Spectral Characteristics

1. Structure:

2. Empirical formula: $C_8H_{11}NO_2$
3. Molecular weight:
   a. Free base: 153.18
   b. Hydrochloride: 189.64
4. Percentage composition:

| | Free base | Hydrochloride |
|---|---|---|
| Carbon | 62.73 | 50.66 |
| Hydrogen | 7.24 | 6.38 |
| Nitrogen | 9.14 | 7.39 |
| Oxygen | 20.89 | 16.87 |
| Chloride | — | 18.70 |

Table 9
ULTRAVIOLET SPECTRAL DATA ON 3- AND 5′-DEOXY COMPOUNDS

| Solvent | $\lambda_{max}$ (nm) | E[a] | $\lambda_{max}$ (nm) | E | Ref. |
|---|---|---|---|---|---|
| 3-Deoxypyridoxine | | | | | |
| 0.1 *N* HCl | 264 | — | — | — | 188 |
| pH 7 | 265 | — | — | — | 188 |
| 3-Deoxypyridoxal phenylhydrazone | | | | | |
| 0.3 *M* $H_2SO_4$ | — | — | 410 | 22,400 | 380 |
| 5′-Deoxypyridoxine | | | | | |
| pH 7 | — | — | 320 | 7360 | 295a |
| pH 7 | 248 | 3790 | 325 | 7580 | 263 |
| 5′-Deoxypyridoxamine | | | | | |
| pH 7 | 250 | 2000 | 325 | 4100 | 263 |
| 5′-Deoxypyridoxal | | | | | |
| Cation | 294 | 6280 | 342 | 1880 | 358 |
| pH 1.1 | 294 | 6200 | — | — | 318 |
| pH 1.4 | 294.5 | 6300 | 342 | 1900 | 218 |
| pH 1.9 | 290 | 6300 | — | — | 195 |
| Zwitterion | 324 | 2970 | 381 | 4220 | 358 |
| pH 6.4 | 323,5 | 3000 | 380.5 | 4200 | 218 |
| pH 6.88 | 324 | 3000 | 381 | 4200 | 318 |
| pH 7.0 | 330 | 990 | 390 | 5100 | 263 |
| Anion | 263 | 3480 | 391 | 6330 | 358 |
| pH 10.0 | — | — | 390 | 6200 | 318 |
| pH 11.0 | — | — | 390 | 6300 | 195 |
| pH 11.0 | 262 | 3500 | 392 | 6300 | 218 |
| 5′-Deoxypyridoxal phenylhydrazone | | | | | |
| 0.1 *N* $H_2SO_4$ | — | — | 410[b] | 31,200 | 407 |
| 0.3 *M* $H_2SO_4$ | — | — | 404 | 18,200 | 380 |

[a] Molar absorbtivity.
[b] The original article lists 340 nm. However, communication with Dr. Reiber[406a] confirms 410 nm is the correct wavelength.

5. Systematic name: 3-hydroxy-2,5-dimethyl-4-pyridinemethanol
6. Melting point:
   a. Free base: 179—181°C Iwata[218]; 181—182°C Heyl et al.[195]
   b. Hydrochloride: 140—142°C Muhlradt and Snell[348]; 143—144°C Heyl et al.,[195] Korytnyk and Ikawa,[251] Korytnyk et al.,[253] and Kuroda.[263]

7. Ionization constants, Metzler et al.[315]

| | $pK_1$ | $pK_2$ | $pK_3$ |
|---|---|---|---|
| 5′-Deoxypyridoxine | 5.48 | 9.46 | — |
| 5′-Deoxypyridoxamine | 3.94 | 8.39 | 9.87 |
| alpha[4]-3-*O*-Isopropylidene-5′-deoxypyridoxine | | 6.10 | |

8. Spectra:

Ultraviolet: See Table 9
Proton magnetic resonance: See Table 2

SCHEME 2 Fragmentation of 5′-deoxpyridoxine[a]

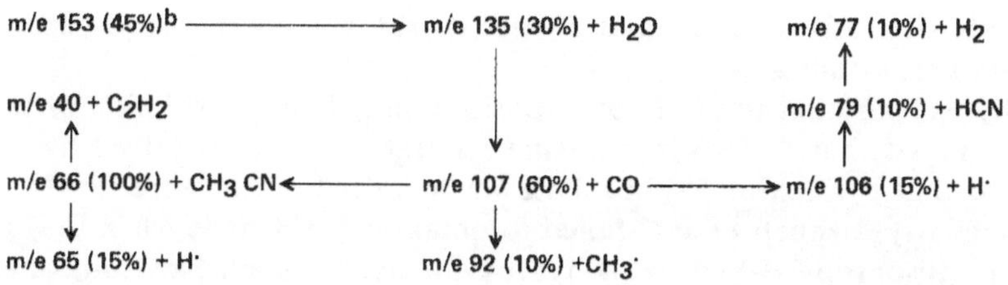


[a] Adapted from DeJongh et al[112]

[b] Values in parentheses are the approximate relative intensities of the peaks.

### Table 10
### $R_f$ VALUES OF 5′-DEOXYPYRIDOXINE (5′-DOP) AND 5′-DEOXYPYRIDOXAL (5′-DPL)

| | | Compound | | |
|---|---|---|---|---|
| Support | Solvent | 5′-DOP | 5′-DPL | Ref. |
| Whatman #1 | A | 0.74 | — | 15 |
| | B | 0.78 | — | 15 |
| | C | 0.57 | — | 15 |
| Silica gel G | D | — | 0.50 | 218 |

*Note:* (A) Ethanol:water:concentrated ammonium hydroxide (80:16:4);(B) l-butanol saturated with water; (C) t-amyl alcohol:acetone:water:concentrated ammonium hydroxide (40:35:20:5); (D) l-butanol:acetic acid:water (4:1:5).

*Note:* For detection reagents see Table 4.

Proton magnetic resonance: See Table 2

Mass fragmentation: See Scheme 2. The fragmentation pattern resembled that of *O*-hydroxybenzyl alcohol.[112]

9. Chromotographic properties: See Table 10

## C. Derivatives

In the course of attempting to extend our work to 5′-deoxypyridoxine, we encountered an additional example of the high activity of the 4′-position. We first encountered this problem in our attempts to synthesize 5′-deoxypyridoxine-4′-sulfate.[94] Every time we started to neutralize the reaction mixture which contained the 4′-sulfate, we obtained a water-insoluble precipitate. We eventually discovered that Singh and Korytnyk,[457] and McCasland et al.[284] had observed similar effects when attempting to prepare the free base of 4-(chloromethyl)-5-hydroxy-6-methyl-3-pyridinemethanol benzoate hydrochloride or 2-methyl-3-hydroxy-4,5-dibromomethyl-pyridine hydrochloride, respectively. It appears that whenever the 4′-position is substituted with a good leaving group, base catalyzed polymerization via a quinone methide mechanism can occur very easily.

We[94] also found some unexpected neighboring group effects which misled Shane and Snell[446] in their examination of the metabolism of 5′-deoxypyridoxine. First, 2,6-

dimethyl substitution around the phenol group almost completely inhibited arylsulfatases from limpet, *Helix pomatia*, abalone entrails, and *Aerobacter aerogenes*. Thus, the 3-sulfate derivatives of 4′- and 5′-deoxypyridoxines were not hydrolyzed even though they are aryl sulfates.

Secondly, substitution in the 3- or 5′-position may inhibit oxidation of the 4′-position. We[94] noted that 5′-deoxypyridoxine-3-sulfate was not oxidized by manganese dioxide and that pyridoxine-5′-phosphate was oxidized much more slowly than pyridoxine. Similarly, Bennett et al.[45] failed to obtain any aldehyde when they attempted to oxidize diisopropyl-3-hydroxy-4-hydroxymethyl-2-methyl-5-pyridine-methylphosphonate hydrochloride with manganese dioxide under conditions which were successful for pyridoxine phosphate. While speaking of pyridoxine phosphate, we would suggest especially careful structure confirmation of commercial samples of pyridoxine phosphate. In one case, we received pyridoxal-5′-phosphate, and in another case it was simply the phosphoric acid salt of pyridoxine rather than pyridoxine-5′-phosphate.

Padyukova et al.[380] synthesized a 5′-deoxypyridoxine derivative, 2,5-dimethyl-3-oxy-4-acetylpyridine, by allowing a mixture of 3.3 g 3-pentene-2-one, 1.8 g 4-methyl-5-propoxyoxazole, 100 mg pyrogallol, 1 m*l* glacial acetic acid, and 5 m*l* anhydrous ether to stand at 20° for 10 days. Subsequently, the mixture was concentrated to an oil and treated with an equimolar amount of concentrated hydrochloric acid in methanol. The hydrochloride crystallized upon addition of acetone and was recrystallized from ethanol yielding 1.25 g (60%); M.P. 185°C.

## III. 5′-DEOXYPYRIDOXAL

### A. Syntheses

#### *1. Unlabeled*

5′-Deoxypyridoxine can readily be converted to 5′-deoxypyridoxal by manganese dioxide oxidation.[218,348] The manganese dioxide must be finely divided and activated. Various procedures for accomplishing this are available in the literature. We have obtained satisfactory results with the following procedure which was compiled[181a,295b] by Bob In-Yu Yang and forwarded to us by Barry Shane.[560a]

Twenty grams of manganous sulfate monohydrate was placed in a 1-liter, 3-neck round bottom flask with a magnet. The flask was seated in a heating mantle on top of a magnetic stirrer. To the flask was added 300 m*l* of water. With stirring, the compound was dissolved and heated to 90°C. Four grams of potassium permanganate dissolved in 100 m*l* of boiling water was transferred to a dropping funnel and allowed to drip into the flask over a period of approximately 2 min. Addition of hot permanganate was repeated four more times, so that a total of 20 g of potassium permanganate in 500 m*l* of water was added.

Following the addition of the last of the permanganate, the content of the flask was allowed to continue stirring at 90°C for 30 min. Then, heating was discontinued while stirring was maintained until the mixture cooled to 60°C (approximately 1½ hr). The content was filtered through a 350 m*l* medium porosity sintered glass funnel. The precipitate was transferred to a 500 m*l* beaker containing 300 m*l* of 2.4 *N* nitric acid, stirred for 5 min and filtered with suction in the cleaned funnel. The acid treated manganese dioxide was removed to a beaker containing approximately 300 m*l* of 50°C water, stirred for 5 min, and filtered with suction in the cleaned funnel. The warm water washing was repeated several times until the washing attained a pH of 6. The brownish black cake was removed from the funnel in as few pieces as possible and dried and stored in a 110 to 130°C oven. The chunks were pulverized just before use. Manganese dioxide thus prepared was suitable for the preparation of pyridoxal sulfate from pyridoxine sulfate.

Manganese dioxide prepared by different methods varies in the content of moisture and permanganate. This may cause differences in reactivity. The oxidation may be carried out in water,[560a] chloroform,[218] or a mixture of both.[348] Iwata[218] used a 10:1 ratio by weight of manganese dioxide:5′-deoxypyridoxine while Yang[560a] used 2:1. The oxidation is conducted with stirring at room temperature. It should be monitored at

390 nm to determine the optimum time. Reported yields range from 58%[348] to 74%.[218] Doktorova et al.[123] also reported the synthesis of 5′-deoxypyridoxal by manganese dioxide oxidation of 5′-deoxypyridoxine. The aldehyde was then converted to an oxime and reduced to the amine. Ultraviolet and proton magnetic resonance spectra were included for both the aldehyde and amine.

While the manganese dioxide technique is very popular for the oxidation of alcohols, we have found the use of pyridinium chlorochromate to be preferable with some derivatives. It seems to be slightly milder than the manganese dioxide and is soluble in organic solvents. However, as a consequence of the increased solubility of the pyridinium chlorochromate, it is sometimes slightly more difficult to purify the desired product. Two other recently described oxidizing agents may also be useful in the conversion of hydroxy groups to aldehydes.[103,140]

Kuroda[263] extended his synthesis of 5′-deoxy compounds by converting 5′-deoxypyridoxamine to the oxime with manganese dioxide and hydroxylamine and then to 5′-deoxypyridoxal with hydrochloric acid and silver nitrite.

Reiber[406] showed that both 5′-deoxypyridoxal and pyridoxal 5′-phosphate would readily undergo photochemical oxidation to the corresponding acids. This action has caused some confusion in studies of $B_6$ metabolism.[100,540]

### *2. Labeled*

Mulligan and Snell[349] used the manganese dioxide procedure to convert labeled 5′-deoxypyridoxine to labeled 5′-deoxypyridoxal. The 5′-deoxypyridoxal and unreacted 5′-deoxypyridoxine were purified by chromatography on silica gel and P11 phosphocellulose.

## B. Physical, Chemical, and Spectral Characteristics

1. Structure:

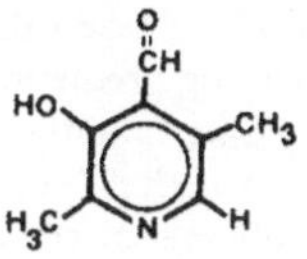


2. Empirical formula: $C_8H_9NO_2$
3. Molecular weight:
   a. Free base: 151.16
   b. Hydrochloride: 187.63
4. Percentage composition:

| | Free base | Hydrochloride |
|---|---|---|
| Carbon | 63.56 | 51.21 |
| Hydrogen | 6.00 | 5.37 |
| Nitrogen | 9.27 | 7.47 |
| Oxygen | 21.17 | 17.05 |
| Chloride | — | 18.90 |

5. Systematic name: 3-hydroxy-2,5-dimethyl-4-pyridinecarboxaldehyde
6. Melting point
   a. Free base: 108—109°C Iwata[218]; 111—113°C Muhlradt and Snell[348]
   b. Hydrochloride: 191—193°C Kuroda[263]
7. Ionization constants: The interaction between the aldehyde group and the adjacent hydroxy groups is of great importance in the mechanism of $B_6$-catalyzed reactions. Eliminating part of this interaction by phosphorylation of the 5′-position or by conversion to 5′-deoxypyridoxal significantly alters the reactivity

of pyridoxal. For example, lack of interference from the intramolecular hemiacetal permits 5′-deoxypyridoxal to react with phenylhydrazine at the same rate as pyridoxal-5-phosphate while pyridoxal alone reacts much more slowly at low pH (Wada and Snell[534]). As a result, while there has been very little work with 5′-deoxypyridoxal in vivo, it has been used extensively for in vitro mechanistic studies.

In the presence of copper ions, 5′-deoxypyridoxal was approximately equal to pyridoxal in its ability to catalyze the nonenzymatic deamination of pyridoxamine and glycine.[214] 5′-Deoxypyridoxal was also the only aldehyde other than pyridoxal phosphate which was able to catalyze the nonenzymatic decarboxylation of alpha-keto-acids to a significant degree, producing 25 to 30% as much activity as pyridoxal phosphate. Pyridoxal, salicylaldehyde, acetaldehyde, and formaldehyde showed less than 10% as much activity as pyridoxal phosphate. Metzler and Snell[318] noted that ionization studies of 5′-deoxypyridoxal are probably applicable to pyridoxal phosphate. At pH 1.10 5′-deoxypyridoxal showed a major peak at 291 nm which was attributed to the hydrated form and a secondary peak at 342 nm which was attributed to the free aldehyde. In the zwitterion form (pH 6.88) the peaks were shifted to 324 nm and 381 nm. The relative height of the peaks was also reversed, reflecting an increase in the stability of the aldehyde form relative to the hydrate. At pH 10.0 the hydrate peak (273 nm) is very weak and the aldehyde peak (391 nm) is strong. These analyses were later expanded.[182,227,315] Nagano and Metzler[358] developed computer programs for calculating equilibrium constants and predicting spectra. With deoxypyridoxal alone, the calculated spectra agreed very well with the observed data. However, when leucine was added, postulation of an additional complex was needed to permit accurate prediction of the spectra.

Ahrens and Maas et al.[6,7,283] examined the hydration of various $B_6$ compounds including 5′-deoxypyridoxal. Data on the dependence of the reciprocal relaxation time on pH were presented. The proton and intramolecular rate constants of 5′-deoxypyridoxal resembled those of pyridoxal but the constant for base catalysis was $5 \times 10^6$ for pyridoxal compared with $1.4 \times 10^3$ for deoxypyridoxal. Intramolecular catalysis was more important than acid or base catalysis. The rate constants of proton catalysis were reduced by at least one order of magnitude compared with benzaldehydes. The authors concluded that this reduction reflected the increased polarization of the carbonyl pi electron systems due to electron withdrawal by the ring nitrogen.

Ahrens[5] presented a detailed consideration of the mechanisms and kinetics of proton transfer between the nitrogen and phenol group of 5′-deoxypyridoxal based on temperature jump techniques. While 5-deoxypyridoxal provides a useful model, Ahrens concluded that attempts to extend this type of analysis to pyridoxal phosphate would be complicated by the additional possible cross-reactions.

Savin et al.[440] calculated electron excitation energies for the first two absorption bands of several $B_6$ compounds including 5-deoxypyridoxal (Table 11). These values showed good agreement with experimentally measured values. They also derived values for predicting the effects of substituents (Table 12). For example, to predict $E_1$ for the cation of pyridoxine, take the value for 3-hydroxypyridine (4.406) and add the corrections for 2-methyl (−0.054), 4-hydroxymethyl (−0.046), and 5-hydroxymethyl (−0.009) yielding 4.297. This compares well with the experimentally measured value of 4.25. Other tabular data in the report included values of pi-electronic charges and bond orders of 3-hydroxypyridine in the ground state and their changes upon transition of the molecule to the first excited singlet and triplet states and changes in the values of the pi-electron charge for various forms of 3-hydroxypyridine on introducing different substituents.

Buell and Hansen[73] used the reaction between pyridoxal phosphate and aminothiols to gain some insights into the interaction with glycogen phosphorylase. They noted that increasing pH produced a greater reduction in the rate of complex formation between cysteine and pyridoxal phosphate than with 5′-deoxypyridoxal presumably because the phosphate group imparted a greater negative charge. On the other hand the ethyl ester of cysteine reacted four times faster with pyridoxal-5′-phosphate than with 5′-deoxypyridoxal suggesting that the phosphate group in some manner activated the aldehyde group.

Thanassi[501] has suggested that since the reactions between 5′-deoxypyridoxal and aminomalonic acid derivatives occur rapidly at low temperature, they might provide better models of $B_6$ reactions than systems involving higher temperatures or slower rates. He has examined this model in detail.[58,59,500-503]

8. Spectra:
   a. Ultraviolet: See Table 9. Buell and Hansen[73] presented spectra of 5-deoxypyridoxal at pH 4, 5, 6, 7, 8, 9, 10 to show the close similarity to the spectra of pyridoxal-5′-phsophate at similar pH. However, they noted two differences. As the pH rises from 5 to 10, the peak at 380 in 5′-deoxypyridoxal shifted 14 nm toward longer wavelengths while the analagous peak in pyridoxal phosphate remained at 380 nm. Also around physiological pH, the ratio of free to hydrated aldehyde was higher with pyridoxal-5′-phosphate than with 5′-deoxypyridoxal. The authors concluded that these differences reflected the inductive effect of the phosphate group. Models suggested that this effect could conceivably be exerted directly through the carbonyl group rather than indirectly through the ring.
   b. See Table 2 for proton magnetic resonance.
   c. See Scheme 3 for mass fragmentation.

### Table 11
### CALCULATED VALUES OF THE ELECTRON EXCITATION ENERGIES (eV) FOR THE FIRST TWO ABSORPTION BANDS OF VARIOUS 3-HYDROXYPYRIDINES

| | Cation | | Bipolar ion | | Neutral molecule | | Anion | |
|---|---|---|---|---|---|---|---|---|
| Compound | $E_1$ | $E_2$ | $E_1$ | $E_2$ | $E_1$ | $E_2$ | $E_1$ | $E_2$ |
| 3-Hydroxypyridine | 4.406 | 5.442 | 3.904 | 4.866 | 4.589 | 5.679 | 4.223 | 5.185 |
| 3-Hydroxypyridine-4-aldehyde hydrate | 4.304 | 5.527 | 3.815 | 5.004 | 4.509 | 5.728 | 4.158 | 5.316 |
| 3-Hydroxypyridine-4-aldehyde | 3.787 | 5.044 | 3.217 | 4.806 | 3.591 | 4.839 | 3.218 | 5.006 |
| Pyridoxine | 4.284 | 5.186 | 3.801 | 4.743 | 4.489 | 5.512 | 4.030 | 5.037 |
| Pyridoxamine | 4.235 | 5.237 | 3.765 | 4.824 | 4.443 | 5.534 | 4.010 | 5.101 |
| Pyridoxal phosphate hydrate | 4.240 | 5.373 | 3.782 | 4.894 | 4.438 | 5.578 | 4.067 | 5.151 |
| Pyridoxal phosphate | 3.715 | 4.789 | 3.180 | 4.709 | 3.516 | 4.712 | 3.190 | 4.846 |
| 5-Deoxypyridoxal hydrate | 4.258 | 5.378 | 3.837 | 4.951 | 4.454 | 5.593 | 4.127 | 5.202 |
| 5-Deoxypyridoxal | 3.742 | 4.707 | 3.240 | 4.725 | 3.511 | 4.623 | 3.236 | 4.838 |

From the English translation of *Mol. Biol. (U.S.S.R.)*, 7, 674, 1973. With permission. Copyright© Plenum Press, New York.

### Table 12
### CHANGES IN $E_1$ AND $E_2$ FOR 3-HYDROXYPYRIDINES ON INTRODUCING VARIOUS SUBSTITUENTS INTO THE MODECULE

| | Cation | | Bipolar ion | | Neutral | | Anion | |
|---|---|---|---|---|---|---|---|---|
| Substituent | $E_1$ | $E_2$ | $E_1$ | $E_2$ | $E_1$ | $E_2$ | $E_1$ | $E_2$ |
| 2-$CH_3$ | −0.054 | −0.046 | −0.023 | −0.042 | −0.052 | −0.038 | −0.052 | −0.036 |
| 4-$CH_2OH$ | −0.046 | −0.165 | −0.055 | −0.006 | −0.075 | −0.058 | −0.093 | +0.038 |
| 4-$CH_2N^+H_3$ | −0.096 | −0.082 | −0.094 | +0.071 | −0.113 | −0.007 | −0.126 | +0.110 |
| 4-$CH(OH)_2$ | −0.102 | +0.085 | −0.089 | +0.138 | −0.080 | +0.049 | −0.035 | +0.131 |
| 4-CHO | −0.619 | −0.398 | −0.687 | −0.060 | −0.998 | −0.840 | −1.005 | −0.179 |
| 5-$CH_3$ | +0.009 | −0.090 | +0.053 | −0.018 | −0.024 | −0.033 | +0.010 | −0.044 |
| 5-$CH_2OH$ | −0.009 | −0.079 | +0.007 | −0.076 | −0.047 | −0.044 | −0.055 | −0.091 |
| 6-$CH_3$ | −0.068 | +0.010 | −0.034 | −0.032 | −0.076 | −0.085 | −0.049 | −0.194 |

From the English translation of *Mol. Biol (U.S.S.R.)*, 7, 674, 1973. With permission. Copyright© Plenum Press, New York.

## IV. 3-DEOXYPYRIDOXINE

### A. Syntheses

Ichiba et al.[213] synthesized 3-deoxypyridoxine by treating 3-cyano-4-(ethoxymethyl)-6-methyl-2-pyridone with phosphorus pentachloride followed by catalytic hydrogenation to produce 2-methyl-4-(ethoxymethyl)-5-(aminomethyl) pyridine. Diazotization led to the 5-hydroxymethyl derivative which in turn was converted to the bis-(bromomethyl) compound by boiling with 40% hydrobromic acid. This was converted to 3-deoxypyridoxine by boiling in water and precipitating the bromide ions with silver.

Jones and Kornfeld[230] obtained 3-deoxypyridoxine from 3-methylisoquinoline. A hot solution of 1422 g (9 m) of potassium permanganate dissolved in the minimum

SCHEME 3 Fragmentation of 5′-deoxypyridoxal[a]

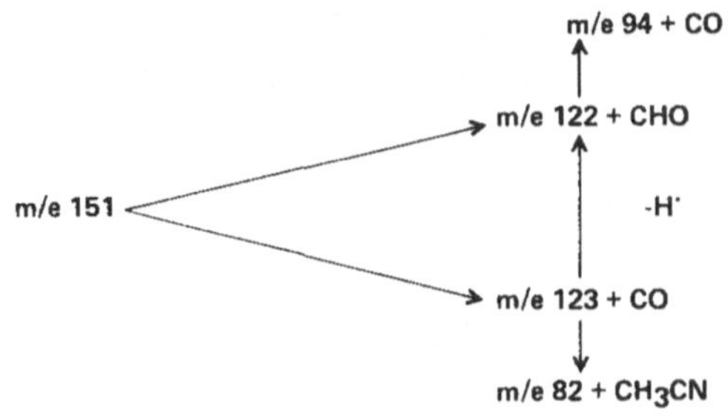


[a] Adapted from Iwata and Metzler[219]

amount of hot water was added with stirring over two hours to 215 g (1.5 moles) of 3-methylisoquinoline in 4 ℓ of water at 90°C. The manganese dioxide was filtered off and washed with 1ℓ of hot water. The filtrate and washings were evaporated to dryness *in vacuo* and the residue was extracted with three 500 mℓ portions of hot ethanol. Evaporation of the ethanol left a red gum which was again dissolved in hot ethanol and again evaporated to dryness *in vacuo*. The residue was dissolved in 2 ℓ of absolute ethanol, saturated with hydrogen chloride, and allowed to stand for 4 days. It was then concentrated *in vacuo* to about 500 mℓ, poured into an excess of sodium carbonate solution, and extracted with ether. The ether was evaporated and the residual liquid distilled at 141 to 142°C (1 torr) yielding about 150 g of a mixture of diethyl phthalate and diethyl 2-methyl-4,5-pyridinedicarboxylate. The mixture was resolved by dissolving the compounds in ether and precipitating the pyridine compound with dry hydrogen chloride yielding 57 g (11%) of the desired product. The diester (41 g) was then reduced to 3-deoxypyridoxine by reduction with 11 g lithium aluminum hydride in ether. Distillation at 170 to 190 °C (7 torr) yielded 16 g (60%) of the free base of 3-deoxypyridoxine which crystallized on standing. It was recrystallized from ethyl acetate (M.P. 95 to 97°C). While we have no experience with this specific compound, we have found with some other pyridine carboxylic acids that lithium aluminum hydride reduction of the free acid sometimes gives better yields than reduction of the esters. We also found that significant amounts of product can be lost in the precipitate in the reaction mixture if it is not thoroughly washed.

Tani[495] started with 2-methyl-4-pyridinecarboxylic acid and built up dimethyl-2-methyl-3,4-pyridinedicarboxylate which was then reduced with lithium aluminum hydride. Harris et al.[188] discovered 3-deoxypyridoxine was formed in significant amounts as a by-product in the synthesis of pyridoxine from 2-methyl-3-amino-4-methoxymethyl-5-aminomethylpyridine using the approach described by Harris and Folkers.[186] The percentage yield of the 3-deoxy compounds is not specified. About 13 g of 3-deoxypyridoxine hydrochloride was obtained from 600 g of crude residue from pyridoxine hydrochloride mother liquor. Therefore, while this report provides further evidence of the complex chemistry of this group of compounds, the procedure described does not appear to be an efficient means of synthesizing 3-deoxypyridoxine. Melius and Marshall[312] prepared 3-cyano-4-ethoxycarbonyl-6-methyl-2-pyridine according to Wuest et al.[555] It was then chlorinated with phosphorus pentachloride and hydrogenated to give 2-methyl-4-ethoxycarbonyl-5-cyanopyridine. Saponification in 15% sodium hydroxide followed by acidification yielded 2-methylpyridine-4,5-dicarboxylic acid which was reduced to 3-deoxypyridoxine with sodium borohydride. No yield was given.

We have had no direct experience with any of these syntheses for 3-deoxypyridoxine. The route used by Jones and Kornfeld is appealing in spite of its low yield because it involves only two steps. The starting material, 3-methylisoquinoline, used to be readily available and inexpensive. Unfortunately, it is now stocked by only a few suppliers, and its cost is currently about $2 per gram. After evaluating the various possibilities, it was our conclusion that the most inexpensive route would be to condense acetone and diethyloxalate to form ethyl acetopyruvate,[299] which could then be condensed with cyanoacetamide,[29,216] hydrolyzed with 50% sulfuric acid to 2-methyl-6-pyridone-4,5-dicarboxylate,[71] chlorinated with thionyl chloride, hydrogenated to 2-methyl-pyridine-4,5-dicarboxylate, and reduced to 3-deoxypyridoxine. Sodium borohydride has been used.[312] It seems likely that lithium aluminum hydride would also be useful. Under the proper conditions,[216,408] the 2-chloro-group can be replaced without reducing the nitrile group to an amine. However, the need for such selectivity in the hydrogenation step could be eliminated by hydrolyzing the nitrile to a carboxyl group at an earlier stage as suggested. We are currently testing this proposed route.

## B. Physical, Chemical, and Spectral Characteristics

1. Structure:

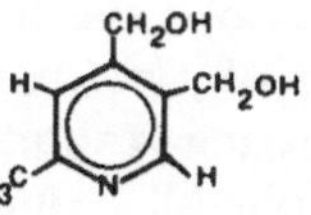


2. Empirical formula: $C_8H_{11}NO_2$
3. Molecular weight:
   a. Free base: 153.18
   b. Hydrochloride: 189.64
4. Percentage composition:

| | Free base | Hydrochloride |
|---|---|---|
| Carbon | 62.73 | 50.66 |
| Hydrogen | 7.24 | 6.38 |
| Nitrogen | 9.14 | 7.39 |
| Oxygen | 20.89 | 16.87 |
| Chloride | — | 18.70 |

5. Systematic name: 6-methyl-3,4-pyridinedimethanol
6. Melting point:
   a. Free base: 99—103° Tani[495]
      95—97° Jones and Kornfeld[230]
   b. Hydrochloride: 202—203° Ichiba et al.[213]
      206—208° Harris et al.[188]
      210—212° Melius and Marshall[312]

7. Spectra:
   a. Ultraviolet: See Table 9
   b. Proton magnetic resonance: See Table 2.
   c. Mass fragmentation: See scheme 4. The largest peak (135) is due to the elimination of water.

SCHEME 4 Fragmentation of 3-deoxypyridoxine[a]

m/e 105 (5%) + $H_2$ ← m/e 107 (25%) + CO → m/e 106 (28%) + H˙

m/e 153 (5%)[b] → m/e 135 (100%) + $H_2O$ (→ m/e 107 (25%) + CO)

m/e 135 (100%) + $H_2O$
↓
m/e 134 (60%) + H˙
↓
m/e 106 (28%) + CO
↓
m/e 79 (25%) + HCN
↓
m/e 77 (25%) + $H_2$

[a] Adapted from DeJongh et al[112]

[b] Values in parentheses are the approximate relative intensities of the peaks.

## V. SPECIAL CHEMICAL PROPERTIES OF THE PYRIDOXINE GROUP

The arrangement of the substituents on the pyridine ring in the $B_6$ compounds imparts some striking characteristics to these molecules including the deoxy-derivatives. For example, while the pK values for pyridine and phenol are about 5.25 and 9.89, respectively, comparison of the ultraviolet spectra of various derivatives has led to the conclusion that in aqueous solutions of pyridoxine compounds, the phenol group has a pK around 5 and the pK of the nitrogen is about 9.[187,227,282,315] This is about a 10,000-fold increase in the acidity of the phenol group compared with phenol itself and a comparable increase in the basicity of the nitrogen compared with pyridine. These analyses also led to the conclusion that at pH 6.8, essentially all of the pyridoxine exists in the zwitterion form with a protonated nitrogen and ionized phenol groups. Metzler, Johnson et al.[227,315] provided a detailed discussion of the techniques of ultraviolet spectral analysis as related to tautomeric equilibria in $B_6$ compounds including some deoxy derivatives. They noted that replacement of the hydroxy group in the 4′- or 5′-position with hydrogen increased $pK_1$ (4.94, 5.36, and 5.54 for pyridoxine, 4′-deoxypyridoxine and 5′-deoxypyridoxine, respectively). Bazhulina et al.[31] presented similar spectral data and used computer analysis to develop individual spectra and equilibrium curves for the various forms of 5′-deoxypyridoxal. Martell[296] presented a similar discussion based on infrared spectroscopy. The 4′-position is very activated. Several examples of the ready loss of substituents from this position have already been mentioned. Neutral aqueous solutions of pyridoxine polymerized at 120°C, but 4′-deoxypyridoxine, which has no good leaving group in the 4′-position, did not polymerize under these conditions.[185] The phenol and hydroxymethyl adjacent to the 4-position provide opportunity for hydrogen bonding, hydride transfer, and quinone methide interactions. The importance of the phenol group is illustrated by the observation that the 4′-methoxy group of 2-methyl-3-chloro-4-methoxymethyl-5-hydroxymethylpyridine is stable to conditions which would hydrolyze the 3-hydroxy derivative.

Chapter 2

# MICROORGANISMS, PLANTS, AND LOWER ANIMALS

Most of the studies to be discussed in this section were simply gross tests for growth inhibition. However, they provide some perspective on the breadth of organisms affected by deoxypyridoxine and in a few cases permit some generalizations.

## I. MICROORGANISMS

The earliest work with deoxypyridoxine involved testing for possible $B_6$ activity. Moeller[330-333] found that compared with pyridoxine, deoxypyridoxine required a 50-fold higher concentration (approximately 50 μg/mℓ) and even then produced only 30 to 40% as much growth in *Streptobacterium plantarum* type 10S. Since Moeller did not specify the synthetic route used, it is impossible to determine the probability of pyridoxine contamination; 2,4,5-trimethyl-3-hydroxypyridine had no growth-promoting effect.

In other studies of deoxypyridoxine as a growth factor, Bohonos et al.[61] reported that taking pyridoxine as 100% deoxypyridoxine was only 3% as effective in promoting the growth of *Lactobacillus casei.* Again, 2,4,5-trimethyl-3-hydroxypyridine had no effect. Robbins and Ma[413] found that deoxypyridoxine supported only about 0.2% as much growth as an equivalent amount of pyridoxine in *Ceratostomella ulmi.* Similarly, Snell[464] reported that deoxypyridoxine was only 5.8, 8.2, and 5.5% as effective as pyridoxine in *Streptococcus faecalis, Lactobacillus casei,* and *Saccharomyces cerevisiae,* respectively.

Under conditions in which pyridoxine synthesis is inadequate to meet growth requirements, 50% inhibition of *Saccharomyces carlsbergensis* was obtained with a deoxypyridoxine concentration of about 1 μg/mℓ.[401] When adequate $B_6$ was synthesized within the cell, the amount of deoxypyridoxine required to produce 50% inhibition of growth increased to about 170 μg/mℓ. Pyridoxine, with an inhibitor:$B_6$ ratio of 113 to 270, was the form of vitamin $B_6$ most effective in reversing deoxypyridoxine inhibition. The ratios for pyridoxal and pyridoxamine ranged from 16 to 47. For all forms of $B_6$, the ratio declined slightly as the $B_6$ concentration was increased from 16.7 to 167 ng/mℓ. The authors felt the decline was insignificant and that the data were consistent with a simple competitive mechanism. However, in four of the six cases, the change in inhibition ratio was greater than 25%. This suggests to us that factors in addition to simple competition were involved. Perhaps the $B_6$ vitamers themselves exert some toxicity at high concentrations of deoxypyridoxine. The authors commented that their results failed to confirm studies which detected slight vitamin activity in deoxypyridoxine. However, their data consistently showed growth stimulation at low concentrations of deoxpyridoxine. In a subsequent paper, Rabinowitz and Snell[402] acknowledged this stimulation, but noted that it is not unusual for toxic substances to be stimulatory at low doses. They reported that 1 μg deoxypyridoxine per milliliter was the lowest concentration which begins to depress growth. Chiao and Peterson[84] also noted that low levels (not further specified) of deoxypyridoxine had a stimulatory effect on *Saccharomyces carlsbergensis* but reported no effect at higher levels (again unspecified).

Rabinowitz and Snell[402] found that 5′-deoxy derivatives were not as effective as 4′-deoxypyridoxine in inhibiting *Saccharomyces carlsbergensis* (Table 1). Tests in combination with two other B antagonists suggested that 4′-deoxypyridoxine and omega-methyl pyridoxine (6-ethyl-5-hydroxy-3,4-pyridinedimethanol) act at the same site; whereas, 5′-deoxypyridoxal acts at a different site.[402] Omega-methyl pyridoxine was

Table 1
COMPARATIVE INHIBITORY POTENCIES OF 5′-DEOXY COMPOUNDS FOR *SACCHAROMYCES CARLSBERGENSIS*[402]

| Inhibitor | Reversing agent (35 n*M*) Pyridoxal | Pyridoxamine | Pyridoxine |
|---|---|---|---|
| 5′-Deoxypyridoxal | 440[a] | 190 | 1,400 |
| 5′-Deoxypyridoxamine | 12,000 | 1,800 | 35,000 |
| 5′-Deoxypyridoxine | 430 | 120 | 6,500 |

[a] Inhibitor: $B_6$ ratio causing 50% growth depression.

Table 2
EFFECTIVENESS OF $B_6$ ANALOGS IN INHIBITING PYRIDOXINE UPTAKE BY *SACCHAROMYCES CARLSBERGENSIS*[447]

| Compound | $K_I$ $\mu M$ |
|---|---|
| 3-Deoxypyridoxine | 2.9 |
| 4′-Deoxypyridoxine | 1.1 |
| 5′-Deoxypyridoxine | 0.18 |
| 5′-Deoxypyridoxal | 0.61 |

about twice as effective as 4′-deoxypyridoxine, which in turn was 8- to 10-fold more effective than 5′-deoxypyridoxal. It is interesting that although Rabinowitz and Snell[402] found 5′-deoxypyridoxine to be 50-fold less effective than 4′-deoxypyridoxine as a pyridoxine antagonist as measured by growth inhibition, Shane and Snell[447] reported that 5′-deoxypyridoxine was about 6-fold more effective than 4′-deoxypyridoxine as an inhibitor of pyridoxine uptake by *Saccharomyces carlsbergensis* (Table 2). Korytnyk et al.[246] also concluded that some 5-homologs of pyridoxine may exert their effect in yeast more through interference with $B_6$ uptake than by enzyme inhibition.

The importance of the phenol group in the actions of 4′-deoxypyridoxine is emphasized by the fact that 3-deoxypyridoxine acts through a different mechanism.[429] Rabinowitz and Snell[401] suggested that 4′-deoxypyridoxine may act by preventing the uptake of pyridoxine from the medium. However, simultaneous addition of both 4′- and 5′-deoxypyridoxines fails to show any additive effect, suggesting that the two compounds act by different routes.[402] This observation combined with the data reported above revealing that 5′-deoxypyridoxine is more effective than 4′-deoxypyridoxine at inhibiting pyridoxine uptake but less effective at inhibiting growth,[447] indicates that the mechanism of action of 4′-deoxypyridoxine is more complex than simple inhibition of uptake.

Sakuragi and Kummerow[432] found that in the presence of 4 ng pyridoxine hydrochloride/m$\ell$, about 0.6 $\mu$g deoxypyridoxine hydrochloride per milliliter produced 50% inhibition. This is a ratio of 150:1 and corresponds with other reports given earlier. However, in these studies, 5′-deoxypyridoxine was 50% as effective as 4′-deoxypyri-

## Table 3
## MOLAR RATIO OF ANTAGONIST NEEDED TO PRODUCE 50% INHIBITION OF *SACCHAROMYCES CARLSBERGENSIS*[429]

| | Pyridoxine (20 n*M*) | Pyridoxal (20 n*M*) | Pyridoxamine (20 n*M*) | Pyridoxine 4′,5′-diacetate (25 n*M*) |
|---|---|---|---|---|
| 4′-Deoxypyridoxine | 110 | 45 | 30 | 55 |
| 4′-Deoxypyridoxine 5′-Monoacetate | 30,000 | 2,600 | 2,600 | 3,600 |
| 3-Deoxypyridoxine | 85 | 7 | 10 | 65 |
| 3-Deoxypyridoxine 4,5 Diacetate | >100,000 | 70,000 | 70,000 | 20,000 |
| 5-Deoxypyridoxine | 250 | 15 | 30 | 20 |
| 4′,5′-bis-Deoxypyridoxine | 900 | 70 | 55 | 35 |

doxine, whereas Rabinowitz and Snell[402] found a 50-fold difference. This may reflect differences in the purity of the deoxypyridoxine preparations.

Sakuragi and Kummerow[432] also found that combining 4′-deoxypyridoxine with isonicotinic acid hydrazide or D-4-amino-3-isoxazolidone enhanced the inhibition of *Saccharomyces carlsbergensis* suggesting that these three antagonists may act by a similar mechanism. 3-Deoxypyridoxine, 5′-deoxypyridoxine, 5′-deoxypyridoxal, 4′,5′-bis-deoxy-pyridoxine, 4′,5′-dithiopyridoxine, pyramin, 5′-deoxypyramin, 5′-ethoxypyramin, and 2-methyl-thiopyramin were additive with each other but not with 4′-deoxypyridoxine, suggesting that 4′-deoxypyridoxine acts by a fairly unique mechanism). It was also at least twice as effective as any of the other compounds listed. Since 4′,5′-bis-deoxypyridoxine was less effective than 4′-deoxypyridoxine and also did not have an additive effect, the authors concluded that the 5′-hydroxy group was involved in the action of 4′-deoxypyridoxine. Similarly, further studies[429] showed that 4′-deoxypyridoxine-5′-acetate was 60- to 300-fold less effective than 4′-deoxypyridoxine alone (Table 3) and in a mixture of the two, the effects were not additive. However, the authors left open the possibility that some of the effect of the acetate could be due to hydrolysis back to 4′-deoxypyridoxine.

Isonicotinic acid hydrazide acts by reacting with the aldehyde group of pyridoxal-5′-phosphate. These workers[429,432] therefore predicted that isonicotinic acid hydrazide should interfere with the effects of antagonists which rely on the 4-aldehyde group. Such interference was noted for 4-nitrosalicylaldehyde[432] and 5′-deoxypyridoxal.[429] However, the effects of isonicotinic acid hydrazide were additive with those of 3- and 5′-deoxypyridoxines. Therefore, the inhibitory action of these two compounds apparently does not depend on the presence of an aldehyde group in the 4′-position.

Korytnyk et al.,[253] found that with a medium containing 1 ng pyridoxine per milliliter 4′-deoxypyridoxine produced 50% inhibition of *Saccharomyces carlsbergensis* at $5 \times 10^{-7}$ *M* (95 ng/m*l*). Derivatives in which one, two, or three methylene groups were inserted between the ring and the 5′-hydroxymethyl group of pyridoxine were about 10-fold more inhibitory than 4′-deoxypyridoxine.

Tuszynska[519] noted that inhibition of *Saccharomyces carlsbergensis* in a medium containing 0.25 to 2.5 mg pyridoxine per milliliter was detectable with a deoxypyridoxine:pyridoxine ratio of 100:1 and was complete with a ratio of 800:1. This appears to be compatible with the ratio of 160:1 for 50% inhibition in a medium containing 167 ng pyridoxine per milliliter (Rabinowitz and Snell).[401]

Ozawa et al,[378,379] reported a 20 to 40% decline in the growth of *Sacchaomyces carlsbergensis* in media containing $0.8 \times 10^{-6}$ or $0.8 \times 10^{-7}$ *M* pyridoxine at concentrations of $0.8 \times 10^{-10}$ *M* deoxypyridoxine. This level of inhibition remained constant with increasing amounts of deoxypyridoxine up to concentrations of about $0.8 \times 10^{-6}$ *M*. About 70% inhibition was achieved at $0.8 \times 10^{-5}$ *M* and total inhibition at $0.8 \times 10^{-4}$ *M;* 3-aminopyridoxine and 3-aminopyridoxal produced similar effects.[379] Except for the initial inhibition at very low concentrations these results are in general agreement with the earlier discussion. The initial decline may result from some specific characteristic of their system.

Wagner-Jauregg and Fischer[536] reported that 2 to 3.5 μg 4′-deoxypyridoxine per milliliter produced 50% inhibition of the growth of *Saccharomyces carlsbergensis.* The $B_6$ content of the medium was not specified.

Kawakita et al.,[236] reported 50% growth inhibition in *Saccharomyces carlsbergensis* with a deoxypyridoxine:pyridoxine ratio of 8:4. Their later observation[235] that $10^{-4}$ *M* deoxypyridoxine caused only 20% inhibition under conditions in which the yeast synthesizes $B_6$ agrees with the observation of Rabinowitz and Snell[401] that about $10^{-3}$ *M* was required to produce 50% inhibition under such conditions.

Deoxypyridoxine partly offset the inhibitory effects of 10 ng thiamine per milliliter, but had little effect against larger amounts of thiamine.[84] The authors noted that although the inhibitory action of thiamine could be offset by pyridoxal, thiamine treatment did not alter the $B_6$ content of the yeast. Thus, they concluded that thiamine interfered with the utilization of vitamin $B_6$ rather than its production.

Some authors failed to find an inhibitory effect. Yamada and Tsuji[558] found that 20 ng of deoxypyridoxine had no effect in the presence of 2 ng of pyridoxine. This is compatible with the reports discussed above where more than a tenfold excess of deoxypyridoxine was needed to produce marked growth inhibition. It is more difficult to explain the results of Struck et al.,[483] who found no inhibition from up to 10 μg of deoxypyridoxine or its phosphonate derivative in the presence of 0.2 to 4 ng pyridoxine per milliliter. The lack of inhibition under these conditions, which apparently involve as much as a 10,000-fold excess of deoxypyridoxine, suggests a possible error in their procedure. The possibility of such an error suggests that it might also be wise to reexamine the effect of the phosphonate (4′-deoxyisopyridoxal did show inhibition in these studies).

Deoxypyridoxine has also been tested in a variety of other organisms, although not as thoroughly as in *Saccharomyces carlsbergensis.* Rabinowitz and Snell[401] noted that none of the four organisms tested (*Clostridium butyricum, Escherichia coli* Y109, *Lactobacillus arabinosus, Torula cremoris* 2512) which grow in the absence of an external source of vitamin $B_6$ were inhibited by deoxypyridoxine (167 μg/mℓ). Several organisms *(Lactobacillus casei, Lactobacillus helveticus, Streptococcus faecalis)* which required external $B_6$ were also unaffected at that concentration. In contrast to the situation in *Saccharomyces carlsbergensis,* where 4′-deoxypyridoxine was much more effective than 5′-deoxypyridoxine, in *Streptococcus faecalis* both 4′- and 5′-deoxypyridoxine were relatively poor inhibitors, possibly because pyridoxine itself cannot serve as a source of vitamin $B_6$ for this organism[402] (Table 4). Deoxypyridoxine and deoxypyridoxine phosphate had no effect at ratios to pyridoxamine of $2.4 \times 10^5$ and $4.8 \times 10^4$, respectively.[369] 5′-Deoxypyridoxamine was the most effective inhibitor.[402]

Under the conditions used, *Streptococcus faecalis* required $B_6$ only for the synthesis of D-alanine. When D-alanine was supplied, $B_6$ was no longer needed and 5′-deoxypyridoxal and 5′-deoxypyridoxamine no longer inhibited growth.[402] 5′-Deoxypyridoxine did inhibit pyridoxamine uptake by *Streptococcus faecalis* ($K_i$ = 0.3 μ*M*) but was not accumulated in the cells.[350] Mora and Snell[335] found that 5′-deoxypyridoxal at low concentrations ($10^{-4}$ *M*) tended to increase glycine accumulation by protoplasts from *Streptococcus faecalis* about 30% but caused 80% inhibition at $10^{-3}$*M.* With whole cells even at $10^{-4}$ *M* about 10% inhibition was noted. The effect appeared to be com-

Table 4
COMPARATIVE POTENCIES OF VITAMIN $B_6$ ANTAGONISTS FOR *STREPTOCOCCUS FAECALIS*[402]

| Inhibitor | Reversing agent: Pyridoxal (8.1 n*M*) | Reversing agent: Pyridoxamine (3.5 n*M*) |
|---|---|---|
| 5-Deoxypyridoxal | 2,700[a] | 48,000 |
| 5-Deoxypyridoxamine | 400 | 5,000 |
| 5-Deoxypyridoxine | 200,000 | 200,000 |

[a] Molar ratio of inhibitor: $B_6$ needed to produce 50% inhibition.

petitive, and the authors suggested that 5′-deoxypyridoxal and glycine formed a complex which was poorly absorbed (5′-deoxypyridoxine had no effect at the highest concentrations tested ($10^{-2}$ *M*).

In the case of *Lactobacillus helveticus,* which requires pyridoxamine phosphate, 5-deoxypyridoxal but not 5′-deoxypyridoxamine or 5′-deoxypyridoxine were inhibitors.[402] Wagner-Jauregg and Fischer[536] also found the growth of *Lactobacillus casei* and *Streptococcus faecalis* unaffected at deoxypyridoxine concentrations up to 100 μg and 1 ng/mℓ, respectively.

Rabinowitz and Snell[401] noted the growth of *Saccharomyces cervesiae* in the presence of 16.7 ng pyridoxine per milliliter was reduced by 98% by 167 μg deoxypyridoxine per milliliter (a 10,000:1 ratio). Loveless et al.[280] obtained a 50% growth reduction in the presence of 1 μg pyridoxine per milliliter with 2 mg deoxypyridoxine per milliliter (a 2000:1 ratio). Martin et al.,[297] tested several analogs against *Saccharomyces cerevisiae.* 2-Acetoxy-3,5-diacetoxymethyltoluene and 2-methyl-3-hydroxy-4-dimethylaminomethylpyridine were inactive. 2-Methyl-3-hydroxy-4-hydroxymethylpyridine was active with an inhibitor:pyridoxine ratio of 250. 2-Ethyl-3-amino-4-ethoxymethyl-5-aminomethylpyridine was a more effective antagonist than deoxypyridoxine.

In contrast to results in *Saccharomyces carlsbergensis* where 5′-deoxypyridoxine was the most potent inhibitor of pyridoxine uptake tested,[447] in *Salmonella typhimurium* LT2, 5′-deoxypyridoxine had no effect on pyridoxine uptake at the concentrations tested (up to 10 μ*M* in the presence of 0.1 μ*M* pyridoxine).[349] 5′-Deoxypyridoxal did inhibit pyridoxine uptake. The authors concluded that the inhibition was due to binding with the transport protein. They also concluded that $B_6$ transport in *Salmonella typhimurium* involved facilitated diffusion, and differed considerably from the active transport mechanism observed in *Saccharomyces carlsbergensis.*

Pflug and Lingens[389] produced a series of $B_6$-deficient mutants of *Bacillus subtilis.* Pyridoxal was the best growth substance for these mutants producing maximum growth at a concentration of 60 to 80 ng/mℓ. Pyridoxine and pyridoxamine were less effective, requiring concentrations of 60 to 100 μg/mℓ for maximum growth. Surprisingly, deoxypyridoxine was almost as effective, requiring 80 to 120 μg/mℓ. While the data show that deoxypyridoxine is not inhibitory at these concentrations, further work would be needed to insure that the growth was the result of deoxypyridoxine itself rather than an impurity. The amounts of pyridoxine, pyridoxamine, and deoxypyridoxine required for maximum growth are so much larger than the amount of pyridoxal, that only 0.1% contamination with pyridoxal would be required to account for the response. Based on the identification of deoxypyridoxine-5′-phosphate in the medium and the ability of isoleucine to substitute for vitamin $B_6$, the authors proposed a scheme for the biosynthesis of $B_6$ which included 4′-deoxypyridoxine as an intermediate. Apparently, the authors did not consider the possibility that rather than participating directly in $B_6$ synthesis, isoleucine might have a sparing effect analogous to the situation in *Streptococcus faecalis* where $B_6$ was required only for alanine synthesis and could be omitted if alanine was supplied.[402] Further study of this interesting group of mutants should help to clarify the pathways involved.

Rabinowitz and Snell[401] found complete inhibition of *Neurospora sitophila* 209 in the presence of 100 ng pyridoxine per milliliter by 100 μg deoxypyridoxine per milliliter. Wagner-Jauregg and Fischer[536] obtained 50% inhibition of *Neurospora sitophila* with 100 μg deoxypyridoxine/mℓ in a medium of unspecified $B_6$ content. Agolini and Caliari[4] also examined the effect of deoxypyridoxine in *Neurospora sitophila*, but we have not yet seen the original article.

*Neurospora crassa* was not affected at deoxypyridoxine concentrations up to 1 mg/mℓ.[536] Carsiotis and Suskind[78] demonstrated that deoxypyridoxine phosphate at about $10^{-4}$ *M* stimulated the aldolytic activity of tryptophan synthetase from *Neurospora crassa* to about the same degree as $10^{-6}$ *M* pyridoxal phosphate. Since most deoxypyridoxine is prepared from pyridoxine, results such as these where there is a 100-fold difference in effectiveness raise the question of whether the activity of the deoxypyridoxine phosphate should be attributed to deoxypyridoxine phosphate itself or to traces of pyridoxine or pyridoxine phosphate which was also effective in stimulating tryptophan synthetase. Neither deoxypyridoxine alone nor pyridoxamine phosphate had any effect, suggesting that the aldehyde group could be replaced by methyl or hydroxymethyl but not by an aminomethyl group.

In studies in thiamineless mutants of *Neurospora*, Harris[183] observed that both pyridoxine and 4′-deoxypyridoxine inhibit thiamine biosynthesis. He concluded that since deoxypyridoxine shows no $B_6$ activity, the effect was due to direct competitive inhibition of pyrimidine utilization and not to any interaction with $B_6$ reactions. He even suggested a possible coupling of thiazole and pyridoxine compounds.

Wooley and Murphy[554] observed little effect of deoxypyridoxine concentrations up to 1 mg/mℓ on the growth of *Escherichia coli*. However, Jannes and Lahikainen[224] found that concentrations above 2 mg/mℓ caused detectable growth inhibition. During such inhibition, two strains showed a decrease in gamma-aminobutyric acid production. One of these strains showed a decrease in glutamic acid. The authors postulated that these changes were due to inhibition of asparatate aminotransferase. Wada and Morino[533] failed to detect much change in aspartate aminotransferase activity but did note about a threefold increase in apoenzyme when the medium contained 10 mg deoxypyridoxine per liter. In later studies, Jannes and Saris[225] reported that under aerobic conditions, anthranilic acid production was increased by deoxypyridoxine. The concentration is listed as 9.7 mg/mℓ in the text and 4.75 mg/mℓ in the summary. Under anaerobic conditions, deoxypyridoxine concentrations above 4.75 mg/mℓ increased production. The data indicated that pyridoxine used at a concentration of 0.25 *M* raised indole production even more. While the authors did not mention growth inhibition in this report, these concentrations of deoxypyridoxine were found to produce growth inhibition their previous report.[224] Jannes and Saris[225] noted that while the accumulation of anthranilic acid might be explained by a block in a subsequent metabolic step, under anaerobic conditions indole is the final product. Thus, we have a paradoxical situation in which growth is inhibited but the metabolism of trpytophan, which is heavily dependent on vitamin $B_6$, seems to be enhanced by both pyridoxine and deoxypyridoxine. Therefore, the authors concluded that the main effect of deoxypyridoxine may be to alter cell permeability to tryptophan and tryptophan metabolites rather than to inhibit tryptophan metabolism. Oya[377] also concluded that deoxypyridoxine exerted its main effect on absorption of $B_6$ into the cell rather than on the conversion of pyridoxine to pyridoxal phosphate. Loveless et al.[280] studied the effect of numerous compounds on *Escherichia coli*. Since he mentions using deoxypyridoxine with *Saccharomyces carlsbergensis*, it is not clear whether his failure to mention it in connection with *Escherichia coli* means it was not tested or whether it was included in the group of unspecified compounds which had no effect.

Scott and Hockney[444] have been investigating the properties of some *Escherichia coli* mutants. When the BL-1 strain was incubated without pyridoxal, pyridoxine was excreted into the medium. In excess pyridoxal (120 ng/m*l*), 50% inhibition was obtained at a deoxypyridoxine concentration of 6.7 mg/m*l*. When pyridoxal was limiting (6 ng/m*l*), the effect of deoxypyridoxine varied with concentration. At 1.5 mg deoxypyridoxine per milliliter growth was about 240% greater than controls. At about 75 μg/m*l* growth about equaled controls. Maximum inhibition (50% of controls) was observed at 5 μg/m*l*. In strain CR63, inhibition of growth was directly proportional to deoxypyridoxine concentration over the range 0 to 10 mg/m*l*. A concentration of 6 mg/m*l* produced 50% inhibition of growth and abolished the uptake of labeled pyridoxine. The failure to observe stimulation in this strain reduces the probability that the increased growth in the BL-1 strain was due to impurities in the deoxypyridoxine. The authors suggested that under certain conditions deoxypyridoxine might be converted to pyridoxal.

Gupta et al.[177] observed that deoxypyridoxine caused a slight stimulation (4 to 21%) of aflatoxin synthesis by *Aspergillus parasiticus* at low concentrations (0.01 m*M*) and slight inhibition (10 to 25%) at higher concentrations (5.0 m*M*). In view of the limited inhibition, the authors conclude pyridoxal phosphate reactions are not involved in aflatoxin synthesis. However, considering the lack of information about deoxypyridoxine metabolism in this species, we do not feel such a conclusion is strongly supported.

Morisawa et al.[340] noted that although other analogs were more potent antivitamins than 4′-deoxypyridoxine in *Saccharomyces carlsbergensis*, deoxypyridoxine was the only one which showed anticoccidial activity against *Eimeria acervulina*. They further state that the anticoccidial activity of the analogs is not proportional to their anti-$B_6$ activity. In a patent, Morisawa et al.[342] stated that feed containing 4′-deoxypyridoxine at 50 to 500 ppm provided protection against coccidiosis and raising the content to 1000 to 2000 ppm was curative. However, in a previous report,[340] they had mentioned that concentrations of 200 to 500 ppm depressed growth in chicks.

Gross et al.[175] examined the effect of deoxypyridoxine on the ability of preloading with glutamate to stimulate increased uptake of amino acids by *Streptomyces hydrogenans*. The authors assumed that deoxypyridoxine administration would inhibit pyridoxal phosphate dependent reactions of amino acids. Therefore, they concluded that since deoxypyridoxine simulated rather than inhibited transstimulation, intermediates of glutamate produced by transamination or decarboxylation could not be involved in transstimulation. However, no data verifying the assumed inhibition were reported, and no references documenting the effects of deoxypyridoxine in *Streptomyces* were cited. Therefore, their assumptions concerning the effects of deoxypyridoxine in *Streptomyces hydrogenans* may be invalid. They also suggested that transstimulation may result from improved coupling between energy metabolism and transport.

Reeves et al.[405] found that deoxypyridoxine hydrochloride concentrations of 0.17 to 2.5 μg/m*l* produced 50% inhibition of varios strains of *Entamoeba histolytica*. The strain differences were constant and reproducible. *Bacteroides symbiosus* was inhibited only slightly by concentrations of 0.05 to 0.5 mg/m*l*.

Cote et al.[104] in conjunction with studies of nicotinamide inhibitors reported that deoxypyridoxine at concentrations of 0.5 to 5 mg/m*l* neither stimulated nor inhibited *Lactobacillus arabinosus* under their experimental conditions. Deoxypyridoxine at 10 mg/mℓ caused 22% inhibition of the growth of *Flavobacterium devorans.*[390] Hiroaka[197] found that deoxypyridoxine failed to inhibit the growth of *Lactobacillaceae.* Gardner et al.[159] found no activity against tuberculosis in mice with a thiosemicarbazide derivative of deoxypyridoxine.

Lanigan and Anderson[265] found that *Desulfovibrio desulfuricans* was inhibited only 15 to 37% by 4′-deoxypyridoxine concentrations up to 500 μg/mℓ but was inhibited over 99% by concentrations of 550 μg or more per milliliter. They attributed this effect primarily to transaminase inhibition. However, they presented no data to support this assumption and were apparently unaware of the many other enzymes which might be affected by deoxypyridoxine.

Bradner and Clark[65] tested the effects of deoxypyridoxine, using filter paper discs, on anaerobic microorganisms as part of a search for a screening test for anticancer drugs. *Bacteroides pseudoramosus* was most susceptible to deoxypyridoxine inhibition showing inhibition at 10 μg deoxypyridoxine per disc. *Clostridium acidi-urici,* and *Clostridium fesseri* were somewhat susceptible showing inhibition at 100 μg/disc but not at 10. *Bacteroides limosus, Butryibacterium rettgeri, Clostridium cylindrosporum, Micrococcus aerogenes* and *Propionibacterium freudenreichii* were inhibited only at the highest concentration of deoxypyridoxine tested (1 mg/disc). *Propionibacterium shermanii* was not inhibited.

In an evaluation of the use of microbiological systems to screen for antitumor activity, Foley et al.[144] reported that more than 1 mg/mℓ was needed to obtain 50% growth inhibition in *Streptococcus faecalis, Escherichia coli, Lactobacillus arabinosus, Leuconostoc citrovorum, Lactobacillus fermenti, Candida albicans, Tetrahymena pyriformis, Glaucoma scintillans, Colpidium campylum* and *Neurospora crassa.* It is not clear whether any growth inhibition at all was noted in these organisms. Growth was reduced by 50% in *Lactobacillus casei* at concentrations of 0.1 to 1 mg/mℓ, in *Saccharomyces carlsbergensis* at concentrations of 0.01 to 0.1 mg/mℓ, and in human tissue cultures at 1 to 10 μg/mℓ. Apparently, no attempt was made to supply pyridoxine-free media.

Sinha and Chatterjee[458] found that a deoxypyridoxine concentration of 750 μg/mℓ was required to inhibit the growth of *Staphlococcus aureus* while only 100 μg/mℓ was required to inhibit a pyrithiamine mutant strain.

Kawakita et al.[236] reported 50% growth inhibition in *Leuconostoc mesenteroides* with a deoxypyridoxine:pyridoxal ratio of 53 and in *Kloeckela apiculata* with a deoxypyridoxine:pyridoxine ratio of 9200. About 50% reduction in the growth of *Vibrio cholera* was observed at 100 μg deoxypyridoxine per milliliter.[81a] Deoxypyridoxine also has some activity against viruses. Wooley and Murphy[554] found that deoxypyridoxine (0.5 mg/mℓ) produced 99% inhibition of multiplication of bacteriophage without affecting the growth of *Escherichia coli* in 5-hr cultures. The inhibition was reversed by as little as 0.01 mg pyridoxine per milliliter. It was also reversed by 2.5 mg glucose-6-phosphate or pyruvic acid per milliliter and by 5 mg formic, acetic, butyric, or valeric acid per milliliter. Glucose alone was not effective suggesting deoxypyridoxine might interfere with glucose utilization.

Cushing and Morgan[109] reported that 0.5 mg deoxypyridoxine per milliliter markedly inhibited the multiplication of mumps and influenza viruses in chick embryo cultures without producing obvious toxic effects on embryonic heart tissue as measured by contractility and fibroblastic proliferation during 4 days of culture. At a concentration of 0.1 mg/mℓ, there was no significant inhibition of viral growth. In a later study, Morgan[336] reported that 0.5 mg deoxypyridoxine per milliliter, but not 0.1 mg/mℓ, produced definite inhibition of psittacosis virus. However, in this case, he noted evidence of toxic effects on the host chick embryo tissue at 0.5 mg/mℓ. However, a later statement that pyridoxine analogs can be used to suppress growth of psittacosis virus "without lethal or permanent toxic effects on the host cells" suggests that the toxic effects were readily reversible. He did not comment on the fact that no toxic effects were noted in the first paper.

Bader and Morgan[26] found that incubation of HeLa and L cells with deoxypyridoxine (0.5 mg/mℓ) for 2 hr before and 24 hr after adding Newcastle disease virus, PR8 influenza virus or mumps virus to the cultures inhibited cytopathic effects except when Newcastle virus was added to HeLa cells. The authors stated that inhibition of cytopathic effects appeared to be correlated with reduced infectivity and production of hemagglutinins. However, the data indicate that in two of the four cases where deoxypyridoxine reduced cytopathic effects there was no decrease in hemagglutinin production. The infectivity also seemed to show relatively little change. The authors suggested that the analogs interfere with viral replication. Jaffe et al.[223] found no inhibitory effect on polio virus from 0.1 mg deoxypyridoxine/mℓ. This is in agreement with the four studies just discussed.

Treatment of mice receiving a normal $B_6$ intake and infected with *Toxoplasma gondii* with a diet containing 0.1% deoxypyridoxine increased survival time 2- to 3-fold (14 to 18 days after infection compared with 5 to 7 days in untreated controls).[488] Platzer et al.,[393] noted that unpublished results from their laboratory show deoxypyridoxine to be an effective antimalarial agent against *Plasmodium lophurae* in ducklings with an $ED_{50}$ of 5.2 mg/kg/day.

## II. PLANTS

We located only six in vivo studies of deoxypyridoxine in plants. In the earliest study, Sundaram et al.[489] demonstrated that deoxypyridoxine (13.3 μg/mℓ) could inhibit the synthesis of niacin and ascorbic acid in germinating pulses *(Phaseolus mungo, Vigna sinensis,* and *Cajanus cajan).* This effect could be overcome by pyridoxine. In addition, the inhibition of ascorbic acid synthesis was counteracted by mannose but not by glucose. This observation suggested a requirement for vitamin $B_6$ in the conversion of glucose to mannose.

As might be expected, there is significant interspecies variation in susceptibility to deoxypyridoxine. Molina and Axelrod[334] reported that 83 μg deoxypyridoxine per milliliter significantly inhibited root growth in *Lotus corniculatus;* whereas 4 mg/mℓ had no significant effect on the elongation of excised roots from *Phaseolus vulgaris.* Nodulation was slightly reduced.

Brucker[72] observed a rapid effect of deoxypyridoxine (50 μg/mℓ) on the root growth of pea seedlings (*Pisum sativum*). The growth depression remained strong for three days after removing the deoxypyridoxine and did not subside rapidly until an additional 4 days had passed. Exposure to deoxypyridoxine for three days, combined with a 100 R dose of radiation, produced slightly greater damage than would have been expected from the sum of the individual treatments. Pretreatment with deoxypyridoxine had no additive effect when the radiation dose was raised to 400 R.

Boll[62] found that both the absolute amount of deoxypyridoxine and the deoxypyridoxine:pyridoxine ratio are factors in determining the degree of inhibition of growth of tomato roots (Table 5). Three combinations gave better growth than any concentration of pyridoxine alone. In general, pyridoxine became relatively less effective in counteracting deoxypyridoxine as the deoxypyridoxine concentration increased. The effect of deoxypyridoxine on the growth of the laterals did not always agree with the data on the main axis. It is likely that a complex interrelationship exists between the growth of the two portions of the root.

Zavenyagina and Bukin[566] reported that 3-deoxypyridoxine (probably $2 \times 10^{-3}$ *M*) markedly inhibited the growth of roots and stems of pea and wheat seedlings. How-

Table 5
EFFECT OF DEOXYPYRIDOXINE AND PYRIDOXINE ON GROWTH OF THE MAIN AXIS OF TOMATO ROOTS[62]

| Deoxypyridoxine ($\mu M$) | Pyridoxine ($\mu M$) | Ratio | Approximate % change from control |
|---|---|---|---|
| 14.7 | 4.9 | 3:1 | 0 |
| 147 | 49 | 3:1 | +10 |
| 4.9 | 1.47 | 3.3:1 | +20 |
| 49 | 14.7 | 3.3:1 | +20 |
| 14.7 | 1.47 | 10:1 | 0 |
| 147 | 14.7 | 10:1 | −20 |
| 14.7 | 0.49 | 30:1 | −60 |
| 147 | 4.9 | 30:1 | −90 |
| 49 | 1.47 | 33.3:1 | −75 |
| 490 | 14.7 | 33.3:1 | −95 |

ever, no specific data were presented. Chlorosis was detected in the distal parts of the leaves. The activities of cytochrome c oxidase and peroxidase in the wheat leaves were not altered but the cytochrome b content of the roots was reduced to 75% of control values. Aspartate aminotransferase activity in the leaves was markedly increased by 3-deoxypyridoxine ($2 \times 10^{-3}$ $M$) and by pyridoxal ($10^{-4}$ $M$).

Finally, Black and SanPietro[54] reported that $B_6$ analogs such as 5′-deoxypyridoxal which have an aldehyde in the 4-position supported photosynthetic phosphorylation in spinach chloroplasts and reduction of exogenous cytochrome c. 4′-Deoxypyridoxine was not active within the concentrations tests ($10^{-2}$ and $10^{-5}$ $M$).

## III. INSECTS

Observation of deoxypyridoxine effects in insects has been restricted to growth and gross morphological studies. Sieburth and McLaren[455] found no decreased growth in the German cockroach *(Periplaneta americana)* when pyridoxine (16 $\mu$g/g diet) was replaced by an equivalent amount of deoxypyridoxine in a purified diet.

Schultz et al.[443] reported that the $LD_{50}$ at various stages of development of Drosophila was

1. $4 \times 10^{-2}$ $M$ for the first instar larvae
2. $4.7 \times 10^{-3}$ $M$ at puparium formation
3. $6.7 \times 10^{-3}$ $M$ at imaginal development

Developmental abnormalities noted were

1. Eyes — rough, seamed at dorsoventral junction, lobulated anterior margin
2. Wings — crumpled, vestigial or dumpy, with posterior crossbain branched
3. Legs — crippled, bent
4. Abdomen — incompletely chitinized, tergital hairs and bristles missing; occasional melanotic tumors, bloated abdomens and rotated genitalia

Shyamala and Bhat[454] showed that the growth of silkworms *(Bombyx mori)* was reduced about 40% when they were fed leaves coated with deoxypyridoxine (250 $\mu$g/leaf). Adding pyridoxine hydrochloride (12.5 $\mu$g/leaf) almost completely restored normal growth. Levinson and Bergmann[275] found that deoxypyridoxine concentrations exceeding 1 mg/g diet severely reduced growth and development of housefly larvae *(Musca vicina)*. The diet contained about 0.6 μg pyridoxine per gram. This amount appeared adequate to support normal development over a 72-hr period. The effects of deoxypyridoxine on growth at 1.5 mg/g diet were almost completely offset by an equivalent amount of pyridoxine. However, the percent of larvae pupating was still somewhat depressed. Pyridoxine itself appeared to be lethal at concentrations of 15 mg/g diet. The authors stated that lack of pyridoxine produced lethargic larvae which moved unusually slowly, lacked appetite, and suffered from heavy diarrhea. They invariably died in a typically paralyzed position. However, since they did not mention use of a $B_6$-deficient diet, these symptoms presumably resulted from deoxypyridoxine treatment rather than dietary lack of pyridoxine. Reproduction of adult flies was inhibited, and longevity was reduced by 0.3% deoxypyridoxine.

Blaustein and Schneiderman[56] found that 2 mg deoxypyridoxine per gram body weight killed 90% of the pupae of the giant silkmoths, *Callosamia promethea* and *Samia cynthia* within 3 days.

Akov and Guggenheim[9] studied $B_6$ interactions in mosquito larvae *(Aedes aegypti)*. As was noted earlier for housefly larvae and tomato roots, pyridoxine hydrochloride at high concentrations (200 μg/mℓ) was inhibitory to mosquito larvae. With a diet containing 0.2 μg pyridoxine hydrochloride per milliliter, a deoxypyridoxine hydrochloride concentration of 10 μg/mℓ produced significant developmental retardation and 100 μg/mℓ was 100% lethal. Raising the pyridoxine hydrochloride concentration to 20 μg/mℓ almost completely offset the effects of 20 μg deoxypyridoxine hydrochloride per milliliter. Whereas a 50-fold excess of deoxypyridoxine over pyridoxine was needed to produce 50% of the maximum inhibition with 0.2 μg pyridoxine hydrochloride per milliliter, only a 10-fold excess was needed with a pyridoxine hydrochloride concentration of 20 μg/mℓ. At 200 μg pyridoxine hydrochloride per milliliter, only a 3-fold excess was needed. Therefore, the authors concluded that the inhibition was reversible but not competitive. Deoxypyridoxine was more toxic than omega-methylpyridoxine or isonicotinic acid hydrazide.

Levinson et al.[274] noted that while pyridoxine deficiency inhibited the development of hide beetle *(Dermestes maculatus)* larvae, in the presence of 11 mg pyridoxine per kilogram diet deoxypyridoxine showed little effect at 1.1 g/kg diet and only 28% inhibition at 2.7 g/kg which is a 250:1 ratio between antagonist and vitamin. The authors commented that silkworms,[454] housefly larvae,[275] and mosquito larvae[9] were more sensitive than the hide beetle to deoxypyridoxine. However, it is questionable whether direct comparisons are valid because of differences in culture techniques and $B_6$ content of the diets.

Gothilf and Waites[172] found that feeding larvae of the Mexican bean beetle leaves dipped for 5 sec into a solution containing 0.1% deoxypyridoxine killed 92% of the larvae within 6 days. Concentrations of 0.0001 and 0.01% had little effect. A 1% solution killed 22% of the larvae by 3 days and 100% by 4 days. Combining 0.1% pyridoxine with 0.1% deoxypyridoxine nullified the inhibitory action of deoxypyridoxine.

## IV. OTHER ORGANISMS

Gustafson and Horstadius[178] observed that in the differentiation of sea urchin *(Paracentrotus lividus)* eggs and larvae, deoxypyridoxine enhanced vegetalizing rather than animalizing effects. Roberts and Mong[415] found that 400 μ*M* deoxypyridoxine inhibited the development of the tapeworm, *Hymenolepsis diminuta*.

Bieber and Hitchings[53] observed no effect on frog embryos when deoxypyridoxine was supplied at 1, 5, or 10 mg/dℓ but noted a toxic effect at 25 mg/dℓ. Yamashita[559] found that deoxypyridoxine administered at 600 to 1500 μg/g to frogs after removal of the nervous system anterior to the diencephalon failed to produce convulsions.

Pappas[382] studied various antimetabolites in the guppy *(Lebistes reticulatus)*. A natural diet of unknown pyridoxine content was used. At a concentration of 500 μg/ℓ deoxypyridoxine reduced survival. Concentrations of 25 and 250 μg/ℓ had little effect. Conflicting data on toxicity are given. His Table 3 notes that 21 days were required to produce 33% mortality while Table 4 states 10% survival at 7 days. Symptoms noted were ataxia, anorexia, rapid and gasping breathing, flexing of opercles, and no weight gain for 6 weeks. Deoxypyridoxine (500 μg/ℓ) produced greater inhibition of weight gain than any other antimetabolite which permitted similar survival. The effects were reversed by an equal concentration of pyridoxine. No other concentration was tested.

Addition of deoxypyridoxine (20 mg/kg) to a $B_6$-deficient diet reduced weight gains in turbot *(Scophthalmus maximus)* to about 66% of those in fish receiving the $B_6$-deficient diet alone.[3] Plasma protein was reduced ($p < 0.05$) about 20%. Packed cell volume in the deoxypyridoxine group was reduced to about 33% of the other groups. It seems likely that failure to indicate this as statistically significant was a misprint. A symbol for $p < 0.01$ is listed as a footnote but no value in the table is so designated. Liver aspartate aminotransferase in the deoxypyridoxine group was about 50% of the activity found with the $B_6$-deficient diet alone. The activity of aspartate aminotransferase in muscle and alanine aminotransferase in liver was scarcely different from activity in the $B_6$-deficient controls. However, the amount of stimulation of aspartate aminotransferase activity from liver and muscle observed in the deoxypyridoxine group after adding pyridoxal phosphate was only about 50% compared with 120 to 250% in the $B_6$-deficient controls.

## V. SUMMARY

Table 6 lists the microorganisms for which dosages have been specified. The dosages should be considered only as a general guide since the response of any given organism may vary greatly depending on the culture conditions. Deoxypyridoxine appears to be most effective under conditions in which the organism requires an external source of vitamin $B_6$.[401] In the case of *Saccharomyces carlsbergensis,* which has received the most attention, it appears well established that in the presence of minimal amounts of pyridoxine, a 4′-deoxypyridoxine:pyridoxine ratio of about 150:1 will produce at least 50% inhibition of growth. Deoxypyridoxine concentrations of 0.5 mg/mℓ produced significant growth inhibition in bacteriophage, mumps, influenza, and psittacosis virus. The evidence concerning toxicity of this dose to the host cells is inconclusive. The data on other organisms is too scanty to warrant any generalizations.

However, special note should be taken of several observations which already suggest mechanisms of deoxypyridoxine metabolism other than simple interference with the commonly recognized pyridoxal phosphate enzymes. Three reports[225,377,401] suggested that deoxypyridoxine may exert its main effect on membrane transport systems rather than intracellular metabolism. However, the transport effect is not the sole mechanism since 5′-deoxypyridoxine was more effective than 4′-deoxypyridoxine for inhibiting $B_6$ uptake by *Saccharomyces carlsbergensis*[447] but less effective at inhibiting growth.[402] Two reports[489,554] raised the possibility that deoxypyridoxine interfered with some aspect of glucose metabolism. The severe effect of deoxypyridoxine on weight gain in guppies[382] and turbot[3] is typical of effects in higher animals.

Perhaps the best indicator of the complexity of deoxypyridoxine action is the obser-

# Table 6
# SUMMARY OF GROWTH INHIBITION STUDIES WITH 4′-DEOXYPYRIDOXINE IN MICROORGANISMS

## Less Than 50% Inhibition at Highest Concentration Tested

| Organism | Highest concentration tested[a] | Ref. |
|---|---|---|
| *Aspergillus parasiticus* | 5 m*M* | 177 |
| *Bacteroides symbiosus* | 0.5 mg/mℓ | 405 |
| *Candida albicans* | 1 mg/mℓ | 144 |
| *Ceratostomella ulmi* | 10 μg/tube | 413 |
| *Clostridium butyricum* | 167 μg/mℓ | 401 |
| *Colpidium campylum* | 1 mg/mℓ | 144 |
| *Glaucoma scintillans* | 1 mg/mℓ | 144 |
| *Lactobacillus arabinosus* | 1 mg/mℓ | 144 |
| *L. casei* | 1 mg/mℓ | 144 |
| *L. fermenti* | 1 mg/mℓ | 144 |
| *L. helveticus* | 167 μg/mℓ | 401 |
| *Leuconostoc citrovorum* | 1 mg/mℓ | 144 |
| *Neurospora crassa* | 1 mg/mℓ | 144 |
| Polio virus | 0.1 mg/mℓ | 223 |
| *Propionibacterium shermanii* | 1 mg/disc | 65 |
| *Streptococcus faecalis* | 1 mg/mℓ | 144 |
| *Tetrahymena pyriformis* | 1 mg/mℓ | 144 |
| *Torula cremoris* 21512 | 167 μg/mℓ | 401 |

## Inhibited Under at Least some Conditions

| Organism | Lowest reported concentration producing at least 50% inhibition | Ref. |
|---|---|---|
| Bacteriophage | 0.5 mg/mℓ | 554 |
| *Bacteroides limosus* | 1 mg/disc | 65 |
| *B. pseudoramosus* | 10 μg/disc | 65 |
| *Butyribacterium rettgeri* | 1 mg/disc | 65 |
| *Clostridium acidi-urici* | 100 μg/disc | 65 |
| *C. cylindrosporum* | 1 mg/disc | 65 |
| *C. feseri* | 100 μg/disc | 65 |
| *Desulfovibrio desulfuricans* | 550 μg/mℓ | 265 |
| *Eimeria acervulina* | | 340 |
| *Entamoeba histolytica* | 0.17 μg/mℓ | 405 |
| *Escherichia coli* | | 224 |
| Influenza virus | 0.5 mg/mℓ | 109 |
| *Kloeckela apiculata* | 18.4 μg/tube | 236 |
| *Leuconostoc mesenteroides* | 2.8 μg/tube | 236 |
| *Micrococcus aerogenes* | 1 mg/disc | 65 |
| Mumps virus | 0.5 mg/mℓ | 109 |
| *Neurospora sitophila* | 100 μg/mℓ | 401 |
| Newcastle virus | 0.5 mg/mℓ | 26 |

**Table 6 (continued)**
**SUMMARY OF GROWTH INHIBITION STUDIES WITH 4′-DEOXYPYRIDOXINE IN MICROORGANISMS**

| Organism | Lowest reported concentration producing at least 50% inhibition | Ref. |
|---|---|---|
| *Propionibacterium freundreichii* | 1 mg/disc | 26 |
| Psittacosis virus | 0.5 mg/mℓ | 336 |
| *Saccharomyces carlsbergensis* | 0.6 μg/mℓ | 432 |
| *S. cerevesiae* | 167 μg/mℓ | 401 |
| *Staphlococcus aureus* | 750 μg/mℓ | 458 |
| *S. aureus* A | 100 μg/mℓ | 458 |
| *Vibrio cholera* | 100 μg/mℓ | 81a |

vation by Rabinowitz and Snell[401] that although Umbreit and Waddell[522] used preparations from *Streptococcus faecalis* to demonstrate the enzymatic phosphorylation of deoxypyridoxine and the inhibition of tyrosine decarboxylase by deoxypyridoxine phosphate, growth of *Streptococcus faecalis* in culture was not inhibited by deoxypyridoxine. This is an excellent example of the need to exercise caution in extrapolating in vitro results to in vivo situations. Also pertinent is the conclusion of Akov and Guggenheim[9] that deoxypyridoxine inhibition was reversible but not competitive.

Finally, it is interesting that under some conditions deoxypyridoxine actually stimulated rather than inhibited growth.[62,84,401,402] It seems likely that this effect is due to an interaction between deoxypyridoxine and $B_6$ rather than $B_6$ contamination of deoxypyridoxine because in the case of the tomato root[62] the combination produced better growth than any amount of pyridoxine alone.

Chapter 3

# GROSS PHYSIOLOGICAL EFFECTS IN HIGHER ANIMALS

## I. GENERALIZED EFFECTS

### A. Rat

Since acrodynia was one of the earliest recognized symptoms of $B_6$ deficiency, cure of this dermatitis was quickly adapted as a test of $B_6$ activity. On the basis of this test, Moeller et al.[333] reported that single doses up to 1 mg of 4′-deoxypyridoxine had no curative effect in rats. This was confirmed by Unna[523] for both 4′-deoxypyridoxine and 2,4,5-trimethyl-3-hydroxypyridine. Emerson[134] was the first worker to extend Ott's report of $B_6$ antagonism in chicks to rats. Using a natural diet a deoxypyridoxine:$B_6$ ratio of 175:1 was needed to produce dermatitis. With a purified diet containing 3.4 μg pyridoxine per gram diet a 50-fold excess of dexoypyridoxine was required. In adult rats the deficient diet alone produced no symptoms in four months. However, addition of 5 μg of deoxypyridoxine per gram produced acrodynia after 55 days and decreased growth. Stoerk[477] reported that Emerson in a personal communication noted that pyridoxamine was not antagonized by 4′-deoxypyridoxine. Emerson[135] also commented in an abstract that deoxypyridoxine was more effective against pyridoxal than against pyridoxine or pyridoxamine. Further study would be needed to clarify whether these differences might be accounted for by losses of pyridoxal due to the reactivity of the aldehyde group.

Woolley[553] presented a review of possible mechanisms of antagonist action. Deoxypyridoxine was mentioned as a $B_6$ antagonist but no original data were presented. Using a $B_6$-deficient diet containing 20 mg deoxypyridoxine per kilogram, Ling et al.[277] observed dermatitis in young rats after 5 weeks and in adult rats after 7 weeks.

Gregoire[173] treated rats with a $B_6$-deficient diet plus 250 μg deoxypyridoxine for 8 days. The $B_6$-deficient diet was continued for another 12 to 16 days at which time the animals were sacrificed. Dermatitis was more severe in the group receiving the deficient diet plus deoxypyridoxine than in the group receiving only the $B_6$-deficient diet. Interestingly, dermatitis did not appear until a few days after deoxypyridoxine administration had ceased.

Delayed responses have also been observed by other workers. Stoerk[477] noted that while rats supplemented with unspecified amounts of both pyridoxine and deoxypyridoxine for 10 days showed no deficiency symptoms, starting about 2 weeks after both supplements were terminated, the animals which had received deoxypyridoxine earlier developed severe dermatitis. Ross and Pike[425,427] treated rats with a $B_6$-deficient ration containing 5 mg deoxypyridoxine per kilogram for 7 to 30 days prior to mating. Weight gains during gestation were reduced even when the deoxypyridoxine treatment was replaced by a normal $B_6$ diet on the day of mating. The appearance of these effects several days or weeks after termination of deoxypyridoxine administration is in sharp contrast to some other effects, such as inhibition of amino acid transport, which are readily reversible.

The most detailed description of the pathology of deoxypyridoxine effects in rats was reported by Stoerk[477] who administered 4′-deoxypyridoxine (0.3 mg/mℓ drinking water) to Sherman rats. Growth retardation, dermatitis, hyperirritability, hunching of the back, prostration, and death were observed in young rats (50 to 80 g body weight) after prolonged feeding of $B_6$-deficient diet but appeared much sooner with simultaneous administration of deoxypyridoxine. Administration of deoxypyridoxine plus a $B_6$-deficient diet to larger rats (100 to 300 g) produced severe deficiency symptoms within 2 to 3 weeks whereas the deficient diet alone had little effect even after several months. All symptoms were alleviated by administration of pyridoxine hydrochloride. Dermatitis and curling of the tail were unique characteristics of the deoxypyridoxine groups. Evidence of this curling also appeared in a photograph published by Pol et al.[395] Histologically, Stoerk found that the dermatitis of the tail did not differ from that reported by Stokes et al.[480] for a dietary $B_6$ deficiency. There was also marked akanthosis, parakeratosis, and hyperkeratosis. In advanced cases superimposed infection with cellulitis extending into the subcutaneous tissues was often noted. Usually, such inflammatory changes appeared to originate in the accessory structures of the skin. The muscles of the tail showed histological evidence of atrophy and fibrosis but not myositis. Some bony lesions were also evident in advanced cases, but these were probably secondary rather than primary effects. The effects of deoxypyridoxine on dermatitis and body weights were comparable whether administration was continuous, or covered only the first ten or last ten days of the three week period.

Sakurai et al.[433] reported that deoxypyridoxine accelerated the appearance of deficiency symptoms in the rat but noted unspecified differences between the symptoms produced by a $B_6$ deficiency and by deoxypyridoxine.

Hawkins and Evans[192] found that a $B_6$-deficient diet plus an oral dose of 100 μg deoxypyridoxine per day in 100 g rats produced rapid weight loss, acrodynia, and some deaths within 31 days.

Caldwell and McHenry[77] observed that supplementing pyridoxine-deficient rats (average initial weight 111 g) with 100 μg deoxypyridoxine hydrochloride per day produced severe dermatitis within four weeks. There was also a marked difference in body weight (95 g for the deoxypyridoxine group vs. 124 g for the $B_6$-deficient controls).

Beaton et al.[36] examined the effect of $B_6$ deficiency on various metabolites in Wistar rats. Each animal received a $B_6$-deficient diet supplemented with 100 μg deoxypyridoxine per day. In addition, control animals also received 50 μg pyridoxine hydrochloride per day. Both ad libitum and pair-fed controls were used. The $B_6$-deficient diet plus deoxypyridoxine produced acrodynia and limited growth to 7 g over 20 days compared with 36 and 42 g for the pair-fed and ad libitum controls, respectively. No statistical analysis of the data was performed. In further studies using a similar design, Beaton et al.[40] usually observed acrodynia in deoxypyridoxine treated rats receiving a 25% casein diet but not with a 20% gelatin diet. In the third paper of this series, Beare, et al.[32] again used the same experimental design. Weight gains were reduced about 75% in animals receiving the $B_6$ deficient diet plus 100 μg deoxypyridoxine per day. Acrodynia appeared in all deficient animals.

In another replication,[39] the deoxypyridoxine group showed only slightly decreased weight gain for about a week and then started losing weight so that by 29 days they were back to their initial weight. Addition of pyridoxine hydrochloride (50 μg/day) to the deoxypyridoxine diet after 16 days restored gains so that by 28 days they were almost equal to rats that never received deoxypyridoxine. Dermatitis appeared only in rats receiving deoxypyridoxine in the 28-day experiment. On a 6-week study without deoxypyridoxine, dermatitis appeared in only two of ten rats on a $B_6$-deficient diet.

With the same treatment, Beare et al.[34] reported that dermatitis appeared within 2 weeks. Adrenalectomy prevented development of dermatitis, but adrenocortical extract did not. Weight gains were markedly reduced even when pair-feeding was employed

Meister et al.[305] noted that adult rats fed a $B_6$-deficient diet and injected with 0.4 mg deoxypyridoxine six times per week developed deficiency symptoms in 73 to 91 days with an average weight loss of 57 g.

Halverson and Hendrick[180] reported data suggesting that feeding young rats 0.01% deoxypyridoxine in conjunction with a diet of unspecified $B_6$ content may have slightly reduced growth but did not produce liver damage and did not protect against selenium toxicity.

Olsen and Martindale[370] used a $B_6$-deficient diet plus 12.5 mg deoxypyridoxine per kilogram diet in the rat. Diarrhea, dermatitis, and convulsions appeared as early as 7 days. After 4 weeks, deoxypyridoxine was removed from most animals because of high mortality. It was restored to some animals after 30 weeks for the final 4 weeks of the experiment. All nine animals left on the deoxypyridoxine treatment died within 10 weeks. Addition of deoxypyridoxine to purified diets (12.5 mg/kg) or to a natural diet (5 mg/day) produced weight losses. Compared with rats receiving a normal $B_6$ intake, deficient animals showed significant ($p<0.01$) increases in the relative size of the adrenals, heart, liver, and kidneys.[371] Exposure to deoxypyridoxine during the final 4 weeks also produced an increase in the relative size of the adrenal glands in animals receiving normal $B_6$. However, addition of deoxypyridoxine to the $B_6$-deficient diet did not produce further increases in relative organ size. Supplementing the $B_6$-deficient diet with 10 mg pyridoxal per kilogram for the last 4 weeks of the experiment did not correct the increased relative size of the adrenals and heart. The data suggest that the kidney size also probably remained significantly higher than in the controls. The authors felt that this increase could not be completely explained by the loss in body weight.

Bianchi-Donnasibilla and Cortesi[49] observed that the dietary deficiency alone had little effect on the animals. Weight remained stable. Clinically the animals appeared normal throughout the study and at autopsy, except for infrequent occurrence of small erythmatose spots on the paws. In contrast, animals receiving 15 $\mu$g pyridoxine plus 150 $\mu$g deoxypyridoxine daily showed noticeable changes even by the end of the first month. Food intake decreased and there was notable weight loss. The hair lost its luster and became sparse and yellow-grey. Erythemato-desquamation appeared in the nose, eyes, and paws. The cornea slowly became opaque, frequently resembling the lesions described by Musini and Tenconi.[357] Changes were also seen at autopsy but are not specifically described.

Boone et al.[63] compared the effects of isonicotinic acid hydrazide (50 mg/kg intraperitoneal [i.p.]) and deoxypyridoxine (100 $\mu$g/day i.p.) in $B_6$-deficient rats (average initial weight 130 g). The isonicotinic acid hydrazide group had a weight loss and the poorest physical appearance. This may have been due to the 50-fold difference in dose. The deoxypyridoxine rats showed decreased weight gains and acrodynia.

Silva[456] reported results of histological examinations in two $B_6$-deficient rats receiving deoxypyridoxine for 6 to 7 weeks and one rat which had received the treatment for 25 days and then had been supplemented with pyridoxine for 4 weeks. The amount of deoxypyridoxine was not specified. Earlier trials used 1 mg/100 g body weight. Photomicrographs of the skin lesions are included. The skin of the paws was covered by a crust formed by desquamated epithelial cells and polymorphonuclear neutrophils. In the dermis, there was infiltration of polymorphonuclear neutrophils, hemorrhages within some hair follicles, and extensive hyperemia. The skin of nose and tail showed similar changes plus some ulceration on the tail. In the animal supplemented with pyridoxine, the histological picture was completely different. The paws and nose were normal but the tail showed slight hyperkeratosis along with dilated follicles and atrophy of the sebaceous glands. Since so few animals were examined, it is impossible to determine the reason for these differences.

Talageri[494] treated male Wistar rats with a $B_6$-deficient diet containing 20 mg deoxypyridoxine hydrochloride per kilogram. Dermatitis developed in deficient animals with and without deoxypyridoxine after 2 weeks. Severe involvement of the nose, paws, and anogenital region was observed within 3 months in the deoxypyridoxine group but not until 5 to 8 months in the group with only dietary deficiency. The dermatitis was relieved within 5 days after restoring normal $B_6$ intake. The deoxypyridox-

Table 1
PERCENTAGE DECREASE IN BODY WEIGHT AND FOOD INTAKE IN THE DEOXYPYRIDOXINE GROUPS COMPARED WITH NORMAL $B_6$ CONTROLS RECEIVING THE SAME LIPID[395]

| | | % Decrease in | | | |
|---|---|---|---|---|---|
| | | Body weight | | Food intake | |
| Lipid | Vitamin A (I.U./kg diet) | M | F | M | F |
| Butter | 8500 | 19 | 10 | 13 | 5 |
| Margarine | 6030 | 24 | 10 | 13 | 6 |
| Corn oil | 1100 | 29 | 18 | 16 | 7 |

ine group showed a progressive weight loss which could not be accounted for by inanition alone.

Gadekar et al.[151] fed weanling (40 g) rats a 30% casein $B_6$-deficient diet for 16 to 17 days. Then deoxypyridoxine (50 to 100 μg/day subcutaneously [s.c.]) was administered. The rats promptly lost weight and died within a week.

Hsu and Kawin[208] found that body weight in rats receiving a $B_6$-deficient diet plus deoxypyridoxine (0.75 mg deoxypyridoxine i.p. three times per week) was only 60% of that of animals receiving normal $B_6$.

Mihich et al.[326] found that in rats fed a pyridoxine-deficient diet for 3 weeks before administering deoxypyridoxine, intraperitoneal doses of 12.5 mg/kg × 18 produced toxic effects. This was only 25% of the dose needed when the $B_6$-deficient diet was started simultaneously with deoxypyridoxine.

Pol et al.[395] studied the interactions between vitamins A, E, and $B_6$ in the rat. A $B_6$-deficient diet (1.75 mg/kg) was used in conjunction with deoxypyridoxine (50 mg/kg). The control diet contained pyridoxine (5.76 mg/kg) in place of deoxypyridoxine. Deoxypyridoxine reduced both food intake and body weight, particularly in males (Table 1). The authors noted that the effects of deoxypyridoxine increased with the increasing unsaturation of the lipid. However, it also increased with the decreasing vitamin A content of these diets. We feel this was more of a factor than the increasing unsaturation.

Similar to the situation in microorganisms[84,401,402] and tomato roots[62] there are at least two reports suggesting that under certain conditions a combination of deoxypyridoxine plus pyridoxine actually improves growth in rats. Beaton et al.,[38] found that the average weight gain in pair-fed animals receiving 100 μg deoxypyridoxine plus 50 μg pyridoxine per day was 33 g compared with 20 g in animals receiving only pyridoxine, 22 g in animals receiving no supplement, and 2 g in animals receiving only deoxypyridoxine. In another replication of that experiment listed in the same paper, the gains were 49, 46, 33, and 29 g, respectively. Iwanowska[217] also reported that 400 μg deoxypyridoxine per day increased growth about 10% in rats receiving 50 μg pyridoxine per day but decreased growth in $B_6$-deficient animals. A dose of 100 μg/day decreased growth about 25% in animals receiving normal $B_6$. It seems rather paradoxical that a low dose would be more detrimental than a high dose. However, Boullin[64] reported a somewhat similar situation. He used a $B_6$-deficient diet with the addition of 70 (36 to 40 days) or 280 μg (26 to 58 days, mean of 32 days) deoxypyridoxine per rat per day. Control animals were pair-fed the same diets (including deoxypyridoxine) plus

**Table 2**
**EFFECT OF DEOXYPYRIDOXINE ON BODY WEIGHT[64]**

| Deoxypyridoxine (μg/rat/day) | Body weight (g) | | | |
|---|---|---|---|---|
| | Maximum | | Final | |
| | D[a] | C | D | C |
| 0 | 86.0 ± 1.8[b] | 103.5 ± 2.2 | 65.5 ± 1.3 | 92.3 ± 3.6 |
| 70 | 75.1 ± 2.1 | 89.7 ± 2.2 | 60.3 ± 2.6 | 82.1 ± 2.5 |
| 280 | 71.0 ± 2.3 | 81.4 ± 3.1 | 56.1 ± 2.3 | 79.2 ± 3.0 |

[a] D = $B_6$-deficient. C = 80 μg pyridoxine hydrochloride per day.
[b] Mean ± standard error.

80 μg pyridoxine hydrochloride per day. Initial weights were 50 to 60 g. Under these conditions the incidence of skin lesions was 90, 100, and 55% in the deficient groups receiving 0, 70, and 280 μg deoxypyridoxine, respectively. Convulsions were observed in 10% of the animals receiving the highest deoxypyridoxine intake. No convulsions were observed in the other two groups. The occurrence of convulsions suggests that the animals were actually receiving a toxic amount of deoxypyridoxine. However, the reduced incidence of dermatitis with that same dosage is puzzling.

The data on weight gains and food intake presented by Boullin[64] are interesting because the addition of deoxypyridoxine to the deficient diet did not alter food intake. In fact, the animals receiving 280 μg deoxypyridoxine ate an average of 5.7 g/day compared with 4.5 g for the group receiving only the deficient diet. No standard deviations were presented for the food intake data. We have found that higher doses of deoxypyridoxine (1 mg/g feed) markedly reduce food intake.

Boullin noted that the maximum body weight of the group receiving only the $B_6$-deficient diet was significantly ($p < 0.01$) less than the weight of its control group but more than the two deoxypyridoxine groups (Table 2). These comparisons are not the only ones which are statistically significant. Deoxypyridoxine exerted noticeable effect on weight gains even in the presence of 80 μg pyridoxine hydrochloride per day. However, since even the pair-fed controls receiving $B_6$ without deoxypyridoxine were losing weight, it is evident that their intake of nutrients other than vitamin $B_6$ was not adequate to sustain growth. It would have been interesting to have seen a comparison between the pair-fed controls and controls fed ad libitum.

Zhdanovich and Udalov[568] reported that deoxypyridoxine (50 mg/kg) tended to increase the toxicity of the insecticide, Dipterex, in rats.

Bayoumi et al.,[30] found a 25% decrease in growth over 10 weeks in rats receiving a normal $B_6$ intake plus 1 mg deoxypyridoxine per milliliter drinking water.

Rasmussen[403] examined the effects of 500 mg deoxypyridoxine per kilogram diet on adult male rats for 17 days. The effects of 200 mg isonicotinic acid hydrazide per kilogram diet were also tested. The rats were fasted for 18 hr prior to drawing blood samples. Deoxypyridoxine had no effect on weight gain when fed with adequate $B_6$ but caused a slight weight loss when fed with a $B_6$-deficient diet. Isonicotinic acid hydrazide was more detrimental than deoxypyridoxine to weight gain both with and without adequate $B_6$. Even though the combination of isonicotinic acid hydrazide plus $B_6$ deficiency caused the greatest weight loss, only deoxypyridoxine produced marked dermatological changes. This observation is further evidence that the dermatitis produced by deoxypyridoxine is a fairly specific effect and is not due to general malnutrition.

Kawakita et al.[236] reported a 25% loss in body weight within 15 days in rats receiving a $B_6$-deficient diet plus 10 μg pyridoxine plus 100 μg deoxypyridoxine per rat per day.

## B. Mouse

Whereas the rat has been studied quite thoroughly, most of the work in mice has been limited to observations of gross toxicity. The first report of deoxypyridoxine use in mice appears to be that of Weir, Heinle, and Welch,[544] who found that subcutaneous administration of more than 1 mg deoxypyridoxine per day to mice receiving a $B_6$-deficient diet was fatal within 10 days. On a dose of 1 mg/day mean survival time was 20 days and weight loss was marked. Higher doses of deoxypyridoxine (4 mg s.c. twice daily) produced high mortality even with a normal $B_6$ diet. A dose of 4 mg/day with a natural diet produced good survival.

Brockman et al.[67] found that 300 mg deoxypyridoxine per kilogram intraperitoneally every other day killed five out of ten mice receiving a normal $B_6$ diet within eight days. Reducing the dose to 150 mg/kg produced little mortality (1/20) in mice receiving normal $B_6$, but killed 13 of 20 animals receiving a $B_6$-deficient diet. These fatalities were observed even though the average weight change was less than 2 g. No symptoms of $B_6$ deficiency were observed. Mickelson and Flippin[320] confirmed that the acute toxic dose was 300 mg/kg.

Skipper et al.,[461] found that intraperitoneal doses of 400 to 600 mg/kg killed all mice within 24 hr regardless of the $B_6$ content of the diet. Doses of 50 to 150 mg/kg/day for 7 days produced no deaths within 14 days regardless of dietary $B_6$ intake. Doses of 200 to 300 mg/kg/day produced more deaths in $B_6$-deficient animals than in controls. The authors also commented that when administered at lethal doses, deoxypyridoxine killed more rapidly than other well-known antitumor agents.

DeRenzo and Cerecedo[114] examined the effects of deoxypyridoxine (20 mg/kg diet) in $B_6$-deficient mice. Acrodynia appeared after average of 23 days of deoxypyridoxine treatment and survival was reduced from 40 days without deoxypyridoxine to 32 days with it. Raising the protein content of the diet to 60% reduced survival to 25 days. The authors reported that 2 μg pyridoxine per mouse per day was generally the minimal amount which would permit some growth. With this $B_6$ intake, a 5:1 deoxypyridoxine yridoxine ratio permitted growth in 5 of 7 mice while a 10:1 ratio permitted growth in only 2 out of 11. Contrasting with the relatively long time required to produce symptoms in this experiment, Summers[488] reported observing deficiency symptoms within 5 or 6 days after mice were placed on a normal $B_6$ diet containing 1000 mg/kg deoxypyridoxine.

Stoerk[477] found that in contrast to his observations in rats, there were no obvious differences in the physical symptoms produced in C and C3H mice by a $B_6$-deficient diet alone and those produced by the deficient diet plus deoxypyridoxine (0.3 mg/mℓ drinking water).

Dietrich and Shapiro[116] observed no weight loss or other external deficiency symptoms in mice receiving a normal $B_6$ intake plus 150 mg deoxypyridoxine per kg per day intraperitoneally for 14 days.

Dubnick et al.,[126] noted that while pyridoxine protected against the toxicity of beta-phenethylhydrazine in mice, deoxypyridoxine (100 mg/kg, i.p.) and pyridoxal actually increased the toxicity. Pyridoxamine did not increase or decrease the toxicity. The authors suggested that the effect of pyridoxal was due to formation of a hydrazone which inhibits pyridoxal kinase. No mechanism was suggested for deoxypyridoxine. The authors reported the $LD_{50}$ of deoxypyridoxine in mice to be about 200 mg/kg intraperitoneally (150 mg/kg killed 0/10, 200 mg/kg killed 6/10, and 250 mg/kg killed 8/10).

Table 3
TOXICITY OF $B_6$ ANALOGS IN MICE IMPLANTED WITH TUMORS[421]

| Compound | Dose (mg/kg/day i.p.) | Wt. change after 8 days (g) | Deaths (%) |
|---|---|---|---|
| | **Normal $B_6$** | | |
| None | 0 | −0.5 | 0 |
| 3-Deoxypyridoxine | 10 | −1.6 | 0 |
| | 20 | −1.6 | 27 |
| | 40 | −1.8 | 90 |
| 4′-Deoxypyridoxine | 50 | −0.5 | 5 |
| | 100 | −1.6 | 0 |
| 5′-Deoxypyridoxal | 25 | −4.4 | 0 |
| | 50 | −3.8 | 13 |
| | 75 | 0 | 93 |
| | **$B_6$ Deficient** | | |
| None | 0 | −0.1 | 0 |
| 3-Deoxypyridoxine | 10 | +0.2 | 10 |
| | 20 | +0.4 | 67 |
| | 40 | — | 100 |
| 4′-Deoxypyridoxine | 50 | +0.5 | 4 |
| | 100 | −0.8 | 0 |
| 5′-Deoxypyridoxine | 200 | −2.8 | 20 |
| | 400 | — | 100 |
| 5′-Deoxypyridoxal | 25 | −1.2 | 0 |
| | 50 | −3.0 | 25 |
| | 75 | −2.8 | 85 |

Melching et al.[310] examined the effect of deoxypyridoxine (4 mg/20g) in mice after whole body irradiation with X-rays. Apparently, a normal $B_6$ diet was used. Treatment with deoxypyridoxine did not increase mortality when administered 15 min before irradiation or two or four days after irradiation. Deoxypyridoxine did increase mortality when administered 15 min or 6 days after irradiation.

Korytnyk et al.[249] found that on a normal diet all five female mice survived a single intraperitoneal dose of 150 mg deoxypyridoxine per kilogram for at least 30 days whereas a dose of 250 mg/kg killed all five. In mice which had received a $B_6$-deficient diet for 6 weeks prior to the treatment all five mice survived a single 10 mg/kg dose but four out of five died after a single 25 mg/kg dose. While their published table simply reports survivors after 30 days, the authors have acknowledged[256a] that the deoxypyridoxine treated animals did die within a day or two after the injection. Since the animals generally died at night, the authors did not know whether convulsions occurred.

Rosen et al.[421] found that 3-deoxypyridoxine and 5′-deoxypyridoxal were more toxic than 4′- or 5′-deoxypyridoxine[421] (Table 3). They suggested that the variations in toxicity might reflect differences in the rates at which the analogs were metabolized and/or excreted.

## C. Chick

Although others had demonstrated that deoxypyridoxine had no $B_6$ activity, Ott's[375] work in chicks was the first evidence of its $B_6$ antagonist action. A dose of 16 μg was fatal but could be conteracted by subsequent administration of pyridoxine. However, two oral doses of 100 μg were fatal even when 16 μg of pyridoxine was administered 15 to 30 min prior to the deoxypyridoxine. Dose response curves suggested that deoxypyridoxine counteracted the effects of about 50% of its weight of pyridoxine. In contrast to the situation with deoxypyridoxine, pyridoxine was generally ineffective in counteracting a lethal dose of methoxypyridoxine.[376]

Cravens and Snell[106,107] studied the response of chick embryos to deoxypyridoxine. Injection of 1 mg prior to setting the eggs was lethal. This effect was almost completely offset by simultaneous injection of a similar amount of pyridoxal hydrochloride. Using 1 mg of deoxypyridoxine the ratios of vitamin to inhibitor which permitted about 50% hatchability were about 1:15 for pyridoxal hydrochloride (the authors state 1:20 but their data show only 30% hatchability at that ratio); 1:50 for pyridoxamine dihydrochloride; and 1:100 for pyridoxine hydrochloride.

At 6 days, 5 mg of deoxypyridoxine was required to produce 100% mortality and this could not be offset by administration of $B_6$. In fact, 5 mg pyridoxal hydrochloride itself produced 80% mortality. Raising the dose to 10 mg produced 100% mortality with pyridoxal hydrochloride and 90% with pyridoxine hydrochloride. Pyridoxamine dihydrochloride seemed to have little toxic effect. Simultaneous injection of niacin or niacinamide along with deoxypyridoxine had no beneficial effect, suggesting that interference with synthesis of niacin from tryptophan was not a major factor in the toxicity of deoxypyridoxine. The authors concluded that the inhibition in the early stage was due to interference with $B_6$ metabolism but in the later stages another mechanism was involved. They suggested that the decreased toxicity of deoxypyridoxine with advancing age might reflect the availability of more enzyme molecules already bound to pyridoxal phosphate and thus less susceptible to deoxypyridoxine inhibition. This work was also summarized later.[105]

Karnofsky et al.,[233] confirmed this work. They reported that the $LD_{50}$ of deoxypyridoxine at 0, 4, and 12 days was 1 to 2, 4, and 12 mg per egg, respectively. In the 4-day embryo, 5 mg of pyridoxine or 3 mg pyridoxal phosphate provided no protection against a 5-mg dose of deoxypyridoxine again suggesting that in these older embryos the effect is not one of simple $B_6$ antagonism. Karnofsky[232] reported that the $LD_{50}$ for deoxypyridoxine in chick embryos was 2 mg per egg and $LD_{100}$ was 3 mg per egg. The embryos were reduced in weight but no developmental abnormalities were noted.

Morisawa et al.[340] reported that incorporation of 0.02 to 0.05% deoxypyridoxine in the feed reduced weight gains in chicks.

Gillette[165] examined the effect of deoxypyridoxine on chick embryo fibroblasts in vitro. A concentration of 20 mg/100 mℓ was sufficient to produce a maximal cellular response with minimal irritation of the fibroblasts. After 48 to 96 hr of perfusion there was a great increase in the relative size and number of nucleolar bodies within the nucleus. This appeared to be accompanied by an elongation of the nucleus itself with a slight decrease in nuclear size. Some small vacuolar bodies appeared in the cytoplasm. As perfusion continued there was a slowing of the movement of the mitochondrial elements in the cells which progressed until there was no apparent activity in the cytoplasm. The cells appeared to be in a cytostatic rather than a dying state. This condition was restricted to those cells which had previously produced multiple nucleoli. The condition was not reversed by the normal perfusion medium or by additional amounts of pyridoxine. Pyridoxal phosphate (20 to 30 mg/100 mℓ) did produce partial reversal. Since mitosis proceeded normally in the early phase of deoxypyridoxine treatment, the author concluded that the decline in mitotic activity reflected a general impairment of cellular function rather than a specific effect on mitosis.

Conti and Milio[98] examined the in vitro effects of deoxypyridoxine and pyridoxine on myocardium from chick embryo. Low doses (40 μg) had no toxic effect but doses of 80 to 160 μg were increasingly toxic as judged by histological examination of the tissue. Pyridoxine was more toxic than deoxypyridoxine. Combining the two substances in various ratios did not prevent the toxicity. Deoxypyridoxine did not affect alkaline phosphatase. Pyridoxine prolonged the time during which high levels of phosphatase could be demonstrated. This effect could be offset by appropriate doses of deoxypyridoxine.

Conti and Milio[99] extended this work to include the intestinal wall from chick embryos. Deoxypyridoxine in doses of 40, 80, 120, and 160 $\mu$g per culture had little effect. Pyridoxine exerted toxic effects at doses of about 40 $\mu$g. Combining the two compounds in a 1:1 ratio did not completely offset the effects of pyridoxine. However, when deoxypyridoxine was present in a 10- or 100-fold excess, the toxic effects of pyridoxine were minimized.

Stoll et al.[481] found that addition of 1 mg deoxypyridoxine did not increase the teratogenic effects of a 0.25-mg dose of semicarbazide on chick embryos. Pyridoxine (2.5 or 5 mg) reduced the teratogenic effects of semicarbazide.

## D. Dog

Mushett et al.,[356] noted that a single large (but unspecified) dose of deoxypyridoxine caused hyperirritability and convulsions in puppies. Puppies receiving a $B_6$ deficient diet plus an injection of 5 mg deoxypyridoxine per kg/day lost weight and died within 2 months.[355]

Forse and Webster[146] used a 5 mg/kg oral dose of deoxypyridoxine with both a normal and $B_6$-deficient diet in dogs. In a group of three dogs, the volume of gastric secretion dropped over 50% within 24 hr after starting the deficient diet plus deoxypyridoxine. There was a slight increase in the pH. Compared with normal $B_6$ controls, the histamine test in the deficient animals showed a marked reduction in the volume of gastric secretion, elevated pH, and a drop in pepsin value. The insulin test showed similar trends but to a lesser degree. The dogs developed a slight microcytic hypochromic anemia, leucopenia, lyphopenia, polynucleosis, and eosinophilia.

Two dogs receiving the $B_6$-deficient diet alone showed a gradual decrease in the volume of gastric secretions over several months. The response to the histamine and insulin tests was altered slightly. In a single female dog fed deoxypyridoxine plus a normal $B_6$ diet, the volume of gastric secretion increased. There was no significant change in the histamine tests. There was a slight leucocytosis, polynucleosis, and lymphopenia. When deoxypyridoxine was discontinued, the volume and pH gradually returned to normal over a 12-day period. When this dog was fed the deoxypyridoxine with the $B_6$-deficient diet, the usual marked reduction in gastric secretion was observed. The authors do not propose any mechanism to explain why the response to deoxypyridoxine should reverse depending on the $B_6$ status.

Williams and Wiegand[550] noted that deoxypyridoxine produces seizures and is synergistic with hydrazide. Intravenous administration of deoxypyridoxine to dogs did not produce a significant acute increase in xanthurenic acid excretion despite the production of severe seizures.

Mihich et al.[326] reported that in dogs, the toxic dose was 40 mg/kg/day × 18. Previous $B_6$ depletion reduced this dose only slightly. Anemia was produced by $B_6$ deficiency and improved by administration of deoxypyridoxine. Lymphopenia developed concurrently. The authors noted that $B_6$ deficiency and deoxypyridoxine have qualitatively different effects. $B_6$ depletion prior to administration of deoxypyridoxine had only quantitative effects on deoxypyridoxine toxicity.

## E. Miscellaneous Animals

Loefer[278] observed no signs of $B_6$ deficiency in hamsters receiving a normal diet plus intraperitoneal injections of 5 mg deoxypyridoxine. The injections were given at one to three day intervals over 59 days and totaled 190 mg of deoxypyridoxine.

Aikawa[8] administered deoxypyridoxine (40 mg/kg/day) intravenously to male rabbits. After the third injection the animals became hyperirritable. After the fifth and sixth injections generalized convulsions lasting several minutes appeared in an unspecified number of animals.

After 50 days of treatment two of nine guinea pigs receiving a $B_6$-deficient diet containing 50 mg deoxypyridoxine per kilogram had died.[524] The average weight increase for the deficient group over the 50 days was 42 g compared with 203 g for the normal $B_6$ controls.

## II. SPECIFIC ORGAN SYSTEMS

### A. Brain

While dermatitis and decreased weight gains are the most commonly reported consequences of deoxypyridoxine administration, convulsions can be produced by some dosages. In mice, levels of 150 mg or more per kilogram seems to be effective even with normal $B_6$ intake. Nozaki[368] reported that injection of 150 to 300 mg deoxypyridoxine per kilogram body weight produced convulsions in 100 of 140 mice and death in 62 others.

Pfeiffer[388] noted that 1.5 mmol deoxypyridoxine per kilogram was a sure convulsant dose in mice. Since no comment was made about $B_6$ status, this presumably applied even in normal mice. The effects of deoxypyridoxine were additive and perhaps synergistic with those of hydrazides. Similarly, Horton and Meldrum[203] reported that the dose needed to produce convulsions in 50% of the mice after intraperitoneal injection of 4′-deoxypyridoxine was 1.1 mmol/kg with a mean latency of 42 min. With a dose of 3.4 mmol/kg the mean latency dropped to 13 min. Mortality was high with five out of six mice dying at a dose of 1.7 mmol/kg.

Bain and Williams[27] and Horton and Meldrum[204] used 250 mg/kg as a convulsive dose in mice. Sawaya et al.,[441] noted that this dose produced convulsions within a mean time of 45 min.

Matano[300] reported that intravenous administration of 300 mg/kg deoxypyridoxine to normal mice produced convulsions with a marked decrease of both cholinesterase activity and potassium content and an increase of sodium content of the cortex. Maynert and Kaji[303] reported data on a compound identified as 3-deoxypyridoxine phosphate. Correspondence with Dr. Maynert[303a] has revealed that the compound used was actually 4′-deoxypyridoxine-5′-phosphate. A dose of 250 mg/kg failed to produce convulsions in mice within 90 min. However, convulsions were observed at a dose of 500 mg/kg.

Deoxypyridoxine may enhance or inhibit the action of other convulsant drugs. For example, Murakami et al.[353,354] found that a subconvulsive dose of 50 mg/kg intraperitoneal dose of deoxypyridoxine protected mice against convulsions induced by castrix (2-chloro-4-dimethyamino-6-methylpyrimidine).

Murakami and Makino[352] noted that the amount of the spectral shift in methylene blue caused by various derivatives of castrix seemed to be correlated with convulsive activity. Pyridoxine, pyridoxamine, pyridoxal, pyridoxal phosphate, and 5′-deoxypyridoxal had little effect. However, 4′-deoxypyridoxine had more effect on the spectrum of methylene blue than several other derivatives (e.g., thiosemicarbozide, toxopyrimidine, and isonicotinic acid hydrazide) which had stronger convulsant effects than 4′-deoxypyridoxine.

While deoxypyridoxine protected against the convulsant action of castrix, Pfeiffer[383] reported that the effects of deoxypyridoxine were additive and perhaps synergistic with those hydrazides. Also, Stone[482] found that a subconvulsive dose of deoxypyridoxine (95 mg/kg) administered interaperitoneally one hour before other convulsants reduced the dose of the second drug needed to produce convulsions in 50% of the mice ($CD_{50}$). The potency ratios ($CD_{50}$ without deoxypyridoxine: $CD_{50}$ with deoxypyridoxine) were 1.45 (95% confidence interval 1.22 to 1.71), 1.82 (1.60 to 2.06), 1.51 (1.29 to 1.77), and 1.30 (1.14 to 1.49) for pentamethylenetetrazol, 3-mercaptopropionic acid, bicuculline, and picrotoxin, respectively. Deoxypyridoxine did not alter the latent periods. The authors assumed that deoxypyridoxine acted primarily by inhibiting glutamic acid decarboxylase and therefore, correctly predicted that deoxypyridoxine would have a greater effect on 3-mercaptopropionic acid than on pentamethylenetetrazol. Since the effect of deoxypyridoxine and other inhibitors of the gamma-aminobutyric acid system on bicuculline and picrotoxin was more similar to their effects on pentamethylenetetrazol than on 3-mercaptopropionic acid, the authors concluded that bicuculline and picrotoxin probably do not act through the gamma-aminobutyric acid system.

Haekkinen and Kulonen[179] also included deoxypyridoxine in their study of the relationship between the gamma-aminobutyric acid system and ethanol intoxication. Treatment of rats receiving a normal $B_6$ intake with 1 mg deoxypyridoxine/mℓ drinking water for 78 days had very little effect while bicuculline diminished the effects of ethanol and amino-oxyacetic acid increased ethanol intoxication. While the total daily intake of deoxypyridoxine in this study may have approached or even exceeded the 95 mg/kg used by Stone,[482] the use of oral administration over many hours would have produced much lower blood concentrations of deoxypyridoxine than in Stone's experiment. On the other hand, the dosage of bicuculline (1.5 to 2 mg/kg s.c.) approached the $CD_{50}$ of 3.25 mg/kg reported by Stone. In fact, Haekkinen and Kulonen noted that the dosage produced intense convulsions and that several animals died. The differences in the timing and route of administration of deoxypyridoxine were probably the major factors in the lack of effect of deoxypyridoxine in the ethanol study.

Yamashita[560] noted that while intracerebral injection of either 25 μg deoxypyridoxine hydrochloride or 30 μg of semicarbazide hydrochloride did not produce convulsions, simultaneous administration of both compounds produced convulsions and death in all five mice tested. While a dose of 60 μg of semicarbazide alone produced convulsions in all animals treated, it killed only 11 out of 20. Therefore, the combination of semicarbazide plus deoxypyridoxine was more lethal than a comparable amount of semicarbazide alone.

While there have been few studies of induction of convulsions by deoxypyridoxine in rats, the limited data available suggest that the convulsive dose is as high as or higher than the dose required in mice. Kosower and Rock[257] studied the effects of deoxypyridoxine in rats made porphyric with allylisopropylacetamide. Deoxypyridoxine (3.5 mmol/kg i.p.) produced convulsions in most rats and killed 70 to 80% of both porphyric and control rats within 60 min. A dose of 3 mmol/kg killed only about 10% of the animals in each group. Snodgrass and Iverson[467] reported some deaths in rats with the much lower intraperitoneal dose of 100 μg/kg while Ebadi and Klangkalya[129] failed to observe convulsions within 2 hr after an intraperitoneal dose of 1000 mg/kg or within 12 hr after a dose of 200 mg/kg. Milholland et al.,[328] did not mention whether they observed convulsions within four hours after a single intraperitoneal dose of 500 mg/kg. There is no obvious explanation for these differences in sensitivity to deoxypyridoxine.

Tsunoo et al.,[518] reported that the brain serotonin level in rats was not altered during convulsions induced by deoxypyridoxine at a dose of 400 mg/kg subcutaneously.

Single injections ranging from 1 to 2 mg/kg did not produce convulsions.[63,151] Neither did daily injections of 100 mg/kg[423] produce convulsions in rats receiving a diet containing 1 mg $B_6$ per kilogram. Like the results of Haekkinen and Kulonen[179] which were discussed earlier, Bayoumi et al.[30] noted no convulsions when deoxypyridoxine was administered at 1 mg/mℓ of drinking water with a normal $B_6$ intake. We[92] have not observed convulsions even when this dosage was accompanied by a $B_6$-deficient diet. Ebadi[130a] failed to observe any convulsions after intracisternal injection of 1.25 μmol per rat. Thus, with a few exceptions, the convulsive dose for normal rats generally appears to lie between 100 and 400 mg/kg body weight. However, lower dosages may be effective when combined with a $B_6$ deficiency. Olsen and Martindale[370] noted typical signs of $B_6$ deficiency including convulsions within a week after placing weanling rats on a $B_6$-deficient diet containing 12.5 mg deoxypyridoxine per kilogram.

When combined with a $B_6$-deficient diet, 3-deoxy derivatives appear to be even more effective than 4′-deoxypyridoxine in producing convulsions. Rosen et al.[423] found that a single intraperitoneal injection of 150 mg 3-deoxypyridoxine per kilogram failed to produce seizures within 4 hr in rats receiving a commercial stock diet. However, after using a purified diet containing only 1 mg $B_6$ per kilogram for 1 week, a single intraperitoneal injection of 3-deoxypyridoxine (100 mg/kg) caused convulsions and death within 3 hr. Injection of a similar dosage of 4′- or 5′-deoxypyridoxine or omega-methylpyridoxine daily for 7 days failed to produce convulsions.

Bukin[75] reported similar results. With normal $B_6$ intake, intraperitoneal doses of 800 mg 3-deoxypyridoxine per kilogram were required to produce convulsions in rats. If the rats were first fed a $B_6$-deficient diet for 5 to 10 days the convulsive dose was reduced to 90 mg/kg. With subarachnoidal injection, 20 mg/kg was not effective, but 30 mg/kg produced convulsions in all four rats tested. When combined with a $B_6$-deficient diet the convulsive dose was reduced to 5 mg/kg.

Gadekar et al.,[151] noted that 3-chloropyridoxine was only 2 to 4% as effective as deoxypyridoxine. However, subcutaneous administration of the chloroderivative produced convulsions in all $B_6$-deficient rats treated with doses of about 12.5 to 31 mg/kg. The convulsions could be avoided by simultaneous administration of pyridoxine (about 0.3 mg/kg). These results suggest that further studies of the interactions between these 3-deoxy derivatives and the central nervous system might provide valuable insights concerning the function of $B_6$ in nerve tissue.

Results with 4′-deoxypyridoxine in primates are similar to those reported for mice and rats. Meldum and Horton[311] noted that intravenous deoxypyridoxine hydrochloride at doses of 40 mg/kg enhanced photosensitivity in baboons. Seizures were observed without photic stimulation at doses about 100 mg/kg. Electroencephalographic characteristics of the seizures were described. Intravenous administration of pyridoxine 10 min before and 30 min after deoxypryidoxine reduced the effect if a 2:1 ratio was used and abolished them in a 4:1 ratio. Subconvulsive doses (40 to 100 mg/kg) produced electrical changes in rhesus monkeys. Convulsive doses (100 to 150 mg/kg) produced effects similar to those seen in baboons. The authors noted that previous unsuccessful attempts to block deoxypyridoxine effects with pyridoxine did not exceed a twofold excess of pyridoxine. Since deoxypyridoxine produced similar effects in both species, the authors concluded that the photosensitivity resulted from abnormal $B_6$ metabolism. The authors postulated an intracortical mechanism involving gamma-aminobutyric acid.

Similarly in a later study, Horton and Meldrum[203] reported that intravenous administration of 0.21 to 0.53 mmol/kg enhanced photosensitivity and 0.53 to 0.87 mmol/kg induced seizures in baboons *(Papio papio)*. The seizures showed features similar to those produced by allylglycine in having a focal origin in one occipital cortex with slow progression to the rest of the cortex.

Depending on the dosage, deoxypyridoxine may also produce morphological alterations. Olsen and Martindale[370] reported no histological changes in the brains of $B_6$-deficient rats. However, Hartsook and Hershberger[191] reported that compared with $B_6$ deficiency alone, addition of deoxypyridoxine (100 mg/kg diet) significantly ($p < 0.01$) decreased gain and increased relative brain weight in rats. With a higher dosage (1 mg/

m*l* drinking water) in rats receiving normal $B_6$, Bayoumi et al.,[30] found no change in brain weight after 29 days but did note a tendency toward lower weight after 48 days.

While Lampert and Schochet[264] produced spongy degeneration of the white matter in ducklings through treatment with isonicotinic acid hydrazide, they were unsuccessful in producing similar effects with deoxypyridoxine. However, the conditions of the deoxypyridoxine studies are not specified.

Klinghardt[238] observed a loss of medullary substance in the exposed sciatic plexus of the rat after local application of deoxypyridoxine. A photograph after 18 days was shown, but no details concerning the experimental procedure were reported. Yonezawa et al.[561,562] examined the effects of deoxypyridoxine ($10^{-3}$ to $10^{-5}$ *M*) on myelinating and matured cultures of newborn cerebellum and fetal spinal ganglia from mice and rats. At $10^{-5}$ *M* there was little effect. However, at $10^{-3}$ *M*, cerebellar cultures showed degeneration of nerve cells and oligodendroglia. "Astrocytes were less affected. The neuron somas generally remained small. The development of Nissl bodies was considerably depressed and granular degeneration of the cytoplasm was often observed. The total oligodendroglial population was decreased as compared to the control nondeficient group." The rate of myelination with $10^{-3}$ *M* deoxypyridoxine was only about 50% of the control value. This decrease was completely offset by equimolar pyridoxine. These changes were produced only when the serum content of the medium was reduced to 15% from its normal 25%. While the authors related this to decreased protein content, serum is also a good source of pyridoxal. The authors described the effects of deoxypyridoxine on mature cultures as follows:

> The effect of deoxypyridoxine on matured myelinated culture was observed principally in nerve cells. With a concentration of $10^{-3}$ mol. initial changes appeared approximately five days after deoxypyridoxine application. Nerve cells began to decrease in size, and their normal rounded form became somewhat angular. In the perikarya, coarse refractile granules appeared. With these changes, chromatolysis and narrowing of axis cylinders were also seen. Alteration of perikarya was followed by Wallerian type of degeneration in the myelinated nerve fibers. It seemed that the neuronal changes were more frequently seen in nerve cells of ganglion cultures than those of cerebellum. On rare occasions, nerve cells were swollen and the cytoplasm developed vacuolar changes. This type of degeneration was seen only with high concentrations of deoxypyridoxine. In addition to this neuronal involvement, alteration of myelin other than Wallerian degeneration was seen, more often in the cultures of peripheral ganglia. The alterations were characterized by swelling and splitting of myelin sheaths close to the Schwannian nuclei. In the swollen area, the axis cylinder remained intact. The swelling was interpreted as being caused by some fluid accumulation between myelin lamellae and the axolemma. Another change in myelin sheaths was seen at Ranvier's nodes, which was characterized by elongation of nodal gaps and clubshaped myelinated stumps. This resulted in gradual shortening of internodal length. Though the shortening of myelin segments took place at both ends, one end may be more severely involved, causing displacement of Schwannian nuclei toward the degenerating end. The shortening of myelin sheaths denuded the axis cylinders, which then disintegrated. This change has been considered as a chronic type of segmental degeneration, suggesting primary involvement of Schwann cells. Wallerian degeneration and primary Schwann cell involvement may occur simultaneously, though the former changes were more frequent. With lower concentration of deoxypyridoxine ($10^{-4}$ to $10^{-5}$ mol), these changes were slower, though the patterns of alteration were identical. Development of deficiency changes was found to occur earlier and more severely in the young cultures. For example, cultures two months *in vitro* required two weeks for the initial change of neutron somas with a concentration of $10^{-3}$ mol of deoxypyridoxine, whereas cultures 20 days *in vitro* might respond with a similar change within a week.*

Comparable doses of isonicotinic acid hydrazide produced less severe changes of a slightly different nature. However, the effects of isonicotinic acid hydrazide were not prevented by pyridoxine. Only slight recovery was possible after deoxypyridoxine treatment even if pyridoxine supplementation was started before Wallerian degeneration

* From Yonezawa, T., Mori, T., and Nakatani, Y., *Ann. N.Y. Acad. Sci.*, 166, 147, 1969. With permission.

occurred. The authors concluded that this suggested severe metabolic changes had occurred even though morphological changes were minimal.

Victor and Adams[528] discussed the neuropathology produced by $B_6$ deficiency in monkeys. One monkey received 100 mg deoxypyridoxine per kilogram subcutaneously for 47 days. The diet was not specified but it seems likely that it was $B_6$ deficient. The axis cylinders were generally intact in the $B_6$-deficient animals. However, in the monkey receiving deoxypyridoxine, many nerve fibers exhibited disintegration involving both the axis cylinder and myelin sheath.

### B. Liver

All reports agree that deoxypyridoxine produces a relative increase in liver weight but no detectable morphological changes. Beare et al.[34] reported a relative increase in the size of the liver in rats receiving a $B_6$-deficient diet plus 100 μg deoxypyridoxine per day compared with animals receiving the same treatment plus $B_6$. Olsen and Martindale[370] found no histological changes in the liver of $B_6$-deficient rats. However, the relative size was significantly ($p < 0.01$) increased.[371] Silva[456] observed no histological lesions in the livers of two rats treated with a $B_6$-deficient diet plus an unspecified amount of deoxypyridoxine for 6 weeks. Hsu and Kawin[208] reported an increase in the relative weight of the liver in $B_6$-deficient rats. Treatment with deoxypyridoxine (0.75 mg deoxypyridoxine hydrochloride i.p. three times per week) produced a further increase. However, it appears that these changes were associated primarily with decreased body weight and that the absolute liver weights of the deficient animals were smaller than those of the controls receiving $B_6$. Easton et al.[128] found no histologic evidence of cirrhosis after 18 days on a $B_6$-deficient diet containing 10 mg deoxypyridoxine per kilogram.

### C. Kidney

The observations on the kidney have been less consistent than those in the liver. Olsen and Martindale[370] found no histological changes in the kidneys of $B_6$-deficient rats. However, their relative size was significantly ($p < 0.01$) increased.[371] Boone et al.[63] reported no change in kidney weight in rats receiving a $B_6$-deficient diet plus 100 μg deoxypyridoxine per day intraperitoneally. Talageri[494] observed no changes in the renal histology of male rats receiving a $B_6$-deficient diet containing 20 mg deoxypyridoxine hydrochloride per kilogram. Hsu and Kawin[208] reported that $B_6$ deficiency produced a significant increase in the relative weights of the kidneys. Deoxypyridoxine treatment produced a further increase. However, it appears that the change was associated more with decreased body weight than with true hypertrophy. Pol et al.[395] reported that oxalate kidney stones occurred almost exclusively in the rats (particularly males) receiving a $B_6$-deficient diet plus deoxypyridoxine (50 mg/kg diet).

### D. Pancreas

Olsen and Martindale[370] observed no histological changes in the pancreas of $B_6$-deficient rats. Injecting 20 mg deoxypyridoxine per kilogram produced a 25% decrease in pancreas weight in $B_6$-deficient rats compared with animals receiving only the deficient diet.[556] Injecting the deficient animals with 10 mg $B_6$ per kg/day tended to increase pancreas weight. Deoxypyridoxine seemed to have little effect in the $B_6$-supplemented animals. The significance of these results is limited by the small size (two rats per treatment) of the groups.

### E. Reproduction

Nelson and Evans[363] noted that dietary $B_6$ deficiency alone even in combination with succinylsulfathiazole to repress intestinal synthesis had little effect on reproduction in rats. Addition of 5 mg deoxypyridoxine per kilogram of diet caused some reproductive impairment if the experimental diet started on the day of breeding. Total reproductive failure occurred if the diet was started 22 days before breeding. However, the effects of 14 days of deoxypyridoxine treatment were completely offset if the normal diet was restored on the day of breeding. Acrodynia often appeared within 15 days after starting deoxypyridoxine treatment.

Nelson and Evans[364] continued their study of deoxypyridoxine effects in reproduction in rats. The animals received a $B_6$-deficient diet either alone or supplemented with 1% succinylsulfathiazole or 0.5 mg% deoxypyridoxine. Control rats received the deoxypyridoxine diet supplemented with either 0.5 or 5.0 mg% pyridoxine hydrochloride. Institution of the deficient diet either 3 weeks before breeding or on the day of breeding had minimal effect on reproductive performance. Neither did the antibiotic treatment aggravate the deficiency. Treatment with deoxypyridoxine for 3 weeks prior to breeding produced 90 to 100% resorption. The deoxypyridoxine groups also showed loss of the estrous cycle and acrodynia. However, despite this severe impairment, these animals showed less than 5% mortality after 60 days of deficiency, and body weight was maintained for 30 to 60 days and then declined gradually. Pair-feeding demonstrated that inanition alone could not explain the results. The growth of young with $B_6$-deficient dams was retarded. Neither succinylsulfathiazole nor deoxypyridoxine accentuated this retardation during lactation.

Goldsmith[170] observed that either partial inanition or deoxypyridoxine in conjunction with a $B_6$-deficient diet reproduced the response of target organs in mice to sex hormones. Paul et al.[385] also reported that treating mice apparently receiving a normal $B_6$ intake with 1.5 mg deoxypyridoxine per day significantly decreased uterine growth following estrogen treatment. Part of this effect may have been due to general poor health.

Ross and Pike[425,427] used a $B_6$-deficient ration containing 5 mg deoxypyridoxine per kilogram for at least 7 days before mating. This depletion period resulted in reduced weight in both dams and young even when a pyridoxine supplement (12 mg/kg diet) was given during gestation. The number of live young was not affected, suggesting that litter size alone is not a completely adequate measure of reproductive performance.

Ross and Pike[425] noted that the average weights of young from the depleted mothers maintained during gestation on diets containing pyridoxine were as low or lower than those of the young from nondepleted animals maintained on the deoxypyridoxine-supplemented diet during gestation. On this basis they stated that the results were in direct contrast to the report by Nelson and Evans[363] that the weight of the young was not affected adversely if pyridoxine was incorporated into the diet of depleted animals on the first day of gestation. Actually, both studies agree that supplementing depleted animals with pyridoxine on the day of breeding improved reproductive performance. The question is whether such treatment completely restored normal reproductive performance. Since Nelson and Evans[363] had no animals which had received a normal $B_6$ intake throughout the entire experiment, they had no basis for assessing maximum performance, and their statement that $B_6$ supplementation completely offset the effects of depletion was unsupported. However, Ross and Pike[425] also erred in claiming that there was a conflict because their results with nondepleted controls did not agree with the results obtained by Nelson and Evans with depleted controls. A more significant difference upon which Ross and Pike did not comment was the fact that Nelson and Evans reported an average fetal weight of over 5 g in their depleted controls whereas Ross and Pike[425] obtained only 4.3 g. Ross and Pike used weight exceeding 5 g as a criterion of normal reproductive performance. Since Nelson and Evans[363] obtained average weights of over 5 g for three of their four groups it is possible that their animals were not quite as deficient as those of Ross and Pike.[425] At any rate, we disagree with the conclusion of Ross and Pike[425] that their data seriously conflict with those of Nelson and Evans.[363]

Pike and Brown[391] increased the deoxypyridoxine content of the $B_6$-deficient diet to 40 mg/kg diet. Treatment of some animals was started on the day of mating and of others 6 to 26 days prior to mating. Estrous cycles became irregular after about a week on the deoxypyridoxine treatment. Some of the depleted animals were restored to a normal diet on the day of mating. Deoxypyridoxine prevented implantation in half of the animals and caused resorption in those that did implant. Reducing the deoxypyridoxine dosage to 20 mg deoxypyridoxine per kilogram diet permitted about 50% of the animals to complete their pregnancies.[69,70] Pike and Kirksey[392] also used a $B_6$-deficient diet containing 40 mg deoxypyridoxine per kilogram and started treatment on the day of mating. The treatment significantly ($p < 0.01$) reduced litter size and increased the number of dead or resorbed fetuses.

In the first of several studies of the effect of deoxypyridoxine on reproduction, Lauro et al.[267] treated adult rats with 2 mg deoxypyridoxine per day intraperitoneally (approximately 5.5 mg/kg body weight) for ten days prior to mating. Another group received 3 mg/day (about 8.5 mg/kg). A third group received 3 mg on days 4 through 8 after mating. A final group received 3 mg/day on days 9 through 12 after mating. The $B_6$ content of diet was not specified. The pretreatment reduced pregnancy and ovulation. Both pretreatment and 4- to 8-day treatment produced about 75% dead ova and a severe fetal weight deficit. The abortion rate did not differ markedly between the treatments. None of the treatments produced malformations.

Lauro et al.[268] administered deoxypyridoxine (2 or 3 mg/day i.p.) to rats for 10 days before mating or between 4 to 8, or 9 to 12 days of pregnancy. No specific data are given. The authors concluded reproduction was affected at several levels but was particularly sensitive at the blastocyst stage. Lauro and Dominici[269] found that deoxypyridoxine damaged rat blastocysts in vitro.

Lauro et al.[270] used a dose of 8.5 mg deoxypyridoxine per kilogram body weight intraperitoneally either alone or followed by an equal dose of pyridoxine 60 min later in pregnant rats, mice, and hamsters. The effect of deoxypyridoxine was much less on the eighth day of pregnancy than on the fourth, fifth, or sixth days. The toxicity was somewhat reduced by injection of pyridoxine. The authors assumed that the eighth day was postnidation and that the results indicated a marked decrease in sensitivity to dosage with nidation. They noted that this prenidation phase is frequently overlooked in embryological studies.

Paul et al.[385] used deoxypyridoxine in an attempt to separate the stimulation of uterine growth from the eosinophilia produced by estrogen. Deoxypyridoxine (1.5 mg/day per mouse) significantly decreased uterine growth following estrogen treatment. Part of this effect may have been due to general poor health. Eosinophilia was noted.

Richter[411] observed no teratogenic effects when pregnant rats receiving a normal diet were treated with 5 mg deoxypyridoxine per day during gestation. However, under conditions of more severe deficiency, teratogenic effects have been observed in both rats and mice.

Davis et al.[111] studied the effect of $B_6$ deficiency during pregnancy. A $B_6$-deficient diet in conjunction with drinking water containing 0.1 mg deoxypyridoxine per m$\ell$ was started the day after mating. Control animals received the same diet and water containing deoxypyridoxine plus 1 mg pyridoxine per m$\ell$ in the water. Pregnant does were sacrificed on day 20. Deficient animals consumed an average of 12.3 g feed plus 2.3 mg deoxypyridoxine. Control animals consumed 19.4 g of feed, 24.5 mg pyridoxine, and 2.5 mg deoxypyridoxine. While both groups averaged 12.6 implantations per rat, there was an average of 5.4 resorptions per rat in the deficient group compared with 0.4 in the controls. Mean fetal length and weight were 1.63 g and 28.3 mm in the deficient animals compared with 3.24 g and 37.0 mm in the controls. While no gross physical anomalies were noted in the controls, the deficient fetuses exhibited digital defects (48), cleft palate (20), omphalocele (12), micrognathia (8), and exencephaly (6). Fetal organ weights were 0.88, 0.19, and 0.03/100 g body weight for the kidney, thymus, and spleen, respectively, in the deficient animals compared with 0.85, 0.24, and 0.14 in the controls. Postnatally, the thymus is usually affected more severely than the spleen. In this experiment the spleen was affected most severely. Since both groups received deoxypyridoxine, it is possible that the thymus was reduced in size in both groups. The authors commented that this was the first example of teratogenicity in $B_6$ deficiency.

In mice, Miller[329] used a $B_6$-deficient diet supplemented with 0.5% succinylsulfathiazole to reduce intestinal flora and 0.1% 4′-deoxypyridoxine. The authors mention preventing coprophagy but apparently assumed that wire floors were adequate for this purpose. Additional precautions are usually necessary. Administration of succinylsulfathiazole on days 10 to 14 did not increase the incidence of cleft palate over that produced by the deficient diet alone. Combining succinylsulfathiazole with deoxypyridoxine during days 10 to 14 caused 61% of the fetuses to have cleft palate with 83% of the litters containing at least one affected fetus. This is a marked increase from 21% affected fetuses with at least one affected pup in 50% of the litters observed with the deficient diet alone. Extending the succinylsulfathiazole treatment to include days 1 to 15 with deoxypyridoxine added on days 10 to 14 produced 100% affected fetuses. Since no group received only the succinylsulfathiazole over days 1 to 15, it is impossible to determine the relative importance of the deoxypyridoxine in producing these last results. The treatments did not change the mean litter size or rate of resorption. The author also found that a dietary pyridoxine deficiency seemed to aggravate the teratogenic action of cortisone.

Helgebostad et al.[194] found that 8 mg deoxypyridoxine per day in conjunction with a $B_6$-deficient diet produced sterility in mink. Histological figures showing testicular hypoplasia were included.

### F. Circulatory System

The response of the components of the circulatory system to deoxypyridoxine varies from minimal to severe. The mechanical aspects are least affected. Olsen and Martindale[370] reported that the relative size of the heart was increased in $B_6$-deficient rats, but Hsu and Kawin[208] did not confirm this observation. No histological changes were detected in the heart and lungs by Olsen and Martindale[370] or by Silva,[456] who used a $B_6$-deficient diet plus an unspecified amount of deoxypyridoxine. Olsen and Martindale[370] found that mean systolic blood pressure was increased when deoxypyridoxine (12.5 mg/kg diet) was added to a normal $B_6$ diet but not with a $B_6$-deficient diet. While these seem to be the only data on blood pressure with a normal $B_6$ intake, the lack of effect on blood pressure with a $B_6$-deficient diet was later confirmed by Brown and Pike[70] using 20 mg deoxypyridoxine per kilogram diet and by Pol et al.[395] using 50 mg deoxypyridoxine per kilogram diet.

Levine et al.,[273] examined the effects of pyridoxal and deoxypyridoxine on the mechanical activity and sodium and potassium content of rabbit atria in vivo and in vitro. Pyridoxal (0.001 *M*), deoxypyridoxine (0.002 *M*), and 0.001 *M* pyridoxal plus 0.001 *M* deoxypyridoxine did not significantly alter the frequency of contraction in vitro. However, 0.001 *M* pyridoxal plus 0.002 *M* deoxypyridoxine and 0.0005 *M* pyridoxal plus 0.001 *M* deoxypyridoxine produced a significant ($p < 0.01$) increase in the frequency of contraction. The lack of response to the 0.001 *M* combination may have been an artifact. Pyridoxal, both alone and with various combinations of deoxypyridoxine, increased the amplitude of atrial contraction. Deoxypyridoxine alone had no effect.

DeLorme et al.[113] studied the effect of $B_6$ on the renin angiotensin system in rats. Deoxypyridoxine (5 mg/100g body weight per day i.p.) was administered in conjunction with a commercial stock diet. Growth was unimpaired and no dermatological effects were noted. Deoxypyridoxine had no effect on the response to angiotensin II, basal blood pressure, plasma angiotensin levels, or plasma renin. However, both deoxypyridoxine and pyridoxine (5 mg/100 g body weight per day i.p.) produced significant decreases in the response to renin. While the data were not sufficient to determine whether deoxypyridoxine and pyridoxine were acting through the same mechanism, the authors did conclude that the use of deoxypyridoxine to produce or accentuate a $B_6$ deficiency may give misleading results in some systems. Dietary $B_6$ deficiency produced a rise in basal blood pressure, plasma angiotensin, and renin with a decreased response to renin and angiotensin II.

After 14 months of treatment with a $B_6$-deficient diet plus intramuscular (i.m.) injection of 5 mg deoxypyridoxine per kilogram body weight every alternate fifth day, adult male monkeys *(Macaca radiata)* showed numerous biochemical symptoms of $B_6$ deficiency but few clinical signs.[260] The authors had anticipated that $B_6$ deficiency would induce homocystinemia followed by development of atherosclerosis. While there were some changes in the lipid composition of the arterial wall, histological studies revealed no changes in the gross or histological characteristics of the organs and tissues including the blood vessels when compared with animals receiving a normal $B_6$ intake and no deoxypyridoxine.

A series of studies by Beaton, Beare et al.[34,36,40] failed to detect any consistent changes in packed cell volume or hemoglobin in rats receiving 100 $\mu$g deoxypyridoxine per day in conjunction with a $B_6$-deficient diet compared with controls receiving the same treatment plus pyridoxine. Bianchi-Donnasibilla and Maggiora[51] detected no significant changes in erythrocyte metabolism as measured by erythrocyte count, reticulocyte count, hemoglobin, hematocrit, or diameter of the erythrocyte in rats after 50 days on a natural diet plus 400 $\mu$g deoxypyridoxine per day intramuscularly. Eight days of treating rats with a $B_6$-deficient diet plus 20 mg deoxypyridoxine per kilogram body weight per day produced a slight increase in hematocrit ($51.74 \pm 0.36$ vs. $49.94 \pm 0.48$ in controls).[206] A dose of 40 mg/kg in rats receiving a commercial stock diet also produced a slight increase in hemoglobin.[122]

More severe effects have been noted in other species. Pyridoxine-deficient puppies showed a progressive decrease in erythrocyte count, hemoglobin, and hematocrit.[355] The resulting anemia was microcytic and hypochromic. Pups receiving the deficient diet plus 5 mg deoxypyridoxine per kilogram body weight per day showed a delayed microcytic anemia, sometimes followed by partial recovery. However, in a dog which had anemia as a result of 12 weeks on a $B_6$-deficient diet, intravenous administration of deoxypyridoxine (10 mg/kg for three days) markedly increased both hemoglobin and hematocrit values.[421] Such rapid changes probably reflect hemoconcentration rather than increased production of erythrocytes.

While monkeys receiving 50 mg deoxypyridoxine per kilogram body weight for 144 days in conjunction with a natural stock diet showed only transient blood changes, animals receiving 100 mg/kg/day developed a microcytic anemia and the femoral bone marrow was pale yellow with a gelatinous consistency.[355]

Benesch et al.[43] have observed some intriguing interactions between vitamin $B_6$ compounds and hemoglobin. Pyridoxal phosphate but not pyridoxal was as effective as 2,3-diphosphoglycerate in reducing the affinity of hemoglobin for oxygen. In further studies[42] they found that 5′-deoxypyridoxal reacted with the terminal amino groups of the alpha-chains of hemoglobin and increased the oxygen affinity of hemoglobin solutions as well as suspensions of normal and sickle erythrocytes. As a result of the increased affinity for oxygen, sickling was reduced. Benesch et al.[42] noted that pyridoxal has a much smaller effect. However, these results do suggest that $B_6$ metabolism might have a role in oxygen transport. It would be interesting to know whether the alterations in the characteristics of erythrocytes in black Americans (e.g., decreased pyridoxal kinase[82a] and decreased fragility[95a]) are correlated with changes in oxygen transport.

In contrast to the limited effects on erythrocytes, leukocytes and lymphoid tissue in general are among the tissues most sensitive to deoxypyridoxine. These effects will be discussed later in conjunction with the immune system (Chapter 5).

## V. SUMMARY

The data reported here illustrate the wide variety of protocols encountered in studies of vitamin $B_6$ deficiency (see Tables 4 to 6). It is also obvious that the effects of deoxypyridoxine are heavily dependent on a variety of factors including the vitamin $B_6$ content of the diet, the source and amount of protein in the diet, and the relative timing of deoxypyridoxine administration and measurement of experimental parameters. Therefore, caution must be used in trying to compare various reports.

The two most consistently reported effects of deoxypyridoxine in rats are dermatitis and weight loss. In both cases, deoxypyridoxine administration appears to produce a significantly more severe effect than simple dietary deficiency. Dosages most commonly used were 100 μg/day or greater. Gadekar et al.[151] reported that 50 μg/day in 40 g rats which had already been receiving a $B_6$-deficient 30% casein diet for 16 days was fatal within a week. These dosages correspond to about 1 mg/kg body weight. The important role of the protein content of the diet is emphasized by the observation that increasing the casein content of the diet significantly decreased weight gain and survival in rats receiving a $B_6$-deficient diet containing 100 mg deoxypyridoxine per kilogram.[191] These workers also reported that addition of 2% glycine to a $B_6$-deficient diet containing 100 mg deoxypyridoxine per kilogram prevented weight loss. These results suggest that increased glycine intake might explain why a 20% gelatin diet prevented acrodynia but a 20% casein diet did not.[40]

The observation that isonicotinic acid hydrazide produced greater weight losses than deoxypyridoxine without producing acrodynia[63,403] indicates that those two effects are not necessarily causally related. Our experience suggests that decreased food consumption is probably the major cause of the marked weight loss which frequently accompanies administration of 4′-deoxypyridoxine. This decreased consumption does not appear to be a taste factor since animals receiving normal $B_6$ in conjunction with deoxypyridoxine maintain normal feed intake. Neither does the decreased food intake appear to be the result of a generalized $B_6$ deficiency. In a recent study in our laboratory, the rats receiving a $B_6$-deficient diet plus deoxypyridoxine (1 mg/g diet) showed a 10% drop in feed intake within 3 days and a 35% drop within 6 days compared to those receiving only the $B_6$-deficient diet. The feed intake of the latter group never decreased below initial levels even though over 5 weeks their plasma pyridoxal phosphate concentrations dropped to about 15% of their original values. No alteration in food intake was noted when deoxypyridoxine was fed in conjunction with normal $B_6$ intake. On the basis of these results, we suspect that the initial rapid effect of deoxypyridoxine on feed intake is due to alteration in the function of some specific organ or system which is very sensitive to the effects of deoxypyridoxine. Boullin's[64] failure to observe a decrease in food intake may indicate that an interaction between several factors is involved.

**Table 4**
**SUMMARY OF GROSS CHANGES IN THE RAT**

| Ref. | $B_6$[a] | Deoxypyridoxine | Decreased growth[b] | Dermatitis[b] | Convulsions[b] | Abnormal hematology[b] |
|---|---|---|---|---|---|---|
| Emerson[134] | − | 5 mg/kg diet | + | + | | |
| Nelson and Evans[363] | − | 5 mg/kg diet | | + | | |
| Nelson and Evans[364] | − | 5 mg/kg diet | + | + | | |
| Ross and Pike[425] | − | 5 mg/kg diet | + | + | | |
| Ross and Pike[427] | − | 5 mg/kg diet | + | | | |
| Hartsook et al.[89] | − | 5 mg/kg diet | + | + | | |
| Gadekar[151] | − | 50 μ/rat | + | | − | |
| Hawkins and Evans[192] | − | 100 μg/day | + | + | | + |
| Caldwell and McHenry[77] | − | 100 μg/day | + | + | | |
| Beaton et al.[36] | − | 100 μg/day | + | + | | − |
| Beare et al.[34] | − | 100 μg/day | + | + | | − |
| Beaton et al.[35] | − | 100 μg/day | + | + | | |
| Beaton and McHenry[39] | − | 100 μg/day | + | + | | |
| Beare et al.[33] | − | 100 μg/day | + | + | | |
| Beare et al.[32] | − | 100 μg/day | + | + | | |
| Beaton et al.[40] | − | 100 μg/day | + | + | | |
| Beaton and Goodwin[37] | − | 100 μg/day | + | + | | |
| Beaton et al.[38] | − | 100 μg/day | + | + | | |
| Boone et al.[63] | − | 100 μg/rat | + | + | − | |
| Silva[456] | − | 10 mg/kg | | + | | |
| Olsen and Martindale[370] | − | 12.5 mg/kg diet | + | + | + | − |
| Iwanowska[217] | − | 200 μg/day | + | | | |
| Gregoire[173] | − | 250 μg/rat | | + | | |
| Ling et al.[277] | − | 20 mg/kg diet | | + | | |
| Brown and Pike[69] | − | 20 mg/kg diet | | | | + |
| Talageri[494] | − | 20 mg/kg diet | + | + | | |
| Hruban et al.[206] | − | 20 mg/kg | − | | | + |
| Meister et al.[305] | — | 0.4 mg/day | + | | | |
| Pike and Kirksey[392] | − | 40 mg/kg diet | + | + | | |
| Pike and Brown[391] | − | 40 mg/kg diet | | | | + |

| | | | | | | |
|---|---|---|---|---|---|---|
| Hsu et al.[207] | – | 0.75 mg 3×/ week | + | | | |
| Hartsook and Hershberger[191] | – | 100 mg/kg diet | + | | | |
| Hruban and Wissler[205] | – | 20 mg/kg | | – | | |
| Stoerk[477] | – | 0.3 mg/mℓ $H_2O$ | + | + | | |
| Rasmussen[403] | – | 500 mg/kg diet | + | + | | |
| Kawakita[236] | 10µg | 100 µg/day | + | | | |
| Bianchi-Donnasibilla[51] | + | 400 µg/day | | | | + |
| DeLorme et al.[113] | + | 50 mg/kg | – | – | | |
| Doctor[122] | + | 100 mg/kg | + | | | + |
| Rosen et al.[423] | + | 100 mg/kg | | | – | |
| Bayoumi et al.[30] | + | 1 mg/mℓ water | + | | – | |
| Halverson and Hendrick[180] | ? | 100 mg/kg diet | ± | | | |
| Anglesio[14] | ? | 75 mg/kg | – | | | – |
| Tsunoo[518] | ? | 400 mg/kg | | | + | |
| Kosower and Rock[257] | ? | 3.5 mmol/kg | | | + | |

*Note:* Blanks indicate that the authors made no comment about that parameter.

[a] + = Normal $B_6$ intake
– = $B_6$ deficient diet
? = The authors did not specify the $B_6$ content of the diet.

[b] + = The abnormality was observed.
– = The authors stated specifically that the abnormality was not observed.

Table 5
EFFECTS OF DEOXYPYRIDOXINE IN ORGANISMS OTHER THAN THE RAT

| Ref. | $B_6$[a] | Deoxypyridoxine | Decreased growth[b] | Death | Dermatitis[b] | Convulsions[b] | Abnormal hematology[b] |
|---|---|---|---|---|---|---|---|
| | | | **Mouse** | | | | |
| DeRenzo and Cerecedo[114] | − | 20 mg/kg diet | | Within 32 days | + | | |
| Korytnyk et al.[249] | − | 25 mg/kg | | Within a few days | | | |
| Weir et al.[544] | − | 1 mg/day | + | Within 10 days | | | |
| Weir et al.[544] | + | 4 mg/day | − | | − | − | + |
| Dietrich and Shapiro[116] | + | 150 mg/kg | − | | − | | |
| Korytnyk et al.[249] | + | 250 mg/kg | | Within a few days | | | |
| Skipper et al.[461] | + or − | 400 mg/kg | | Within 24 hours | | | |
| Weir and Heinle[543] | + | 500 mg/kg diet | + | | + | | + |
| Mueller et al.[347] | + | 1000 mg/kg | | | | | + |
| Nozaki[368] | ? | 150 mg/kg | | | | | + |
| Dubnick et al.[126] | ? | 200 mg/kg | | 6 of 10 animals | | | |
| Horton and Meldrum[203] | ? | 1.1 mmol/kg | | | | + | |
| Bain and Williams[27] | ? | 250 mg/kg | | | | + | |
| Horton and Meldrum[204] | ? | 250 mg/kg | | | | + | |
| Matano[301] | + | 300 mg/kg | | | | + | |
| Pfeiffer[388] | ? | 1.5 mmol/kg | | | | + | |
| | | | **Dog** | | | | |
| Forse and Webster[146] | + or − | 5 mg/kg | | | | | + |
| Mushett et al.[355] | − | 5 mg/kg | + | Within 60 days | + | + | + |
| Hawkins and Evans[192] | − | 5 mg/kg | + | Within 60 days | + | + | + |
| Mihich et al.[326] | + or − | 40 mg/kg | | | | | + |
| Rosen et al.[421] | + | 500 mg/kg diet | | | | | + |
| | | | **Chick** | | | | |
| Ott[375] | + | 100 μg | | Yes | | | |
| Cravens and Snell[106] | | 1 mg/egg at 0 days of age | | Yes | | | |

| | | | | | |
|---|---|---|---|---|---|
| Cravens and Snell[106] | | 5 mg/egg at 6 days of age | | Yes | |
| Karnofsky[232] | | 3 mg/egg | + | Yes | |
| | | | **Baboon** | | |
| Moldrum and Horton[311] | ? | 100 mg/kg | | | + |
| Horton and Meldrum[203] | ? | 0.53 mmol/kg | | | + |
| | | | **Hamster** | | |
| Shwartzman and Hift[453] | + | 50 μg 3×/week | − | | |
| Shwartzman and Hift[453] | − | 50 μg 3×/week | + | | |

*Note:* Blanks indicate that the authors made no comment about that parameter.

[a] + = Normal $B_6$ intake
− = $B_6$ deficient diet
? = The authors did not specify the $B_6$ content of the diet.

[b] + = The abnormality was observed.
− = The authors stated specifically that the abnormality was not observed.

## Table 6
## EFFECTS OF DEOXYPYRIDOXINE ON ORGAN WEIGHTS AND HISTOLOGY

| Ref. | $B_6$ | DOP | Adrenal | | Liver | | Kidney | | Brain | | Spleen | | Thymus | | Testes | | Lungs | | Heart | |
|---|---|---|---|---|---|---|---|---|---|---|---|---|---|---|---|---|---|---|---|---|
| | | | Wt | Hist | Wt | Hist | Wt | Hist | Wt | Hist | Wt | Hist | Wt | Hist | Wt | Hist | Wt | Hist | Wt | Hist |
| **Rat** | | | | | | | | | | | | | | | | | | | | |
| Ross and Pike[425] | – | 5 mg/kg diet | | | + | | | | | | | | | | | | | | | |
| Ross and Pike[426] | – | 5 mg/kg diet | | | +(R) | | | | | | | | | | | | | | | |
| Ross and Pike[427] | – | 5 mg/kg diet | | | +(R) | | | | | | | | | | | | | | | |
| Easton et al.[128] | – | 10 mg/kg diet | | | | N | | | | | | | | | | | | | | |
| Hawkins and Evans[192] | – | 100 μg/day | | N | | | | | | | + | N | + | A | | | | | | |
| Beare et al.[33] | – | 100 μg/day | +(R) | | | | | | | | −(R) | | – | | | | | | | |
| Beare et al.[34] | – | 100 μg/day | +(R) | | +(R) | | | | | | | | −(R) | | | | | | | |
| Boone et al.[63] | – | 100 μg/day | | | | | 0 | | | | 0 | | – | | 0 | | | | | |
| Gregoire[173] | – | 250 μg/rat | | | | | | | | | | N | – | A | | | | | | |
| Talageri[494] | – | 20 mg/kg diet | | | | | | N | | | | | | | | | | | | |
| Silva[456] | – | 10 mg/kg (?) | | N | | N | | N | | | | | | | | | | N | | N |
| Hartsook and Hershberger[191] | – | 100 mg/kg diet | | | | | | | +(R) | | | | | | | | | | | |
| Hruban et al.[206] | – | 20 mg/kg | | | 0 | | 0 | | | | 0 | | | | | | | | | |
| Yonezawa et al.[562] | – | $10^{-3}$ *M* | | | | | | | | A | | | | | | | | | | |
| Helgebostad et al.[194] | – | 8 mg/day | | | | | | | | | | | | | A | | | | | |

| | | | | | | | | | | | | | |
|---|---|---|---|---|---|---|---|---|---|---|---|---|---|
| Bayoumi et al.[30] | + | 1 mg/ml $H_2O$ | | | | | – | | | | | | |
| Hruban and Wissler[205] | – | 20 mg/day | | | – | 0 | | 0 | | – | | | |
| Hsu and Kawin[208] | – | ? | +(R) | | +(R) | +(R) | | +(R) | | | – | +(R) | O(R) |
| **Mouse** | | | | | | | | | | | | | |
| Weir et al.[544] | + | 4 mg/day | | N | | | | | N | | N | | |
| Weir and Heinle[542] | + | 500 mg/kg diet | | | | | | | | 0 | N | | |
| **Chicks** | | | | | | | | | | | | | |
| Mushett et al.[355] | – | 2 mg/kg diet | | | | | | – | | | | | |
| **Dog** | | | | | | | | | | | | | |
| Mushett et al.[255] | – | 5 mg/kg | | A | | | | – | A | – | A | | |
| **Monkeys** | | | | | | | | | | | | | |
| Mushett et al.[355] | + | 100 mg/kg | | A | | | | | A | | | | |

*Note:* Organ weight: + = enlarged in the deoxypyridoxine group; 0 = no change; – = decreased; R = data based on organ weight relative to body weight rather than absolute organ weights. Histology: N = normal; A = abnormal.

Interpreting changes in organ weights is difficult, particularly in animals which are losing weight. In the data discussed in this chapter there was little change in absolute organ weight but relative weights of several organs were significantly increased due to the smaller body size of the animals treated with deoxypyridoxine. Does this represent true hypertrophy? In the case of animals that are losing weight, it seems logical that organs such as the adrenals and kidneys will retain their original size. In the case of growing animals, it appears that development of these organs is not impeded as much as development of body mass. Therefore, it is our opinion that the increased relative weight of several organs represents normal development and not a compensatory hypertrophy.

The possible interaction between deoxypyridoxine and steroid hormones is an interesting question. There have been no consistently reported histological changes in the adrenal glands and, as discussed above, we question whether the observed increase in relative weight represents true hypertrophy, especially in animals that are losing weight. However, dermatitis did not occur in rats that had been adrenalectomized, indicating that the adrenal gland does influence these processes.[34] Both deoxypyridoxine and cortisone tend to promote granulocytosis and lymphopenia. However, the observation that $B_6$ does not prevent the cortisone effect, and adrenalectomy does not prevent the deoxypyridoxine effect, suggests that these two compounds act via different mechanisms.

While it is well documented that deoxypyridoxine has hematologic effects, there seems to be a significant variation between species. In rats, doses as low as 1 mg/kg in conjunction with a $B_6$-deficient diet produce leukopenia with lymphopenia and increased neutrophils. Atrophy of lymphoid elements in the thymus is consistently observed. Histological changes in the spleen and lymph nodes are not always reported. It is interesting that circulating lymphocytes doubled within two days after restoration of a normal diet and returned to normal within two weeks. However, if deoxypyridoxine was withdrawn but the deficient diet continued, the leukocyte count returned to normal but the granulocyte:lymphocyte ratio did not.[421] Despite the dramatic changes in leukocytes, no abnormalities in circulating erythrocytes have been detected although changes in bone marrow have been noted.[51]

Considering the complexity of the reproductive processes and the metabolic demands associated with pregnancy, it is not surprising that deoxypyridoxine should diminish reproductive performance. However, the data provide a good example of the widely differing sensitivity of various processes to deoxypyridoxine. For example, deoxypyridoxine treatment caused essentially total reproductive failure even under conditions in which the general physical condition of the animals was not severely altered.[364] Does the marked deterioration of reproductive performance with deoxypyridoxine treatment reflect a more severe generalized deficiency state or is it due to an impairment of certain systems (perhaps hormonal) which are especially sensitive to deoxypyridoxine? We tend to favor the latter possibility.

Among the most intriguing observations are those where deoxypyridoxine treatment caused an effect several days or even weeks after treatment ceased. In some cases the effect occurred even in the presence of adequate $B_6$. For example, Ross and Pike[425] found that treatment with deoxypyridoxine prior to mating reduced litter weight even though adequate $B_6$ was supplied during gestation. Such results might be rationalized by postulating that the body of the dam is competing with the embryo for the available $B_6$. However, the most readily available measures of $B_6$ status such as blood pyridoxal phosphate concentrations and pyridoxic acid excretion return to normal very rapidly. Therefore, it seems likely that the embryo has access to an adequate number of $B_6$ molecules. Another explanation might be that although the dam has access to adequate $B_6$ it takes time to replenish her body stores, and this process takes precedence over the needs of the embryo. However, to us, the most attractive explanation is that the preliminary exposure to deoxypyridoxine induces metabolic or hormonal changes which are not immediately corrected upon restoring $B_6$ and thus affect the course of the pregnancy. A similar mechanism might also explain the delayed development of dermatitis.[179,477]

The final major effect of severe $B_6$ deficiency is disturbances of neurological function, most commonly evidenced by convulsions. In this regard deoxypyridoxine is not as potent as some other $B_6$ antagonists. Whereas injection of 3.5 mmol deoxypyridoxine per kilogram killed 80% of the rats within 60 min, reducing the dose to 3 mmol/kg reduced the mortality to 10%. Even a 3 mmol/kg dose is much higher than usually employed but is of the same order of magnitude as the amount needed to affect nerve cultures in vitro.[562] Such convulsions are not associated with altered brain serotinin levels.[518]

Data on species other than rats are quite incomplete. In mice, doses above 150 mg deoxypyridoxine per kilogram are likely to cause convulsions. The $LD_{50}$ is about 200 mg/kg. Doses above 50 mg/kg in conjunction with a $B_6$-deficient diet may be lethal within 3 to 4 weeks. As we observed in rats, increasing the protein content of the diet increased deficiency symptoms. While large doses of deoxypyridoxine are obviously toxic to mice, there has been little detailed study of the pathology involved. Although deoxypyridoxine produces a reversal of the granulocyte:lymphocyte ratio similar to that observed in rats, no histological changes have been observed consistently, probably because several of the studies used a normal $B_6$ diet.

Mushett et al.[355] provided the most extensive description of the pathological effects of deoxypyridoxine in dogs. They also presented the most thorough description of the pathology in monkeys although this was obtained in conjunction with a normal $B_6$ diet. The results of Mushett et al.,[355] Meldrum and Horton,[311] and Horton and Meldrum,[203] all indicate that doses above 100 mg/kg are convulsive in monkeys.

Chapter 4

# BIOCHEMICAL EFFECTS IN ANIMALS

## I. VITAMINS

### A. Vitamin $B_6$

#### *1. Tissue Concentrations*

One of the events which stimulated our interest in the mechanism of deoxypyridoxine action was the report by Johnston et al.[229] that pyridoxal phosphate concentrations in the liver of rats receiving a $B_6$-deficient diet containing 1 g deoxypyridoxine per kilogram were significantly ($p < 0.02$) higher ($6.9 \pm 0.4$ $\mu$g/g wet tissue) than in pair-fed controls ($5.3 \pm 0.2$ $\mu$g/g) receiving only the $B_6$-deficient diet. We subsequently discovered that Stoerk[477] had also found increased pyridoxal phosphate concentrations in the livers of rats receiving deoxypyridoxine (0.3 mg/m*l* drinking water) compared with $B_6$-deficient controls (2.4 vs. 1.6 $\mu$g/g, respectively) and that Rosen, et al.[421] reported a slight increase in liver pyridoxal phosphate after deoxypyridoxine treatment. The decreases reported by Bhagavan and Stewart[48] do not conflict with these earlier reports because Bhagavan and Stewart compared normal $B_6$ controls with $B_6$-deficient animals receiving deoxypyridoxine. They reported no values for rats receiving only the $B_6$-deficient diet. The data of Johnston et al.[229] led us to question the common assumption that deoxypyridoxine exerts its effect primarily as a result of being phosphorylated and then inhibiting enzymes requiring pyridoxal phosphate as a cofactor. Any deoxypyridoxine phosphate in the tissues would presumably be isolated along with pyridoxal phosphate in the tissue hydrolysates and would be expected to inhibit the enzymes used to assay pyridoxal phosphate. Therefore, failure to find decreased concentrations of pyridoxal phosphate suggested that there was no marked interference with the formation of pyridoxal phosphate and also that the concentration of deoxypyridoxine in the tissues was not high enough to produce inhibition of the in vitro assay procedures.

Because of the questions raised by these observations, we repeated the experiment of Johnston et al.[229] and examined muscle and kidney as well as liver. While we found no significant increases in pyridoxal phosphate concentrations when deoxypyridoxine was added to the deficient diet, neither was there any significant decrease (Table 1) compared with animals receiving only the deficient diet for 3 weeks.

In a more recent study using the same diet for 5 weeks, the pyridoxal phosphate concentrations in the plasma, liver, adrenals, and thymus were significantly ($p < 0.05$) increased in the deoxypyridoxine group compared with those receiving only the deficient diet (Table 2). The manufacturer acknowledged that the deoxypyridoxine was produced by reduction of pyridoxine, and we calculated that less than 0.1% contamination with pyridoxine would be adequate to account for the observed increases in pyridoxal phosphate. Therefore, in order to insure that these increases were not due to contamination of the deoxypyridoxine by traces of pyridoxine, we included a group treated with deoxypyridoxine synthesized in our laboratory as described in Chapter 1. There was no difference between the effect of our deoxypyridoxine and the commercial material. Note that the pyridoxal phosphate concentration in the thymus, which shows marked atrophy in response to deoxypyridoxine, was higher in the deficient deoxypyridoxine groups than in the thymus of any other group. This suggests that the normal metabolism of pyridoxal phosphate was seriously altered in the thymus. Several lines of evidence have suggested possible interaction between deoxypyridoxine and the adrenal glands. As in the thymus, the pyridoxal phosphate concentrations of the adrenals were doubled by deoxypyridoxine compared with the deficient diet alone. Plasma con-

Table 1
EFFECT OF DEOXYPYRIDOXINE (1 g/kg DIET) ON TISSUE PYRIDOXAL PHOSPHATE (μg/g)[92]

| $B_6$ | Normal | | Deficient | |
|---|---|---|---|---|
| Deoxypyridoxine | No | Yes | No | Yes |
| Liver | 4.9 ± 1.1 | 4.2 ± 1.3 | 3.5 ± 1.1 | 3.5 ± 1.6 |
| Kidney | 3.0 ± 0.5 | 2.0 ± 0.6 | 2.0 ± 0.9 | 2.0 ± 1.2 |
| Muscle | 4.7 ± 0.9 | 4.0 ± 1.0 | 3.9 ± 1.2 | 3.6 ± 1.0 |

Table 2
EFFECT OF DEOXYPYRIDOXINE ON TISSUE PYRIDOXAL PHOSPHATE CONCENTRATIONS[a,93]

| $B_6$ content | Deficient | | | 22 mg/kg diet | |
|---|---|---|---|---|---|
| Deoxypyridoxine source[b] | None | Sigma Chemical Co. | Synthetic | None | Sigma Chemical Co. |
| Plasma | 3.8 ± 0.7 | 10.1 ± 4.8 | 9.3 ± 4.2 | 87.2 ± 13.0 | 47.4 ± 6.5 |
| Liver | 2.36 ± 0.11 | 3.18 ± 0.19 | 3.04 ± 0.43 | 3.78 ± 0.26 | 4.49 ± 0.63 |
| Adrenals | 0.18 ± 0.06 | 0.42 ± 0.16 | 0.38 ± 0.09 | 0.73 ± 0.12 | 0.64 ± 0.12 |
| Kidney | 0.32 ± 0.08 | 0.33 ± 0.05 | 0.39 ± 0.21 | 0.70 ± 0.28 | 0.84 ± 0.24 |
| Muscle | 3.53 ± 0.61 | 4.61 ± 0.46 | 4.02 ± 0.43 | 5.84 ± 0.37 | 6.18 ± 1.46 |
| Thymus | 0.12 ± 0.04 | 0.24 ± 0.05 | 0.20 ± 0.07 | 0.14 ± 0.05 | 0.18 ± 0.04 |
| Large intestine | 0.07 ± 0.03 | 0.12 ± 0.11 | 0.14 ± 0.05 | 0.11 ± 0.02 | 0.11 ± 0.03 |

[a] Plasma values are ng/mℓ. All others are μg/g wet tissue.
[b] Deoxypyridoxine was used at 1 g/kg diet.

centrations also showed less of a decline in the deficient group receiving deoxypyridoxine than in those receiving only the deficient diet. This difference developed during the fourth and fifth weeks of treatment (Table 3).

The complexity of deoxypyridoxine effects is evident from the difference in response depending on the amount of $B_6$ available. Deoxypyridoxine reduced the fall in plasma pyridoxal phosphate in deficient animals but caused a fall in the normal $B_6$ animals (Table 2). With the exception of the liver, other tissues from the rats receiving normal $B_6$ showed little response to deoxypyridoxine. In analyzing tissues the possible contribution of the blood contained in the tissues must be considered. While we have no direct data concerning this effect, the fact that pyridoxal phosphate concentrations in most of the tissues from the normal $B_6$ animals treated with deoxypyridoxine showed little change or a tendency to increase even though the pyridoxal phosphate concentration of the plasma dropped almost 50% suggests that the blood content of the tissues was not a major factor in the results.

While moderate (50 mg/kg diet) oral doses failed to reduce pyridoxine concentrations in brain,[395] injection of higher doses may reduce pyridoxal phosphate concentrations. Bain and Williams[27] reported a 90% drop in pyridoxal phosphate in mouse brain sacrificed during convulsions within 90 min after an intraperitoneal dose of 250 mg deoxypyridoxine per kilogram. Similar results have been reported in rabbits. An intramuscular dose of 100 mg/kg in rabbits receiving a normal $B_6$ diet reduced brain pyridoxal phosphate concentrations by 80% ($p < 0.05$) within 24 hr.[130] Continuing the treatment for 7 days produced no further effect. Pyridoxal phosphate concentrations in the treated group fluctuated between 15 to 34% of control values. In a more recent

### Table 3
### CHANGES IN PLASMA PYRIDOXAL PHOSPHATE (ng/mℓ) as $B_6$ DEFICIENCY DEVELOPS[90a]

| $B_6$ content | Deficient | | | 22 mg/kg diet | |
|---|---|---|---|---|---|
| Deoxypyridoxine source[a] | None | Sigma Chemical Co. | Synthetic | None | Sigma Chemical Co. |
| Days | | | | | |
| 0 | 36.1 ± 30.6[b] | 34.0 ± 10.8 | 43.1 ± 17.0 | 42.3 ± 11.1 | 96.2 ± 73.1 |
| 7 | 15.3 | 20.0 | 19.6 | 143 | 89.0 |
| 13 | 15.8 | 14.1 | 17.7 | 78.3 | 40.7 |
| 22 | 10.9 | 10.5 | 10.7 | 84.9 | 59.0 |
| 35 — 37 | 3.8 ± 0.7 | 10.1 ± 4.8 | 9.3 ± 4.2 | 87.2 ± 13.0 | 47.4 ± 6.5 |

[a] Deoxypyridoxine was used at 1 g/kg diet.
[b] Initial and final values represent 4 or 5 animals. Intermediate values are an average of just 2 animals. Initial samples were obtained immediately after animals were received and there were large individual variations. However, the values became much more uniform under controlled experimental conditions.

study, Ebadi and Klangkalya[129] measured pyridoxal phosphate in the brains of rats receiving a normal $B_6$ diet 2 hr after intraperitoneal injection of deoxypyridoxine at 0, 25, 50, 100, and 200 mg/kg. Only the 100 and 200 mg/kg dose had a statistically significant ($p < 0.01$) effect, lowering pyridoxal phosphate from the control value of 1.27 μg/g wet tissue to about 0.9. Spector and Shikuma[470] found that subcutaneous administration of deoxypyridoxine (120 mg/kg) to rabbits 18 hr before death produced a 57% decline in pyridoxal phosphate concentrations in the cortex and a 42% decrease in whole brain. There was little effect on the concentrations of pyridoxal.

Certain tumor cells also showed reduced pyridoxal phosphate concentrations after deoxypyridoxine treatment. Bain and Williams[27] reported that deoxypyridoxine (75 mg/kg, route of administration, and $B_6$ content of diet not specified) decreased the total $B_6$ content of mouse ascites fluid but increased the pyridoxal phosphate concentration slightly above that of the controls. Isonicotinic acid hydrazide (75 mg/kg) had a similar effect. However, simultaneous administration of both produced an 80% decrease in total $B_6$ and a 65% decrease in pyridoxal phosphate. Within the Ehrlich ascites tumor cells, the concentration of pyridoxal phosphate dropped 44% with deoxypyridoxine treatment and 62% when deoxypyridoxine was combined with isonicotinic hydrazide. In solid mouse 755 adenocarcinoma, deoxypyridoxine produced a 28% drop in pyridoxal phosphate. This effect was not enhanced by adding isonicotinic acid hydrazide. Since the diet used was not specified, presumably all animals were receiving adequate $B_6$.

Based on the foregoing discussion, we concluded that clinical symptoms produced by addition of moderate doses of deoxypyridoxine to a $B_6$-deficient diet are not associated with decreased pyridoxal phosphate concentrations in most tissues when compared with animals receiving only the $B_6$-deficient diet. Deoxypyridoxine may reduce pyridoxal phosphate concentrations in some tissues when administered to animals receiving a normal $B_6$ intake.

Data on the total $B_6$ content of the tissue are not as consistent as those on pyridoxal phosphate. In two reports, total $B_6$ values increased following deoxypyridoxine treatment. Stoerk[477] reported that liver $B_6$ as measured by *Neurospora* averaged about 11.7 μg/g in rats and mice receiving the $B_6$-deficient diet plus deoxypyridoxine (0.3 mg/mℓ drinking water) compared with 8.4 μg/g in animals receiving only the deficient diet. Talageri[494] reported vitamin $B_6$ concentrations (measured by *Saccharomyces carlsbergensis*) of 4.25 ± 0.36 μg/dℓ in the blood and 3.68 ± 0.24 μg/g in the liver of rats receiving a $B_6$-deficient diet. In rats receiving the same diet containing 20 mg deoxypyridoxine per kilogram, the $B_6$ concentrations were 6.13 ± 0.25 μg/dℓ in the blood and 4.58 μg/g in the liver. Even though $B_6$ concentrations were increased in the deoxypyridoxine group, these rats lost an average of 68 g during the experiment while the $B_6$-deficient rats gained an average of 5 g.

However, Beaton and McHenry[39] found no differences in total liver $B_6$ between rats receiving only the deficient diet and those receiving the deficient diet plus 100 μg deoxypyridoxine per day. Olsen and Martindale[371] reported that treating rats receiving a commercial stock diet with an oral dose of 5 mg deoxypyridoxine per day or $B_6$-deficient rats with 12.5 mg deoxypyridoxine per kilogram diet for 4 weeks produced slight but not statistically significant decreases in kidney and liver $B_6$ as measured by *Saccharomyces carlsbergensis.*

Beaton and McHenry[39] found that liver $B_6$ rose to 11 μg/g 12 days after the $B_6$-deficient diet was supplemented with pyridoxine. Similar supplementation of the group receiving the deficient diet plus 100 μg deoxypyridoxine per day raised liver $B_6$ to only 9.6 μg/g. No statistical analysis was performed, thus complicating the interpretation of this reduced response. However, it is compatible with the previous discussion of a decrease in pyridoxal phosphate produced by combining deoxypyridoxine with normal $B_6$ intake.

Sakuragi and Kummerow[430] demonstrated that the interaction between pyridoxine and deoxypyridoxine is affected by the form of administration in rats. They used a $B_6$-deficient diet without corn oil for 7 weeks and then supplemented with a single dose 5 mg deoxypyridoxine hydrochloride. When administered with 70 mg pyridoxine hydrochloride, weight loss was greater than when no supplements were given. Use of palmitate derivatives delayed the effects of either supplement indicating that the esters were stored. Weight gains were observed when either member of the pair was administered as the palmitate, but the largest weight losses occurred when both were administered as the palmitate.

Since the foregoing discussion suggests that pyridoxal phosphate concentrations can be altered with little change in total $B_6$, there must be compensating changes in the concentrations of other forms of $B_6$. Murakami and Matsuda[353] have documented such shifts. In mice apparently receiving normal $B_6$, the intraperitoneal injection of deoxypyridoxine (50 mg/kg) caused little change in the concentration of total $B_6$ or pyridoxamine phosphate in the brain within 60 min. However, the concentration of pyridoxal phosphate was significantly ($p < 0.005$) reduced while the concentrations of pyridoxal and pyridoxamine were significantly ($p < 0.01$) increased (Table 4). The authors proposed that the convulsive action of castrix is related to decreased synthesis of pyridoxal phosphate from pyridoxamine phosphate. Combining castrix with either deoxypyridoxine or toxopyrimidine prevented the increase in pyridoxamine phosphate produced by castrix alone and also prevented convulsions. The authors also noted that there may be some interaction between the effects of castrix and deoxypyridoxine on pyridoxal phosphate. This report is one of the first attempts to evaluate the significance of shifts among the various forms of $B_6$. We feel that this approach will be the key to providing more detailed information about the functions of vitamin $B_6$ in the future.

### *2. Transport*

The possibility that deoxypyridoxine may interfere with $B_6$ transport has already been mentioned in connection with microorganisms. Alterations in $B_6$ transport have also been noted in various mammalian tissues.

A 1:1 ratio of deoxypyridoxine:pyridoxine or pyridoxamine produced about 66% inhibition of $B_6$ uptake by human erythrocytes. Uptake of pyridoxal or pyridoxal phos-

Table 4
**EFFECT OF $B_6$ ANTAGONISTS ON THE DISTRIBUTION OF $B_6$ COMPOUNDS IN MOUSE BRAIN[353]**

| Compound | Dose (mg/kg) | Total $B_6$ | PLP | PMP | PL | PM |
|---|---|---|---|---|---|---|
| None | — | 15.18 ± 0.83[a] | 7.11 ± 0.28 | 8.12 ± 0.37 | 0.36 ± 0.05 | 0.07 ± 0.01 |
| Deoxypyridoxine (DOP) | 50 | 93 ± 5[b] | 65 ± 5 | 103 ± 3 | 241 ± 15 | 257 ± 11 |
| Castrix | 2.5 | 101 ± 5 | 89 ± 6 | 110 ± 3 | 78 ± 14 | 143 ± 43 |
| Toxopyrimidine (TP) | 25 | 92 ± 3 | 89 ± 2 | 101 ± 1 | 139 ± 6 | 44 ± 44 |
| Castrix & DOP | 2.5 + 50 | 96 ± 3 | 83 ± 2 | 94 ± 2 | 237 ± 13 | 343 ± 71 |
| Castrix & TP | 2.5 + 25 | 99 ± 4 | 86 ± 6 | 99 ± 3 | 120 ± 27 | 47 ± 18 |

[a] nmol/g wet tissue.
[b] % of control value.

phate was not affected under the same conditions[510,557,558] However, an ever-present question in uptake studies is whether the material is actually within the cell or merely bound on outer surface. The widespread use of pyridoxal to label proteins suggests that interpretation of pyridoxal uptake studies might be very complex.

Yamada and Tsuji[557] noted that deoxypyridoxine decreased the concentrations of pyridoxal phosphate found after incubation of erythrocytes with the free forms of $B_6$ by about 50%. Pyridoxine phosphate and pyridoxamine phosphate were transported into the cell much less readily than the free forms. Similarly, deoxypyridoxine phosphate did not interefere with the uptake of vitamin $B_6$. It did seem to cause a slight but consistent decrease in the concentration of pyridoxal phosphate.

Deoxypyridoxine also reduced $B_6$ uptake in human leukocytes[92] and platelets.[160] In leukocytes, the maximum concentration of pyridoxal phosphate obtained was about 37 ng/$10^8$ cells. This value was obtained with pyridoxine uptakes ranging from 31 to 191 ng/$10^8$ cells suggesting that it is an equilibrium value. Deoxypyridoxine caused a detectable decrease (15%) in pyridoxine uptake even at a 1:1 deoxypyridoxine: pyridoxine ratio. This is only about 25% of the inhibition reported above for a 1:1 ratio in erythrocytes. While Yamada and Tsuji did not provide cell counts, assuming an average of $5 \times 10^6$ erythrocytes per cubic millimeter in normal blood, their procedure probably produced deoxypyridoxine concentrations of about 1 μg/2 × $10^9$ cells per milliliter, which is a lower concentration per cell than the 0.05 μg/21 × $10^6$ cells per milliliter used for leukocytes. While a comparison of these two studies suggests that erythrocytes may be more susceptible than leukocytes to deoxypyridoxine effects, the data are very limited, and differences in methodology make comparisons difficult. Since we have obtained preliminary evidence that at least some aspects of $B_6$ metabolism differ considerably between erythrocytes and leukoyctes (see Chapter 9), it would not be surprising if susceptibility to deoxypyridoxine also differed. However, further data are needed to clarify these points.

Gaut and Solomon[160] reported that the accumulation of labeled pyridoxine by human platelets was dependent on conversion to pyridoxine phosphate since the concentration of intracellular pyridoxine never rose above that of the medium. The maximum intraplatelet concentration of pyridoxine phosphate was about 26 μ*M* regardless of the external pyridoxine concentration. This is similar to the situation reported above for leukocytes and suggests product inhibition of pyridoxine kinase. With a pyridoxine concentration of 3 μ*M*, 50% inhibition of pyridoxine uptake was achieved with 2.1 μ*M* deoxypyridoxine. While it is unusual to achieve such a large effect with less than a 1:1 ratio of deoxypyridoxine:pyridoxine, the results are comparable to those reported above for erythrocytes. Also, the platelets are normally exposed to pyridoxal, rather than pyridoxine, and were unable to convert pyridoxine phosphate to pyridoxal phosphate. Therefore, the sensitivity to deoxypyridoxine shown under these conditions may not be representative of its behavior under more natural circumstances.

Tsuji et al.[513,517] concluded that uptake of pyridoxine by rat intestine also involves diffusion followed by phosphorylation. At low concentrations of pyridoxine (e.g., 0.3 μ*M*) the diffusion step is rate limiting. However, whereas 60 min incubation of platelets with 3 μ*M* pyridoxine produced 26 μ*M* pyridoxine phosphate, similar incubation in intestine produced only a 2 μ*M* concentration of $B_6$ phosphates. Also, this system was much less sensitive to deoxypyridoxine than the platelets. With 2 μ*M* pyridoxine in the medium, 0.2 m*M* deoxypyridoxine produced only 53% inhibition of pyridoxine uptake after 60 min. Similarly, in the presence of 0.3 μ*M* pyridoxine, 20 μ*M* deoxypyridoxine reduced uptake of label 61%. The ratio of pyridoxine:$B_6$ phosphates was increased from 1:194 in the absence of deoxypyridoxine to 1:1.3 in its presence. With 300 μ*M* deoxypyridoxine, total uptake was decreased 80% and the pyridoxine:phosphate ratio was 5:1. Therefore, the authors concluded that the primary effect of deoxypyridoxine was to inhibit the phosphorylation rather than the diffusion step.

In further studies using everted sacs rather than rings, Tsuji and Yamada[514] found that addition of 300 $\mu M$ deoxypyridoxine to 0.3 $\mu M$ pyridoxine actually increased the transport of pyridoxine through the tissue but markedly reduced the accumulation of pyridoxine within the tissue. Again the authors concluded that at low concentrations, the movement of pyridoxine into the tissue is controlled by diffusion. Furthermore, the accumulation of pyridoxine within the cell seems to have no role in the initial uptake but does reduce the release of $B_6$ compounds to the serosal side.

Middleton[321] obtained similar results with jejunal sacs from rats. In the presence of 2 $\mu M$ pyridoxine, 10 $\mu M$ deoxypyridoxine did not reduce $B_6$ uptake during the first 4 min after preincubation with the inhibitor for 15 min. However, the total amount of labeled $B_6$ phosphates in the tissue was reduced 63.6% ($p < 0.05$) within that time.[322] After 30 min, tissue accumulations of labeled $B_6$ and $B_6$ phosphate were reduced 19.6% ($p < 0.005$) and 47.2% ($p < 0.0125$), respectively. When similar studies were conducted in vivo, deoxypridoxine (100 $\mu M$) did not alter the loss of label (0.2 $\mu M$ pyridoxine) from the lumen but facilitated the loss from the tissue.[323] High concentrations (1 m$M$) of deoxypyridoxine or unlabeled pyridoxine also showed no effect on loss of label from the lumen, but did reduce both the fraction of the absorbed label which remained in the tissue and the portion which appeared as labeled $B_6$ phosphate. The fact that pyridoxine itself tended to have a greater effect on the relative distribution of label than deoxypyridoxine, suggests to us that this is another case where deoxypyridoxine was acting as a substrate rather than a true inhibitor. Since deoxypyridoxine is not quite as good a substrate as pyridoxine, it did not have quite as much effect as the same amount of pyridoxine. On the basis of these studies, Middleton concluded that pyridoxine absorption was nonsaturable at the concentrations studied, and involved passive diffusion rather than a carrier mechanism. Phosporylation of pyridoxine in the jejunum was saturable. Phosphorylation did not affect uptake but did delay the release of some $B_6$ from the tissue.

When everted sacs were used with pyridoxal,[515] 300 $\mu M$ deoxypyridoxine added to 0.3 $\mu M$ pyridoxal did not affect the time course of the ratio of serosal to mucosal concentration of pyridoxal. However, pyridoxal uptake was affected, presumably decreased. Deoxypyridoxine also failed to affect the serosal to mucosal concentration ratios when pyridoxamine was used.[516] The abstract did not mention effects on uptake.

In isolated, vascularly perfused intestine, addition of 20 $\mu$mol deoxypyridoxine to 0.02 $\mu$mol labeled pyridoxal significantly ($p < 0.05$) increased the fraction of label remaining in the lumen, decreased ($p < 0.05$) the fraction detected in the mucosa, and produced no change in the fraction found in the perfusate.[303a] Adding 20 $\mu$mol unlabeled pyridoxal to 0.02 $\mu$mol labeled pyridoxal tended to increase the fraction of label in the lumen and mucosa and significantly ($p < 0.01$) decreased the fraction of label appearing in the perfusate compared with 0.02 $\mu$mol labeled pyridoxal alone. Thus, with pyridoxal as with pyridoxine, deoxypyridoxine seems to facilitate movement through the tissue rather than accumulation within the cells. The same concentrations of deoxypyridoxine and unlabeled pyridoxal which were used above also reduced by over 50% $p < 0.001$) the fraction of label appearing as pyridoxal phosphate in the mucosa. As was seen with pyridoxine, unlabeled pyridoxal tended to produce a slightly greater decrease than deoxypyridoxine. Neither deoxypyridoxine nor unlabeled pyridoxal had a statistically significant effect on the relative amounts of label in the various forms of $B_6$ which appeared in the lumen or perfusate.

In studies of $B_6$ transport in the central nervous system, Spector and Shikuma[470] treated 17 rabbits receiving a normal $B_6$ diet with a single subcutaneous injection of deoxypyridoxine (120 mg/kg) and sacrificed them 18 hr later. The treatment had no effect on the pyridoxal content of the cortex or whole brain. However, the pyridoxal phosphate content was reduced by 57% in the cortex and by 42% in the whole brain ($p < 0.05$). The treatment had no effect on the in vitro accumulation of labeled $B_6$ by choroid plexus. These results agreed with Ebadi et al.[130] However, Spector and Shikuma[470] failed to find the inverse correlation between pyridoxal kinase activity and the level of biogenic amines reported by Ebadi et al.[130] Spector and Shikuma[470] suggested that interactions between biogenic amines and pyridoxal phosphate described by Neary et al.[362] may have contributed to the discrepancy by interfering with the use of pyridoxal phosphate as an indicator of pyridoxal kinase activity.

Spector[468] examined vitamin $B_6$ transport by brain slices and choroid plexuses from rabbits in vitro. Deoxypyridoxine (1.0 m*M*) caused an 88% reduction in the accumulation of labeled $B_6$ by choroid plexus. However, 0.2 m*M* pyridoxine had a similar effect. In brain slices, 20 μ*M* deoxypyridoxine produced an 80% decrease compared with 90% for 20 μ*M* pyridoxine. Unfortunately, no data on the phosphorylated derivatives are presented. These results suggest that deoxypyridoxine and pyridoxine are about equally effective in competing with the labeled molecules for transport and possibly phosphorylation. As was found in other tissues discussed above, the concentration of phosphorylated forms tended to plateau at about 30 μ*M* although it did apparently rise to 80 μ*M* when the tissues were incubated with 200 μ*M* pyridoxine.

In further studies Spector and Greenwald[469] commented that they used pyridoxine rather than pyridoxal because it was not bound to proteins. Since tissues other than the gastrointestinal tract will rarely encounter pyridoxine, we question whether the kinetics of pyridoxine metabolism necessarily reflect normal in vivo rates for pyridoxal derivatives. Their results show that with an incubation mixture containing 48 n*M* pyridoxine, 50% inhibition of pyridoxine uptake was produced by about 2 μ*M* deoxypyridoxine in brain slices, and about 10 μ*M* in choroid plexus. Unlabeled pyridoxine gave similar results. The authors did not comment on the rather marked difference between the tissues in their susceptibility to deoxypyridoxine or pyridoxine inhibition. They also reported data showing a strong correlation between the inhibition of pyridoxine uptake and inhibition of pyridoxal kinase. However, the pyridoxal kinase assays were conducted with cell-free homogenates containing only 50 nCi of pyridoxine but the same concentration of deoxypyridoxine as was used with the intact cells. Since the deoxypyridoxine:pyridoxine ratio was much higher in the kinase assays than in the uptake studies, and since the uptakes were measured in intact cells while the kinase was measured in homogenates, we question whether it is appropriate to attempt a direct correlation between the two. While no data are provided, we suspect that a similar degree of kinase inhibition would be obtained using equivalent concentrations of unlabeled pyridoxine. Our conclusion from these data would be that both the uptake and kinase systems can be saturated with either pyridoxine or deoxypyridoxine, thus reducing the uptake and phosphorylation of labeled pyridoxine. 5′-Deoxypyridoxine was not accumulated and did not inhibit pyridoxine uptake.

These authors used phosphate to inhibit alkaline phosphatase thus preventing the hydrolysis of extracellular $B_6$ phosphates. They did not comment on the possibility that this treatment might also affect intracellular metabolism. Similar treatment significantly increased pyridoxal phosphate concentrations in erythrocytes[13] and liver.[276] However, unpublished results from our laboratory[91a] suggest that phosphatase inhibition has little effect on pyridoxal phosphate concentrations in normal leukocytes. Since Spector's data are restricted to the disposition of labeled molecules, the effect of phosphatase inhibition on total $B_6$ metabolism in brain and choroid cannot be evaluated.

Tsuji et al.[510,511] found that the intracellular concentrations of pyridoxal and pyridoxal phosphate in ascites sarcoma cells after incubation with pyridoxine, pyridoxal, or pyridoxal phosphate were reduced by deoxypyridoxine especially when glucose was included in the medium.

In the preface we discussed the problems of defining the term, inhibitor, in relation to enzyme function. A similar problem arises in the case of deoxypyridoxine effects on membrane transport. Other than use of labeled deoxypyridoxine there is no easy way to assess the amount of deoxypyridoxine which enters the cells or tissues. Very limited data from a single experiment in our laboratory[92] suggested that in leukocytes the uptake of deoxypyridoxine might be 2- to 3-fold greater than the uptake of pyridoxine itself under the same conditions. It seems likely that in most cases deoxypyridoxine will enter cells at least as readily as pyridoxine. The degree to which deoxypyridoxine interferes with $B_6$ uptake probably depends both on the susceptibility of the membrane process to saturation and on the subsequent interactions between deoxypyridoxine and $B_6$ in intracellular metabolism. These factors must be considered when interpreting uptake studies.

### 3. $B_6$ Excretion

Vaughn and Winders[526,527] examined the effect of deoxypyridoxine on the excretion of $B_6$. In one study, deoxypyridoxine (10 mg/rat i.p.) was injected either 4 hr before or 4 hr after a dose of pyridoxine. The results suggested that deoxypyridoxine tended to increase the percentage of label lost in the urine. However, the interpretation is complicated by the fact that the group which did not receive deoxypyridoxine until 4 hr after the injection of pyridoxine still excreted 25% more label than controls during the first 4 hr when no deoxypyridoxine was present. Use of deoxypyridoxine (200 μg/g diet) in a $B_6$-deficient diet for 2 weeks seemed to cause about a 16% increase in excretion of label, but there was no difference if deoxypyridoxine administration was terminated 2 weeks before testing. Neither did deoxypyridoxine treatment increase excretion when the injected dose of pyridoxine was increased 40-fold to 500 μg per rat. Under these conditions, 76 to 79% of the dose was excreted compared with 30 to 40% at the lower level of pyridoxine. Unpublished results from our laboratory[91a] demonstrated a 3-fold increase in the excretion of label in $B_6$-deficient rats receiving drinking water containing 80 ng labeled pyridoxine plus 1 mg deoxypyridoxine per milliliter compared with rats receiving only the $B_6$-deficient diet. Similar results were obtained more recently using a $B_6$-deficient diet containing 1 g deoxypyridoxine per kilogram (Table 5). The effect of deoxypyridoxine in markedly increasing pyridoxic acid excretion in the deficient animals is particularly striking considering that these animals were smaller and were eating less than those receiving only the deficient diet. We estimate that the deoxypyridoxine group was excreting about five times its $B_6$ intake. Yet, as was shown earlier, pyridoxal phosphate levels were not decreased. This is further evidence that one of the major effects of deoxypyridoxine is to push the metabolism of $B_6$ toward pyridoxal phosphate. We suspect that the additional pyridoxal phosphate arises either from Shiff base or pyridoxamine forms. Vaughn and Winders[527] suggested that deoxypyridoxine increased urinary losses of pyridoxine because it blocked some of the tissue binding sites for pyridoxine. Considering that deoxypyridoxine markedly inhibits $B_6$ uptake in many situations and is very rapidly excreted, we feel that interference with reabsorption in the renal tubules is also likely to be a factor in the increased urinary losses of $B_6$.

Iwanowska[217] concluded that deoxypyridoxine decreased rather than increased $B_6$ excretion in $B_6$-deficient rats and had no effect on $B_6$ excretion in rats receiving a normal $B_6$ intake (Table 6). However, the data suggest that vitamin $B_6$ excretion was

Table 5
EFFECT OF DEOXYPYRIDOXINE ON PYRIDOXIC ACID EXCRETION[90a]

| $B_6$ | Deficient | | 22 mg/kg diet | |
|---|---|---|---|---|
| Deoxpypridoxine[a] | No | Yes | No | Yes |
| Pyridoxic acid excretion | | | | |
| μg/day | 2.2 ± 1.0 | 13.5 ± 7.7 | 115.9 ± 27.3 | 103.0 ± 31.5 |
| μg/kg/day | 5.2 ± 2.0 | 46.5 ± 29.1 | 236.6 ± 57.9 | 221.0 ± 78.6 |

[a] 1 g deoxypyridoxine HCl/kg diet

Table 6
URINARY EXCRETION OF VITAMIN $B_6$[217]

| $B_6$ (μg/day) | Deoxypyridoxine (μg/day) | Vitamin $B_6$ excretion (μg/day) |
|---|---|---|
| 0 | 0 | 0.07 ± 0.001 |
| 0 | 100 | 0.05 ± 0.001 |
| 0 | 200 | Trace |
| 0 | 400 | Trace |
| 50 | 0 | 20.0 ± 0.28 |
| 50 | 100 | 24.0 ± 0.30 |
| 50 | 200 | 27.8 ± 0.24 |
| 50 | 400 | 28.6 ± 0.26 |

actually significantly increased by deoxypyridoxine in the normal $B_6$ group by each increase in deoxypyridoxine dose. Since deoxypyridoxine is excreted very rapidly,[92] the deoxypyridoxine:$B_6$ ratio in the urine of the $B_6$-deficient animals was probably at least 1000:1. We[91a] have noted that the *Saccharomyces carlsbergensis* assay for vitamin $B_6$ is inhibited at ratios above 100:1. Therefore, the apparent decline in $B_6$ excretion in deficient animals treated with increasing amounts of deoxypyridoxine may actually reflect increasing inhibition of the assay. Another point of confusion is that Iwanowska's Table 1 shows the pyridoxine dose as 0.5 mg/day whereas the text and other tables show 50 μg. It seems likely that 50 μg is the correct figure.

### *4. Pyridoxal Kinase*

Pyridoxal kinase (E.C. 2.7.1.35) has a broad specificity and may be referred by various names depending on the substrate used in a particular experiment. For consistency, we will use the term, pyridoxal kinase, throughout this discussion even though the original authors may have used an alternate name (e.g., pyridoxine kinase). Also, in the following review of the interaction between deoxypyridoxines and pyridoxal kinases, the reader should keep in mind the problems discussed in the preface in distinguishing between a substrate and a competitive inhibitor. The term competitive inhibitor is used below as used in the original articles and does not necessarily mean that the inhibitor was not a substrate for the enzyme. It is very likely that 4′-deoxypyridoxine was a substrate in every case.

The effect of deoxypyridoxine on pyridoxal kinase has been examined both in vitro and in vivo. Phosphorylation of deoxypyridoxine has been demonstrated in vitro using kinases from *Streptococcus faecalis*,[522] brewer's yeast,[210] and mouse brain.[508,509] $K_m$ values for pyridoxal, pyridoxine, and 4′-deoxypyridoxine were $0.86 \times 10^{-4}$, $1.07 \times 10^{-4}$,

Table 7
COMPARATIVE AFFINITIES OF SEVERAL COMPOUNDS FOR VARIOUS PYRIDOXAL KINASES[285]

| Compound | Source of kinase | | | | |
|---|---|---|---|---|---|
| | *L. casei* | *S. faecalis* | *S. carlsbergensis* K[a] | Rat liver | Beef brain |
| Pyridoxal | 0.30 | 0.15 | 1.5 | 0.15 | 0.50 |
| 5-Deoxypyridoxal | 0.45 | 0.20 | 2.5 | 1.5 | 0.35 |
| Pyridoxine | 40 | 15 | 0.25 | 0.25 | 2.0 |
| 4'-Deoxypyridoxine | 20 | 45 | 0.40 | 0.15 | 0.90 |

[a] All values expressed as mol/ℓ × $10^4$. $K_s$ for pyridoxal and pyridoxine. $K_i$ for other derivatives were determined using various amounts of pyridoxal with the analog at 2 × $10^{-4}$ *M*.

and 0.35 × $10^{-4}$ *M*, respectively.[509] The values are not directly comparable since a slightly different reaction mixture was employed for each substance. However, it does appear that the $K_m$ values for these three substances are at least generally similar.

Surprisingly, Tsubosaka,[508] Tsubosaka and Makino,[509] and Korytnyk et al.[250] all detected phosporylation of 5'-deoxy derivatives. McCormick and Snell[285] observed that the affinity of 4'-deoxypyridoxine for the pyridoxal kinases from *Lactobacillus casei* and *Streptococcus faecalis* was relatively low. Affinity for the kinases from *Saccharomyces carlsbergensis,* rat liver, and beef brain was much higher. This is compatible with the observation that *Lactobacillus casei* and *Streptococcus faecalis* are less susceptible than *Saccharomyces carlsbergensis* to deoxypyridoxine inhibition (Table 7). The affinity of 5'-deoxypyridoxal was much higher than that of 4'-deoxypyridoxine for the enzymes from *Lactobacillus casei* and *Streptococcus faecalis* and very slightly higher for the enzyme from beef brain. However, 4'-deoxypyridoxine had slightly higher affinity than 5'-deoxpyridoxal for the enzymes from *Saccharomyces carlsbergensis* and rat liver. In fact, in the case of *S. carlsbergensis,* the affinity of 4'-deoxypyridoxine for the kinase was greater than that of pyridoxal. The pyridoxal kinase from *Escherichia coli* had a $K_m$ of 3 × $10^{-4}$ *M* for pyridoxal, and $K_i$ of 8 × $10^{-6}$, 5 × $10^{-4}$, and 2 × $10^{-4}$ *M* for pyridoxine, pyridoxamine, and 5'-deoxypyridoxal, respectively.[548]

There may be differences in enzyme-substrate affinities even within individual cells. Nishiwaki[366] noted that deoxypyridoxine inhibited the pyridoxal kinase activity in the soluble fraction of tumor cell homogenate but not in the membrane fraction.

3-Deoxypyridoxine (0.2 m*M*) produced about a 20% reduction in pyridoxal phosphate production by the rat liver kinase at a substrate concentration of 0.3 m*M*.[285] McCormick and Snell[285] commented that they did not determine whether 3-deoxypyridoxine itself was phosphorylated. However, Hurwitz[210,211] did note phosphorylation of 3-deoxypyridoxine by kinases from *Streptococcus faecalis* R and yeast. Of particular interest to us was his observation[210] that 2-ethyl-3-amino-4-ethoxymethyl-5-aminomethylpyridine was a good inhibitor of the kinase reaction but did not inhibit tyrosine decarboxylase probably because it was not phosphorylated. Such compounds might be useful for isolating the effects of kinase inhibition.

Chern and Beutler[82] found that deoxypyridoxine would inhibit pyridoxal kinase from human erythrocytes. The inhibitor:substrate ratio is not specifically stated. Assuming that the same substrate concentration was used throughout the paper, a deoxypyridoxine:pyridoxine ratio of 2:1 produced 48% inhibition. Platzer et al.[393] reported $K_i$ values of 1.5 × $10^{-6}$ and 8.6 × $10^{-6}$ *M* for pyridoxal kinase from duckling erythrocytes and the malarial parasite, *Plasmodium lophurea*, respectively. Korytnyk et al.[250] found that deoxypyridoxine was a competitive inhibitor of pyridoxal kinase from rat liver ($K_i$ = 360 μM).

Two groups have reported in vivo data from brain. No effect was noted in brain kinase activity two hours after deoxypyridoxine treatment in either rabbits (100 mg/kg i.m.)[130] or rats (25 to 200 mg/kg i.p.).[129] However, by 12 to 18 hr, deoxypyridoxine (100 mg/kg i.m.) has produced about a 4-fold increase in pyridoxal kinase activity in rabbit brain. This difference remained constant even when deoxypyridoxine administration was continued for seven days. Administration of actinomycin D or puromycin prevented the increased activity suggesting that nucleic acids and protein synthesis were involved in the increase. The authors concluded that the turnover of pyridoxal phosphate and the kinase in rabbit brain was very rapid and that the availability of pyridoxal phosphate determined the activity of pyridoxal kinase. No data were available in the rat for periods longer than 2 hr. Spector and Shikuma[470] reported that a single subcutaneous injection of deoxypyridoxine (120 mg/kg) 18 hr prior to death had no effect on the kinetic constants ($K_m$ or $V_{max}$) of pyridoxal kinase from rabbit brain. Ebadi et at.[130] noted that the cellular particulate fractions of the brain contain substances which mask kinase activity and therefore they removed the particulate fraction from their homogenates. This might have contributed to the fact that Ebadi et al. found a greater effect than Spector and Shikuma in vivo. However, when Spector and Shikuma added 10 μM 4′-deoxypyridoxine directly to the assay mixture they obtained 75% inhibition. In contrast, 5′-deoxypyridoxine (100 μM) had no effect in vitro.

Lower oral doses (1 g/kg feed) seem to have little effect on pyridoxal kinase activity in vivo (Table 8). Considering recent suggestions[76] that pyridoxal phosphate may have a role in regulating steroid hormone action, it is interesting to note that the kinase activity of the adrenals is about twice as high as any other tissue.[93] The similarity of the effect of deoxypyridoxine on pyridoxal phosphate concentrations in the thymus and adrenals has already been mentioned. The thymus also has a fairly high kinase activity. Since the thymus is known to be very susceptible to deoxypyridoxine, these additional similarities to the adrenals further strengthen the possibility that deoxypyridoxine may alter adrenal function. While the relative kinase values are not affected, the absolute values listed in Table 8 for the kidneys and intestines are much lower than the values of 1 to 6 ng/mg/hr we had obtained in an earlier phase of that experiment.

*5. Oxidases*

Morino et al[338] demonstrated that deoxypyridoxine acted as a competitive inhibitor ($K_i$ = 3.33.6 × $10^{-3}$ M) for a pyridoxine oxidizing system from rabbit liver. Deoxypyridoxine was unusually effective (67% inhibition with a 1.25:1 ratio of deoxypyridoxine yridoxine) in blocking oxidation of pyridoxine but had no effect on the oxidation of pyridoxal. The system was active only when aldehyde oxidase was added. The pyridoxine oxidase preparation alone did not produce pyridoxal. Neither did aldehyde oxidase oxidize pyridoxine. Together, the enzymes converted pyridoxine but not pyridoxine phosphate to pyridoxic acid. Shane and Snell[445] later concluded that the conversion of pyridoxine to pyridoxal in this system was a nonenzymatic oxidation probably resulting from the hydrogen peroxide produced by aldehyde oxidase. We also were unable to demonstrate any oxidation of 4′-deoxypyridoxine in this system or in crude liver homogenates even though 4′-deoxy-5-pyridoxine acid is a major urinary metabolite of 4′-deoxypyridoxine in the rabbit. However, 4′-deoxyisopyridoxal was oxidized even more readily than pyridoxal. The enzymes involved in forming the deoxyisopyridoxal remain to be identified.

In a subsequent paper, Morisue et al.[343] isolated from rabbit liver a system capable of oxidizing pyridoxine, pyridoxine-5′-phosphate, or pyridoxamine-5′-phosphate to the corresponding aldehyde. Since both pyridoxine and pyridoxine phosphate were

### Table 8
### PYRIDOXAL KINASE ACTIVITY (ng PYRIDOXINE PHOSPHATE/mg TISSUE/hr)[90a,93]

| $B_6$ content | Deficient | | | 22 mg/kg diet | |
|---|---|---|---|---|---|
| Deoxypyridoxine source[a] | None | Sigma Chemical Co. | Synthetic | None | Sigma Chemical Co. |
| Liver | 27.4 ± 7.6 | 29.1 ± 13.9 | 32.4 ± 7.2 | 43.6 ± 15.8 | 54.5 ± 9.6 |
| Adrenal | 73.2 ± 42.6 | 86.3 ± 17.5 | 80.5 ± 14.5 | 92.0 ± 10.8 | 94.2 ± 27.4 |
| Muscle | 0.83 ± 0.05 | 1.23 ± 0.28 | 1.25 ± 0.12 | 0.95 ± 0.14 | 0.90 ± 0.27 |
| Thymus | 19.3 ± 3.7 | 15.5 ± 2.7 | 20.4 ± 4.8 | 16.5 ± 3.3 | 16.6 ± 2.8 |
| Kidney | 0.06 ± 0.03 | 0.06 ± 0.02 | 0.04 ± 0.01 | 0.04 ± 0.02 | 0.03 ± 0.01 |
| Small intestine | | | | | |
| 1st half | 0.29 ± 0.20 | 0.03 ± 0.03 | 0.05 ± 0.02 | 0.10 ± 0.03 | 0.13 ± 0.05 |
| 3rd quarter | 0.10 ± 0.04 | 0.09 ± 0.03 | 0.09 ± 0.03 | 0.07 ± 0.03 | 0.07 ± 0.03 |
| Large intestine | 1.42 ± 0.47 | 3.27 ± 3.37 | 3.11 ± 3.06 | 1.58 ± 1.21 | 1.14 ± 0.36 |

[a] Administered at 1 g/kg diet.

active as substrates, it is surprising that deoxypyridoxine phosphate but not deoxypyridoxine acted as an inhibitor. Wada and Snell[534] noted that the unphosphorylated forms of $B_6$ and deoxypyridoxine were not inhibitors of pyridoxine phosphate oxidase from rabbit liver. However, they did not state whether pyridoxine was a substrate. Deoxypyridoxine phosphate was an effective inhibitor. Korytnyk et al.[250] also reported that deoxypyridoxine phosphate was a competitive inhibitor ($K_i$ = 1.8 $\mu M$). Pyridoxine phosphate oxidase was found in highest concentration in rabbit liver. Kidney and erythrocytes had only 20% as much activity. Brain, spleen, heart, bone marrow, lung, and pancreas had less than 10% of the activity found in liver. Rat liver showed about 67% of the activity found in rabbit liver.[343]

Melius and Marshall[312] reported that pyridoxine dehydrogenase from baker's yeast was completely inhibited by 15 m$M$ 3-deoxypyridoxine or 3-deoxypyridoxamine. Korytnyk et al.[246] found that 5′-deoxypyridoxine and pyridoxine had the same $K_m$ (14.3 × $10^{-3}$ $M$) for this enzyme.

## B. Summary

Deoxypyridoxine can be phosphorylated in vitro. Deoxypyridoxine does inhibit the accumulation of phosphorylated forms of $B_6$ in tissues in vitro. Deoxypyridoxine will inhibit "pyridoxine oxidase" and deoxypyridoxine phosphate will inhibit pyridoxine phosphate oxidase in vitro. Based on these and related observations Korytnyk et al.[246] concluded that two conditions are necessary for a $B_6$ analog to produce significant growth inhibition. First, it must be phosphorylated by pyridoxal phosphokinase. Then it must efficiently inhibit pyridoxine phosphate oxidase.

In spite of these in vitro observations it appears that the deteriorating physical condition of $B_6$-deficient animals treated with deoxypyridoxine is not usually correlated with a detectable change in tissue pyridoxal phosphate concentrations when compared with animals receiving only the $B_6$-deficient diet. Beaton and McHenry[39] observed that the concentration of $B_6$ in liver seemed to reach a saturation point at a $B_6$ intake which was not adequate to produce maximum growth. These observations indicate that normal tissue concentrations of $B_6$ do not necessarily guarantee optimum $B_6$ status.

## C. Other Vitamins

Yamada et al.[556] examined the effect of $B_6$ deficiency and $B_6$ antagonists on the coenzyme A content of the liver in rats. Isonicotinic acid hydrazide, and 4′-deoxypyridoxine were injected daily at 20 mg/kg. In some groups, the $B_6$-deficient diet was supplemented with daily injections of 10 mg $B_6$/kg during the three week study. There were two rats in each of the six treatment and two control groups. Supplementing the $B_6$-deficient diet with $B_6$ increased liver coenzyme A about 30%. This increase was not observed when the $B_6$ supplement was combined with deoxypyridoxine. Injecting deoxypyridoxine into the rats receiving the deficient diet had no effect. The isonicotinic acid hydrazide treatment increased coenzyme A concentrations about 40% in the $B_6$-deficient animals resulting in concentrations slightly higher than those found in the $B_6$-supplemented animals. Combining the isonicotinic acid hydrazide treatment with the $B_6$ supplement had only a slightly greater effect than the $B_6$ supplement alone. Thus the effects of deoxypyridoxine on coenzyme A are essentially opposite to those of isonicotinoyl hydrazide.

## II. Protein Metabolism

Because vitamin $B_6$ is a cofactor for many reactions involving amino acids, measurements of various aspects of amino acid metabolism have commonly been incorporated into studies involving deoxypyridoxine. In the following section we will look first at some of the gross indicators of protein metabolism and then at specific amino acids and enzymes.

### A. Gross Indicators

The most extensive studies of general protein metabolism in normal rats were reported in a series of papers by Beare, Beaton, and others.[32-38,40] Unless otherwise specified, the work of Beare and Beaton was based on a design in which all rats including controls received a $B_6$-deficient diet plus 100 μg deoxypyridoxine. The controls received both pyridoxine (50 μg/day) and deoxypyridoxine. This design does not distinguish between the effects of a dietary deficiency alone and those of deoxypyridoxine. However, the results still provide some valuable information. There was no marked difference in plasma amino nitrogen between the groups with and without a pyridoxine supplement,[34,36,40] neither did serum proteins show a marked change.[36] Although weight gains were markedly reduced,[35,38] the group without pyridoxine remained in positive nitrogen balance. However, the balance was less positive than in the controls.[36]

Similarly, Ross and Pike[425,427] found that exposure to a $B_6$-deficient diet containing 5 mg deoxypyridoxine per kilogram for 7 to 30 days prior to mating caused significantly reduced nitrogen retention during pregnancy even if the deoxypyridoxine treatment was terminated and pyridoxine supplementation (12 mg/kg diet) was started on the day of mating. In a later report[392] the deoxypyridoxine content of the $B_6$-deficient diet was increased to 40 mg/kg and treatment started on the day of mating. Under these conditions the animals were in negative nitrogen balance by the third week.

Data reported by Iwanowska[217] suggested that in $B_6$-deficient animals, increasing doses of deoxypyridoxine tended to decrease urinary excretion of alpha amino nitrogen. In rats receiving 50 μg $B_6$ per day, deoxypyridoxine (100 to 400 μg/day) doubled urinary alpha amino acid excretion making it comparable to the excretion observed with a $B_6$-deficient diet.

Blood urea was markedly increased in the group without pyridoxine.[34,36] Beaton et al.[36] did not feel the increase was due to impaired excretion. Intraperitoneal injection of glutamic acid or lysine increased blood urea concentrations of this group, indicating that $B_6$ was not needed for urea formation.[40] Blood urea was also increased by histidine. Methionine had little effect. Administration of adrenocortical extract produced a larger increase in blood urea in rats without pyridoxine than in controls.[34]

$B_6$ deficiency alone significantly ($p < 0.02$) increased urea formation by liver slices 28% compared with pair-fed pyridoxine supplemented controls.[38] Adding deoxypyridoxine (100 μg/day) increased urea formation about 11% in $B_6$-deficient animals compared with $B_6$-deficient controls and 18% in animals receiving pyridoxine (50 μg/day) compared with animals receiving pyridoxine alone. Apparently, the fact that urea formation between the two deoxypyridoxine groups was not significantly different led the authors to conclude that deoxypyridoxine reduced the effect of $B_6$ deficiency on urea formation. However, we feel that this conclusion fails to recognize that the animals receiving the $B_6$-deficient diet plus deoxypyridoxine had the highest urea formation of any group and even those receiving deoxypyridoxine plus pyridoxine were almost as high as those receiving only the deficient diet. This suggests to us that deoxypyridoxine increases urea formation even in the presence of pyridoxine.

Beaton et al.[38] found that $B_6$ deficiency was associated with a tendency toward increased liver arginase activity, particularly in males. When deoxypyridoxine treatment was combined with the deficient diet, the arginase values returned to normal. The authors again commented that deoxypyridoxine accentuated the external deficiency symptoms even though it did not affect liver arginase activity. They concluded that " . . . although deoxypyridoxine accentuates external manifestations of deficiency, it is not a true antivitamin in accentuating biochemical abnormalities." They also concluded that the dermitatis produced by deoxypyridoxine was not related to disturbances in nitrogen metabolism.

Talageri[494]confirmed that compared with rats receiving only a $B_6$-deficient diet, animals receiving a deficient diet containing 20 mg deoxypyridoxine per kilogram showed a 60% increase in blood urea. Urea concentrations in the livers of the $B_6$-deficient group were 16% lower than in the normal $B_6$ group. In animals receiving the deficient diet plus deoxypyridoxine, liver urea concentrations were 2.16 mg/g compared with 2.20 mg/g in normal $B_6$ controls.

Iwanowska[217] also noted that rats receiving only a $B_6$-deficient diet had blood urea levels of 52 ± 5.5 mg urea per deciliter compared with 69 ± 46, 99 ± 4.7, and 112 ± 4.5 mg/dℓ in $B_6$-deficient rats supplemented with 100, 200, or 400 μg deoxypyridoxine per day, respectively. In rats receiving 50 μg $B_6$ per day, deoxypyridoxine doses of 0, 100, 200, and 400 μg/day produced blood urea levels of 46 ± 4.7, 85 ± 4.8, 47 ± 4.58, and a trace, respectively. The almost total absence of urea at the 400 μg dose is unexpected. It may be an artifact.

In contrast to the increased blood urea levels reported above, Ross and Pike[426] found that pregnant animals receiving a $B_6$-deficient diet containing 5 mg deoxypyridoxine per kilogram had a lower serum concentration of nonprotein nitrogen than animals receiving only the $B_6$-deficient diet.

While Beaton et al.[34,36,40] noted no changes in serum porteins or alpha amino nitrogen, all of their animals received deoxypyridoxine. Ross and Pike[426] stated that the lowest serum globulin values were found in depleted animals maintained on deoxypyridoxine-supplemented and pyridoxine-free ratios. However, while $B_6$ deficiency alone reduced serum globulin, Table 8 of their data showed that the globulin levels in the deoxypyridoxine group were comparable to or higher than those in the pyridoxine-supplemented animals.

Ross and Pike[427] noted that treatment of female rats with a $B_6$-deficient diet containing 5 mg deoxypyridoxine per kilogram for at least 7 days before mating resulted in significantly ($p < 0.01$) reduced serum globulin levels compared with nondepleted animals regardless of the diet used during gestation. The depleted animals which could not be bred had the highest globulin levels of the depleted groups even though they continued to receive deoxypyridoxine throughout the experiment. This suggests that the fall in serum globulin results from an interaction with the stress of pregnancy. The lowest serum globulin level (2.12 ± 0.31) was observed in a depleted group which received only the $B_6$-deficient diet during gestation. The globulin concentration (2.51 ± 0.30) in the group which continued to receive the diet containing deoxypyridoxine was comparable to those (2.72 ± 0.12, 2.51 ± 0.25, and 2.50 ± 0.17) found in depleted animals treated with pyridoxine. Brown and Pike[69] increased the deoxypyridoxine content of the diet to 20 mg/kg. Gamma globulin was decreased in pregnant animals and in nonpregnant animals receiving deoxypyridoxine.

Bergamini et al.[47] reported that intraperitoneal administration of 20 mg deoxypyridoxine per day to rabbits apparently receiving a normal $B_6$ diet produced about an 80% rise in serum alpha-amino-nitrogen. Administration of 1 g cholesterol per day seemed to produce a slight rise in alpha-amino nitrogen compared with the basal diet alone. No further increase was produced by combining the deoxypyridoxine and cholesterol treatments. Iwanowska[217] reported that in rats with normal $B_6$ intake deoxypyridoxine (200 or 400 μg/day) increased both growth and amino-nitrogen excretion. $B_6$ deficiency also increased amino nitrogen excretion. Addition of deoxypyridoxine to the deficient diet produced no further effect.

Heidelberger and Keller[193] intraperitoneally administered 200 mg deoxypyridoxine per kilogram to rats bearing Flexner-Jobling carcinoma at 27 and 3 hr before sacrifice. They observed a 22% reduction in protein synthesis in the tumors.

While Hruban et al.[206] concluded that a $B_6$-deficient diet plus deoxypyridoxine (20 mg/kg body weight per day) for 8 days had no effect on serum proteins in tumor-free rats, their data showed a 17% decrease in the alpha$_1$-fraction (1.43 ± 0.03 in controls vs. 1.18 ± 0.02 g/dℓ) and a 27% increase in the alpha$_2$-fraction (0.63 ± 0.03 in controls vs. 0.80 ± 0.03) of the deoxypyridoxine group. It appears that the differences would be statistically significant. In all four experiments with Murphy-Sturm lymphosarcoma, the rats receiving a $B_6$-deficient diet plus deoxypyridoxine showed a 10 to 25% increase in alpha$_1$-globulin and a 20 to 80% increase in beta-globulin compared with normal $B_6$ controls. Two of the four tumor studies showed increased alpha$_2$-globulins in the deoxypyridoxine group. However, the design did not separate the effects of $B_6$ deficiency from those of deoxypyridoxine.

Several workers have studied the effects of deoxypyridoxine on hemoglobulin synthesis. Jensen et al.[226] reported that deoxypyridoxine at concentrations greater than 0.005 *M* suppressed the incorporation of iron into heme by erythrocytes from white Pekin ducks. The inhibition was not reversed by pyridoxine hydrochloride at up to a 5-fold excess. However, it might have been more appropriate to use pyridoxal or pyridoxal phosphate in the reversal studies. Deoxypyridoxine did not alter the uptake of iron into the nonheme fraction. The authors noted that in this species the erythrocytes are nucleated and that plasma iron does not seem to be protein bound. Therefore, the results may not be directly transferable to man.

Klein[237] found that deoxypyridoxine inhibited both heme and porphyrin formation in duck erythrocytes in vitro. At $3 \times 10^{-5}$ *M* deoxypyridoxine heme formation was reduced to less than 50% of control values. Addition of $9 \times 10^{-3}$ *M* glycine stimulated both heme and free porphyrin synthesis. At $3 \times 10^{-4}$ *M* deoxypyridoxine porphyrin synthesis was completely inhibited while heme synthesis still proceeded at about 8% of its control rate. Addition of glycine had little effect.

Pike and Brown[391] treated rats with a $B_6$-deficient diet containing 40 mg deoxypyridoxine per kilogram. They felt that failure of hemoglobin and hematocrit values to return to normal after drawing a blood sample suggested impaired ability to regenerate hemoglobin and cells. In a subsequent experiment, Brown and Pike[69] reduced the deoxypyridoxine content of the diet to 20 mg/kg. Treatment was initiated at least six days before mating and in some groups was continued during gestation. Again the decline in hematocrit and hemoglobin values in both pregnant and nonpregnant females over 21 days of deoxypyridoxine treatment led the authors to conclude that hemoglobin and erythrocyte formation were impaired.

Five adult male monkeys *(Macaca radiata)* who received a $B_6$-deficient diet plus deoxypyridoxine (5 mg/kg intramuscularly on alternate days) for 14 months showed a slight but statistically significant ($p < 0.01$) reduction in hemoglobin (12.52 g/dℓ) compared with animals receiving normal $B_6$ and no deoxypyridoxine (13.98 g/dℓ).[260]

### B. Glutamic Acid-Glutamine

In addition to the marked increase in blood urea, Beaton et al.[36] observed that compared with rats receiving pyridoxine plus 100 μg deoxypyridoxine per day, those receiving only the $B_y$-deficient diet plus deoxypyridoxine showed a slight increase in plasma glutamic acid and a marked decrease in glutamine. Beare et al.[34] found no consistent change in glutamic acid but confirmed the decrease in plasma glutamine. Plasma glutamic acid seemed to increase more in $B_6$-deficient rats receiving deoxypyridoxine than in rats receiving a normal $B_6$ intake plus deoxypyridoxine after injection with glutamic acid or lysine[40] or alanine.[33] Beaton et al.[40] noted decreased formation or increased utilization of glutamine after treatment with glutamic acid.

Rats receiving deoxypyridoxine plus either a normal or deficient $B_6$ intake responded similarly to supplementing the 20% gelatin diet with either methionine or histidine. However, when both amino acids were added simultaneously, the animals receiving deoxypyridoxine plus the $B_6$-deficient diet showed a 2-fold increase in plasma glutamic and a 4-fold increase in glutamine compared with the animals receiving deoxypyridoxine plus a normal $B_6$ intake.[40]

Beaton and Goodwin[37] attempted to elucidate the effect of deoxypyridoxine on renal glutaminase in rats. In 21 days, deoxypyridoxine (100 μg/day) treatment plus a $B_6$-deficient diet produced a significant decline in renal glutaminase activity. Compared with pair-fed controls, $B_6$ deficiency alone had little effect on glutaminase within 21 days but did produce a significant decline with 35 days. Pyridoxine supplementation prevented the effect of deoxypyridoxine. Restoration of activity by supplementation of the assay with pyridoxal phosphate in animals with dietary deficiency indicated the decreased activity was due to lack of cofactor rather than decreased apoenzyme. However, similar data from the deoxypyridoxine group were not presented.

Using semiquantitative paper chromatography, Bergamini et al.[47] noted no detectable change in the glutamic acid concentration of the serum in rabbits receiving 20 mg deoxypyridoxine per day intraperitoneally apparently in conjunction with a normal $B_6$ diet.

In mice, either a $B_6$-deficient diet or injection of normal mice with deoxypyridoxine produced an increase in the glutamic acid content of the cerebral cortex.[301] Braunstein et al.[66] reported that intraperitoneal administration of deoxypyridoxine (10 mg/kg body weight) reduced the uptake of glutamine by rat intestine perfused *in situ* by about 50%. Intravenous administration of pyridoxine at the end of the first hour of perfusion produced a 150% increase in glutamine uptake thus returning it to the normal range.

Rasmussen[403] found that deoxypyridoxine treatment (500 mg/kg diet) in conjunction with either normal or deficient $B_6$ intake had no effect on the concentration of glutamic acid in plasma. No data on glutamine were presented.

### C. Glycine-Serine

The first hint that glycine metabolism might be involved in deoxypyridoxine action was the observation by Beaton et al.,[40] that changing from casein to gelatin as a protein source prevented acrodynia in $B_6$-deficient rats receiving 100 μg deoxypyridoxine per day. However, using paper chromatography Bergamini et al.[47] observed no gross changes in the concentration of glycine in the plasma of rabbits receiving 20 mg deoxypyridoxine per day apparently in conjunction with a normal $B_6$ intake. Serine seemed to be slightly reduced.

The first specific study of glycine-serine interactions was conducted by Sakami[428a] who injected chicks maintained on a $B_6$-deficient diet with 16 μg deoxypyridoxine 2 hr before sacrifice. $B_6$ deficiency alone had no effect on the incorporation of formate into serine, but liver extracts from the deoxypyridoxine group showed a 50% reduction. Much of the reduction could be overcome by adding pyridoxal phosphate in vitro. Similar inhibition of formate incorporation was obtained by adding 2 $\mu M$ deoxypyridoxine plus adenosine triphosphate to liver extracts from $B_6$-deficient chicks. Again, much of the inhibition was overcome by adding pyridoxal phosphate (1.8 mg/2 mℓ). Gershoff and Faragalla[163] examined the relationships between endogenous oxalate synthesis, glycine, serine, and deoxypyridoxine (500 mg/kg diet) in $B_6$-deficient rats. Addition of deoxypyridoxine to the $B_6$-deficient diet reduced weight gains and caused some deaths. Addition of 3% glycine to the $B_6$-deficient diet reduced weight gains by 33 to 50% compared to the deficient diet alone. However, adding both deoxypyridoxine and glycine to the deficient diet produced gains comparable to the deficient diet alone and prevented the deaths. The increased oxalate excretion associated with $B_6$ deficiency was further increased by deoxypyridoxine and still further by glycine. On a molar basis, serine was even more effective than glycine in preventing the weight loss associated with deoxypyridoxine. The authors suggested that deoxypyridoxine may inhibit the formation of serine and glycine or enhance their degradation.

Hartsook and Harshberger[191] reported that addition of 2% glycine to a $B_6$-deficient diet containing 20% casein and 100 μg deoxypyridoxine per gram prevented weight loss and reduced liver weight. Adding 25% casein to the deoxypyridoxine treatment significantly ($p < 0.01$) decreased weight gain and survival. Supplementation of the deoxypyridoxine diet with gamma-aminobutyric acid or lysine had no significant effect on survival, gain, or liver or brain weight.

Addition of glycine in vitro did not overcome the inhibitory effect of deoxypyridoxine on heme and porphyrin formation in duck erythrocytes.[237]

Matano[301] reported that injection of normal mice with deoxypyridoxine produced an increase in glycine in the cerebral cortex. This was not observed with dietary $B_6$ deficiency.

Rasmussen[403] found that compared with a normal $B_6$ intake, $B_6$ deficiency produced a rise in the glycine concentration of the plasma ($p < 0.01$), muscle ($p < 0.01$), and liver ($p < 0.10$). Compared with a diet containing normal $B_6$ and 500 mg deoxypyridoxine per kilogram, a $B_6$-deficient diet containing a similar amount of deoxypyridoxine caused no change in plasma glycine but a significant decrease in the glycine content of muscle ($p < 0.01$) and liver ($p < 0.05$) tissue. Thus the effect of low $B_6$ intake on tissue glycine is completely reversed by the presence of deoxypyridoxine.

There have been several in vitro studies which provide further evidence for an interaction between deoxypyridoxine and glycine. Christensen et al.[87] concluded that loss of potassium is not necessarily associated with glycine uptake because incubation of Ehrlich mouse ascites tumor cells with deoxypyridoxine caused both loss of potassium and decreased glycine uptake. However, we feel that perhaps the potassium part of the transport system was still active but the glycine part of the cycle was blocked by deoxypyridoxine. Christensen and Riggs[85] found that 5′-deoxypyridoxal and 4-nitroalicylaldehyde were as effective as pyridoxal in stimulating the concentration of glycine by Ehrlich ascites tumor cells. While peak values achieved by the two analogs were slightly higher than those obtained with pyridoxal, the analogs became inhibitory at about 10% of the concentration required to produce inhibitory effects with pyridoxal

(5 to 10 m*M*). 3-Deoxypyridoxine, salicylaldehyde, and 4′-deoxypyridoxine showed no stimulatory effect.[87] However, 4′-deoxypyridoxine was inhibitory at high concentrations (unspecified). The authors concluded that the aldehyde group, the phenol group, and the nitro group or ring nitrogen were essential for the molecule to stimulate the concentration process but that phosphorylation of the 5′-position was not required. This work was summarized by Christensen et al.[86]

Braunstein et al.[66] measured the uptake of amino acids by rat intestine by measuring the loss of amino acid from a solution which perfused the lumen of an intestinal segment with an intact blood supply to the donor animal. Intraperitoneal administration of deoxypyridoxine (1 mg/100 g body weight) at the start of the experiment inhibited glycine absorption. Intravenous administration of pyridoxine (probably 1 mg/100 g body weight) at the start of the second hour of perfusion caused a 110% increase in glycine uptake during the second hour. Thus the intestinal transport system apparently responds very rapidly. However, these data seem to be based on only one rat.

Runyan and Gershoff[428] used a $B_6$-deficient diet containing 125 mg deoxypyridoxine per kilogram. After 8 to 11 days on the diet, rats receiving only the $B_6$-deficient diet tended to have increased tissue concentrations of glycine, particularly in liver, kidney, and muscle. However, in rats receiving the deficient diet plus deoxypyridoxine, glycine concentrations were lower than control values except in muscle. The lack of a decrease in muscle conflicts with the data of Rasmussen.[403] Conversion of glycine, calcium glycolate, and sodium glyoxylate to hippuric acid was reduced after 5 to 9 days of $B_6$ deficiency and further reduced by addition of deoxypyridoxine. Conversion of glycine to respiratory $CO_2$ was reduced equally by $B_6$-deficient diets with and without deoxypyridoxine. In vitro conversion of glycine to $CO_2$ by liver homogenates was decreased in the deoxypyridoxine groups. Upon adding pyridoxal phosphate, all groups increased. The activity in the control and deficient groups was about equal. Both were higher than the activity in deoxypyridoxine group. This suggests that the deficient diet alone reduced the cofactor but had little effect on the amount of apoenzyme. The decreased response of the deoxypyridoxine group to pyridoxal phosphate indicates loss of apoenzyme either through decreased synthesis or perhaps by permanent inactivation by deoxypyridoxine phosphate. In kidney homogenates, the conversion of glycine to $CO_2$ was markedly reduced by $B_6$ deficiency. The deoxypyridoxine group was slightly higher than the deficient group. After addition of pyridoxal phosphate, there was little change in the controls, a 100% increase in the $B_6$-deficient tissues, and a 50% increase in the deoxypyridoxine group. The authors suggested that the decreased liver glycine in the deoxypyridoxine group might be a consequence of increased glycine oxidation. The authors suggested that the enzymes for oxidizing glycine in liver and kidney may be quite different. Alternatively, perhaps it is simply the degree of saturation which differs.

Metabolism of serine and glycine is closely interrelated. Alexander and Greenberg[10] examined the synthesis of serine in rat liver preparations. The authors noted that their system was identical to the one used for methylating nicotinamide. Deoxypyridoxine (25 μmol) preincubated with an equimolar amount of ATP produced 50% inhibition. Addition of 5 μmol pyridoxal phosphate almost eliminated the inhibition. Five μmol of pyridoxal, pyridoxine, or pyridoxamine did not reverse the inhibition. While the authors commented that they anticipated phosphorylation of pyridoxal, our experience suggests that the deoxypyridoxine may have interfered with the kinase reaction. A deoxypyridoxine:pyridoxal phosphate ratio of 150:1 caused 60% inhibition of serine synthesis. However, it appears that the ratios were varied by holding the deoxypyridoxine at 25 μmol and varying the pyridoxal phosphate. It would have been interesting to keep the pyridoxal phosphate at 5 μmol and vary the deoxypyridoxine.

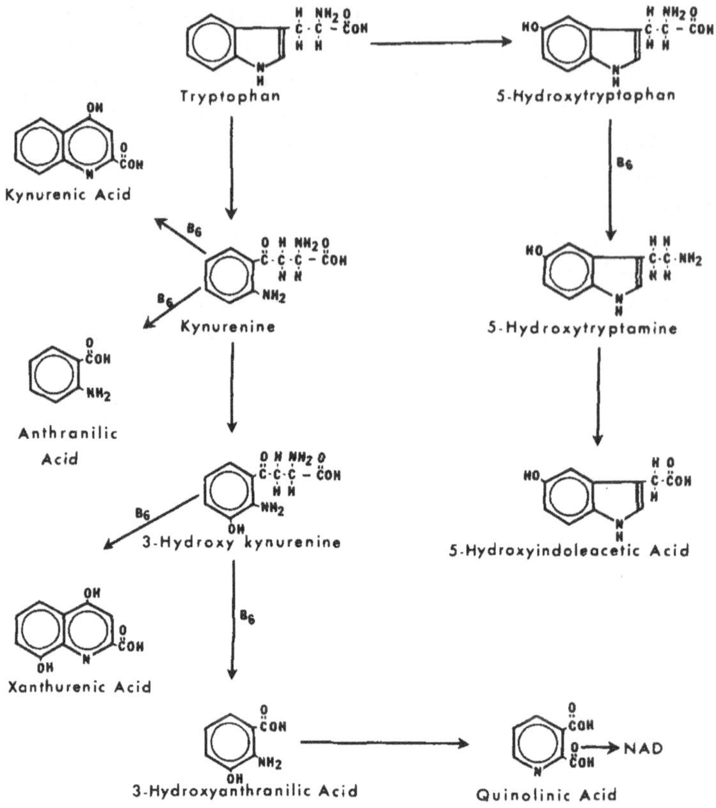


FIGURE 1. Tryptophan metabolism.

Nakao and Greenberg[361] showed that incubation of a sheep liver preparation with deoxypyridoxine reduced the incorporation of carbon from serine and formaldehyde into methionine. The effect of 12 μmol of deoxypyridoxine was partially alleviated by 2 μmol pyridoxal phosphate. Since the system was simply a dialyzed liver homogenate, its complexity makes it impossible to determine the point of deoxypyridoxine effect.

Blakley[55] observed that the in vitro synthesis of serine from glycine by a rabbit liver system was more sensitive to inhibition by deoxypyridoxine phosphate than by pyridoxine phosphate whereas Meister et al.[306] had reported that aspartate aminotransferase showed similar sensitivity to both substances.

## D. Tryptophan

As can be seen from Figure 1, vitamin $B_6$ has an important role in tryptophan metabolism. Although vitamin $B_6$ is a cofactor for the conversion of 3-hydroxykynurenine to both 3-hydroxyanthanilic acid and xanthurenic acid, the former step appears to be more sensitive to $B_6$ deficiency than the latter. Detection of increased xanthurenic acid excretion after a tryptophan load is a commonly used criterion for $B_6$ deficiency.

Porter et al.[396] examined the effect of deoxypyridoxine on tryptophan metabolism in rats. The design consisted of a 1.5 hr feeding period, a dose of 1 mg of deoxypyridoxine administered by stomach tube and a 22.5 hr urine collection. In $B_6$-deficient rats, administration of deoxypyridoxine increased xanthurenic acid excretion from a control level of 1 mg to 3 mg. There was little change in kynurenine or pyridoxic acid. In $B_6$-deficient animals, a tryptophan load caused a marked increase in both kynurenine and xanthurenic acid excretion. Administration of deoxypyridoxine caused a slight

Table 9
EXCRETION OF XANTHURENIC ACID IN $B_6$-DEFICIENT RATS TREATED WITH DEOXYPYRIDOXINE OR ISONICOTINIC ACID HYDRAZIDE

| Treatment | Xanthurenic acid (μg/24 hr) Before tryptophan load | After tryptophan load | Ref. |
|---|---|---|---|
| $B_6$ deficient | 210 | 600 | 63 |
| | 135 | 237 | 420 |
| + 100 μg deoxypyridoxine/day | 260 | 12100 | 63 |
| + 500 mg deoxypyridoxine/kg diet | — | 875 | 420 |
| + 50 mg INH/kg | 270 | 5700 | 63 |
| + 10 mg INH/rat | 43 | 80 | 420 |

additional increase (30%) in kynurenine excretion and about a 75% increase in xanthurenic acid excretion. Deoxypyridoxine even exerted some effect on animals receiving a normal $B_6$ intake.

Ling et al.[277] examined the effect of $B_6$ deficiency plus deoxypyridoxine (20 mg/kg diet) on the conversion of tryptophan into pyridine nucleotides in the erythrocytes of rats. Deoxypyridoxine was removed from the diet after 7 weeks because of the poor condition of the animals. It is not clear how long after 7 weeks the samples were collected. A very crude method of determining pyridine nucleotides was employed. Intravenous injection of 50 mg of tryptophan produced a 50% increase in erythrocyte pyridine nucleotides in normal rats but not in $B_6$-deficient animals. However, subcutaneous injection of 1 mg pyridoxine hydrochloride one hour before the tryptophan injection permitted a normal response in the deficient animals. $B_6$ deficiency did not diminish the increase in pyridine nucleotides seen after niacin injection.

Since simultaneous injection of niacin did not protect chick embryos from the effects of deoxypyridoxine, Cravens and Snell[106] concluded that interference in the synthesis of niacin from tryptophan was not a major factor in the toxicity of deoxypyridoxine.

Kring et al.[259] used deoxypyridoxine in their work with pyridine nucleotides in rats because in earlier studies they were unable to obtain an effect with a simple $B_6$ deficiency. Sulfasuxidine was used to minimize the intestinal flora. A $B_6$-deficient diet containing 20 mg deoxypyridoxine per kilogram was used. All animals were fed ad libitum. Deoxypyridoxine reduced the conversion of tryptophan but not niacin to pyridine nucleotides.

Chiancone et al.[83] examined the effect of riboflavin on xanthurenic acid metabolism in rats. Administration of pyridoxine (100 μg s.c. or 150 μg orally) with a tryptophan load corrected the increased xanthurenic acid excretion induced by deoxypyridoxine or a $B_6$-deficient diet. Riboflavin had no effect.

Olsen and Martindale[370] found that addition of deoxypyridoxine (12.5 mg/kg) to normal or $B_6$-deficient diets increased xanthurenic acid excretion after a tryptophan load in rats.

Comparisons of the effects of isonicotinic acid hydrazide and deoxypyridoxine on xanthurenic acid excretion produced conflicting results (Table 9). Boone et al.[63] and Rosen[420] obtained comparable excretions before a tryptophan load using only the $B_6$-deficient diet. However, Boone et al. administered an oral load of 200 mg DL-tryptophan for three days while Rosen use a single intraperitoneal dose of 20 or 25 mg L-tryptophan and collected urine for 48 hr. It is quite likely that xanthurenic acid excretion dropped markedly during the second 24 hr thus reducing the average 24 hr excretion. The fact that xanthurenic acid excretion in the $B_6$-deficient group did not even double after the tryptophan load suggests that the small load used by Rosen did not stress the animals very much. Therefore, we suspect that the decreased xanthurenic acid excretion observed with isonicotinic acid hydrazide treatment may have resulted from the small load and extended collection period rather than any marked difference between the effects of deoxypyridoxine and isonicotinic acid hydrazide. These same factors may also explain why xanthurenic acid excretion after a tryptophan load in Rosen's deoxypyridoxine group was 15-fold lower than in Boone's study even though Rosen's dosage of deoxypyridoxine was probably 5- to 25-fold higher than Boone's. Yamada et al.[556] also compared deoxypyridoxine and isonicotinic acid hydrazide. Unfortunately, their xanthurenic acid data appear to represent only one rat per treatment. There is no statement that a tryptophan load was used. In the $B_6$-deficient rats, injection of 20 mg deoxypyridoxine per kilogram per day increased the excretion of xanthurenic acid by 80% whereas a similar dose of isonicotinic acid hydrazide reduced excretion by 30% compared with $B_6$-deficient controls. In rats receiving 10 mg $B_6$ per kilogram per day, xanthurenic acid excretion was increased 47% by deoxypyridoxine and 12% by isonicotinic acid hydrazide. Although the effect of isonicotinic acid hydrazide on xanthurenic acid excretion needs further clarification, all three reports agree that deoxypyridoxine produced a greater increase in xanthurenic acid excretion than was produced by isonicotinic acid hydrazide. In view of this, Rosen concluded that vitamin antagonists may affect only selected enzymes and need not inhibit all biochemical functions of the vitamin.

Yamada et al.[556] included data showing that hydroxymethylpyrimidine (20 mg/kg/day) increased xanthurenic acid excretion two- to threefold in rats both with and without $B_6$. Hydroxymethylpyrimidine also had a greater effect than deoxypyridoxine or isonicotinoyl hydrazide on growth. Although the structure was not specified, it seems likely that the term hydroxymethylpyrimidine actually referred to toxopyrimidine (2-methyl-4-amino-5-pyrimidinemethanol). The abstract of another paper[149] stated that the thiaminase of *Bacillus aneurinolyticus* split thiamine into pyrimidinemethanol and a thiazole moiety while a third paper[215] reported the same enzyme as yielding the thiazole derivative plus toxopyrimidine.

Druyan and Haeger-Aronsen[125] administered 5 mg deoxypyridoxine per kilogram body weight per day, via an unspecified route, to rabbits receiving a presumably nutritionally complete natural diet. Excretion of kynurenine and 3-hydroxykynurenine was significantly increased before a tryptophan load. Compared with controls, deoxypyridoxine treatment decreased kynurenine excretion after a tryptophan load but did not significantly alter excretion of 3-hydroxykynurenine. Deoxypyridoxine significantly decreased delta-aminolevulinic acid excretion. Deoxypyridoxine also tended to reduce excretion of these substances in experimental porphyria. Administration of a single 25 mg dose of pyridoxine tended to reverse these effects.

Bianchi-Donnasibilla and Mainardi[52] evaluated the effect of $B_6$ and deoxypyridoxine on the xanthurenic acid index in the rat. They found that with an intake of 30 μg pyridoxine per day a single dose of up to 1.6 mg of deoxypyridoxine failed to change the xanthurenic acid index. After six weeks on a $B_6$-deficient diet, 6 days of treatment with 100 μg pyridoxine and 200 μg deoxypyridoxine produced a normal xanthurenic index whereas the index remained abnormal in animals receiving only 25 μg pyridoxine with 200 μg deoxypyridoxine. Administering 4 μg pyridoxine in conjunction with 40 μg deoxypyridoxine produced a change in the xanthurenic acid index within 4 days which progressively increased for about 5 weeks and then seemed to stabilize.

Amer et al.[11] compared the effects of deoxypyridoxine and *Schistosoma* infection on tryptophan metabolism in mouse liver homogenates. Both deoxypyridoxine and deoxypyridoxine phosphate at 10 $\mu g/m\ell$ inhibited kynurenine metabolism. Since formation of kynurenic acid seemed to be reduced more than anthranilic acid, the authors concluded that kynurenine aminotransferase was more susceptible to inhibition than kynureninase. The presence of an equal concentration of pyridoxal partially reversed the inhibition of kynurenic acid synthesis and appeared to stimulate synthesis of anthranilic acid above control values. Pyridoxal phosphate completely reversed the inhibition of kynurenic acid synthesis and again stimulated production of anthranilic acid.

In connection with studies of bladder cancer in bilharziasis, Amer et al.[12] examined the effect of deoxypyridoxine phosphate (10 $\mu g/m\ell$) on in vitro kynurenine metabolism by homogenates of normal mouse liver. The authors again concluded kynurenine aminotransferase was more susceptible to inhibition than kynureninase. They also suggested that antimony can form an inactive chelate with both pyridoxal phosphate and deoxypyridoxine phosphate.

Koerner and Nowak[239] used the ratio of xanthurenic acid excretion following a tryptophan load before and after antagonist administration to evaluate the action of deoxypyridoxine. Male rats (150 g) were treated with a $B_6$-deficient diet for the first 10 days. For the next 10 days they received 30 $\mu g$ pyridoxine hydrochloride per day. This dose just sufficed to prevent an increase in xanthurenic acid excretion after a tryptophan load of 200 mg/kg. After this 20-day preparation period, the rats received an oral dose of 200 mg tryptophan per kilogram body weight per day in addition to the pyridoxine supplement for 5 days. During the last 5 days of the experiment, the animals received intraperitoneal injections of deoxypyridoxine (0.1 to 100 mg/kg body weight) in addition to the pyridoxine and tryptophan supplements.

$$Q_{xs} = \frac{\text{Average xanthurenic acid excretion before deoxypyridoxine } (\mu g/100 \text{ g body weight/day})}{\text{Average xanthurenic acid excretion during deoxypyridoxine administration}}$$

$Q_{xs}$ was slightly over 1.0 for saline-injected controls. It decreased with increasing dosages to about 0.3 at 10 mg deoxypyridoxine per kilogram and then seemed to level off. The authors concluded that this procedure provides a faster method of evaluating $B_6$ antagonism than the usual growth studies.

Probably the most revealing data on the interaction between tryptophan and deoxypyridoxine comes from Milholland and Rosen.[327] Using a $B_6$-deficient diet containing 100 mg deoxypyridoxine hydrochloride per kilogram, they found that compared with the $B_6$-deficient diet alone, deoxypyridoxine markedly increased urinary xanthurenic acid excretion after a tryptophan load even after just 1 day on the diet. Increasing the activity of tryptophan pyrrolase threefold with cortisone produced little change in xanthurenic acid excretion either before or after a tryptophan load. The excretion of xanthurenic acid was correlated with the dose of tryptophan. Using cyclohexamide also failed to reduce xanthurenic acid excretion. Treatment with the deficient diet alone or with deoxypyridoxine was equally effective in reducing liver kynureninase but the deoxypyridoxine group excreted five to ten times more xanthurenic acid than the group receiving only the $B_6$-deficient diet. Therefore, the authors noted that changes in enzyme activity detected in vitro may not have significant effects in vivo. They also agreed with Spiera and Vallarino[471] that spontaneous increases in urinary excretion of tryptophan metabolites without tryptophan load may result from altered renal clearance. In deoxypyridoxine-treated rats, intraperitoneal administration of 1 mg pyridoxine HCl 1 hr after the tryptophan load prevented the excretion of xanthurenic acid.

In other studies relating to the tryptophan-niacin pathway, Umbreit and Waddell[522] noted that deoxypyridoxine at concentrations up to 900 $\mu g/m\ell$ showed no effect on tryptophanase from *Escherichia coli.* Kawakita et al.[234] also reported that deoxypyridoxine phosphate:pyridoxal phosphate ratios as high as 1000 had no effect on that enzyme.

The serotonin pathway is also susceptible to deoxypyridoxine inhibition. Huidobro et al.[209] obtained some indirect evidence of this when they observed that deoxypyridoxine (18 mg/kg body weight intraperitoneally) decreased the intensity of the abstinence syndrome induced in morphine treated mice by nalorphine. The authors assumed that $B_6$ deficiency decreased conversion of tryptophan to 5-hydroxytryptamine. Therefore, the effect of deoxypyridoxine might conflict with the hypothesis that increasing 5-hydroxytryptamine concentrations attenuates the abstinence syndrome. In an attempt to deplete serotonin more completely than could be accomplished with drugs, Boullin[64] studied the effects of a $B_6$-deficient diet with and without added deoxypyridoxine (70 or 280 μg/day per rat). Controls were pair-fed and received the same diets (including deoxypyridoxine) plus 80 μg pyridoxine hydrochloride. Since there were no statistically significant differences between deficient groups and their controls, he concluded that there was no depletion. However, the averages (Table 10) suggest that there may well have been depletion which was obscured by the wide variability between animals in some groups. The deficient animals receiving 180 μg deoxypyridoxine clearly had the lowest average value and minimal variation between animals. We have experienced similar instances where the variation between animals is markedly reduced as the deficiency becomes more severe. Also, as was noted in Chapter 2, the pair-fed controls were losing weight even though they were receiving 80 μg pyridoxine hydrochloride per day. No serotonin values were presented for animals receiving either the $B_6$-deficient diet alone or a complete diet ad libitum. It seems likely that comparing such treatments with the deoxypyridoxine treatments might have revealed more obvious depletion of serotonin in the deoxypyridoxine groups. For example, Sano et al.[435] found that intraperitoneal administration of 43 mg deoxypyridoxine per kilogram per day caused a 45% reduction in the concentration of serotonin in tumor tissue in mice. 3-Aminopyridoxine had little effect. Also, Zervas et al.[567] noted that daily oral administration of 8 mg deoxypyridoxine per kilogram body weight to four rhesus monkeys for 18 to 27 days decreased blood 5-hydroxytryptamine concentrations by about 50%.

In summary, the marked increase in the urinary excretion of xanthurenic acid after a tryptophan load is one of the most consistently observed effects of deoxypyridoxine. The rapidity with which xanthurenic acid excretion responds to deoxypyridoxine or pyridoxine administration, the lack of effect of cyclohexamide, and the poor correlation between xanthurenic acid excretion and kynureninase activity all suggest that deoxypyridoxine acts in a fairly simple and direct manner. One hypothesis might be that deoxypyridoxine facilitates the movement of xanthurenic acid into the blood and/or the urine. Another possibility is that deoxypyridoxine inhibits the further metabolism of xanthurenic acid thus causing a build-up of xanthurenic acid.

### E. Other Amino Acids

Beaton[40] compared rats receiving only $B_6$-deficient diet plus 100 μg deoxypyridoxine per day with those receiving the deficient diet plus deoxypyridoxine plus pyridoxine. Using gelatin as the protein source seemed to produce decreased blood sugar in the animals not supplemented with deoxypyridoxine. Although histidine supplementation had a marked effect on increasing blood urea and methionine had little effect, the situation was reversed with respect to blood sugar. Histidine alone had little effect, but histidine plus methionine decreased blood sugar.

In another study using the same design with casein as the protein source, blood sugar was increased in every group of rats receiving the pyridoxine supplement one hour after an alanine load.[33] This contrasted with the results in the animals not supple-

**Table 10**
**EFFECT OF $B_6$ DEFICIENCY ON THE SEROTONIN CONTENT OF THE INTESTINAL MUCOSA OF RATS[64]**

| Deoxypyridoxine (μg/day) | Serotonin (μg/g) | |
|---|---|---|
| | Deficient | Control |
| 70 | 3.86 ± 0.73[a] | 4.28 ± 0.64 |
| 180 | 2.52 ± 0.23 | 3.95 ± 0.84 |

[a] Mean ± standard error

mented with pyridoxine where in three out of four comparisons they had higher blood sugar than the controls at 0 time, showed a decrease after 1 hr, and a peak after 3 hr.

Fridhandler and Quastel[147] found that 0.01 *M* deoxypyridoxine inhibited the absorption of DL-alanine from isolated guinea pig intestine. The inhibition was not reversed by 0.005 *M* pyridoxal, 0.01 *M* pyridoxine, 0.003 *M* pyridoxamine phosphate, or 0.012 *M* pyridoxal phosphate (calcium salt). Neither was the inhibition prevented by exposing the tissue to 0.001 *M* pyridoxal phosphate (calcium salt) for 15 min prior to adding deoxypyridoxine. Deoxypyridoxine also inhibited the uptake of glucose and the conversion of fructose into glucose. The authors noted that failure of $B_6$ compounds to reverse the inhibition does not rule out the possibility that deoxypyridoxine acts by interfering with $B_6$ activity. They suggested that deoxypyridoxine may interfere with an aspect of active transport that is common to both amino acids and sugars.

Bergamini et al.[47] administered 20 mg deoxypyridoxine per day intraperitoneally for 30 days to rabbits apparently receiving a normal $B_6$ diet. Two-dimensional paper chromatography of serum revealed that in controls, the major spots were alanine, valine, and leucine-phenylalanine. After treatment with deoxypyridoxine, the concentrations tyrosine, lysine, and arginine seemed to increase. Threonine was markedly increased in three of the four animals examined. Methionine was detected in three of the four treated animals but in neither of the two controls. Proline and serine seemed to be reduced. Aspartic acid, glutamic acid, glycine, tryptophan, alanine, valine, leucine, phenylalanine, and histidine were unchanged. Similar trends were found in the animals receiving deoxypyridoxine plus cholesterol. This differs from the results of Matano[301] who reported that injection of normal mice with deoxypyridoxine produced an increase in glutamic acid, alanine, glycine, and valine in the cerebral cortex.

Jacobs and Hillman[221] examined the effect of deoxypyridoxine (0.2 or 0.4 mg/day s.c. or i.p.) on intestinal absorption of methionine in rats receiving a commercial stock diet. Administration for 5 to 10 days had no greater effect than a single dose administered 30 to 90 min before initiating perfusion tests. However, administration of deoxypyridoxine over 5 days had little effect when the last injection was 5 hr before perfusion. This is compatible with our observation of rapid excretion of deoxypyridoxine. In contrast to some effects of deoxypyridoxine which seem to persist long after deoxypyridoxine administration ceases, the effect on methionine absorption occurs and disappears very rapidly, suggesting that deoxypyridoxine probably interferes directly with the transport system.

Jacobs et al.[220] continued their studies again using a commercial stock diet. Deoxypyridoxine (0.5 mg administered i.p. 1 hr before perfusion *in situ*) produced methionine uptakes intermediate between those of the controls and a group which had received a $B_6$-deficient diet for 5 to 6 weeks. When injection of deoxypyridoxine was followed 30 min later by an equimolar amount of pyridoxine, methionine uptake remained at control levels. However, pyridoxine alone stimulated uptake above control levels. The authors noted that pyridoxal phosphate did not stimulate increased uptake but we feel this might be due to poor uptake by tissues compared with the uptake of pyridoxine. Therefore, it is difficult to determine whether the pyridoxine completely offset the effects of deoxypyridoxine. Pyridoxal phosphate but not pyridoxine alleviated the inhibition produced by dinitrophenol.

In a third group of experiments,[222] this time using tyrosine, deoxypyridoxine (0.5 mg i.p.) was again injected 60 min before perfusion. It was observed that while the amount of radioactivity removed from the lumen was reduced, the amount in the intestinal wall was not affected. Therefore, the authors concluded that deoxypyridoxine did not interfere with the initial movement from the lumen into the mucosal cells but did interfere with the movement from the mucosa to the blood.

Szmelcman and Guggenheim[491] used a $B_6$-deficient diet plus 20 mg deoxypyridoxine per kilogram diet. Some rats received 5 mg deoxypyridoxine per 100 g body weight with the protein meal. Deoxypyridoxine, whether administered to normal rats with the protein meal or administered chronically with the $B_6$-deficient diet, reduced the concentrations of lysine and tryptophan but not methionine in the portal blood. Administration of pyridoxine to the deficient animals at an unspecified time (probably with the protein meal) and in an unspecified amount did not correct the depressed uptake of tryptophan but did partially restore lysine uptake. When used, deoxypyridoxine either orally or intraperitoneally seemed to reduce the uptake of all three but did not effect the baseline concentrations. The difference in methionine effects may simply be that there was little increase in methionine concentrations after the protein meals; whereas, there was an increase of the amino acid load. Again the timing and dosage are not specified. It was shown that including starch (0.5 g/100 g body weight) with the tryptophan dose counteracted the inhibitory effect of deoxypyridoxine. The route, dose, and timing of deoxypyridoxine administration are not specified.

Braunstein et al.[66] after reviewing earlier work in their own laboratory and in the literature, concluded that ascites tumor cells did not provide a good model for studying amino acid transport. They therefore adopted an intestinal perfusion technique. Deoxypyridoxine (1 mg/100 g body weight i.p.) produced a 25 to 50% decrease in uptake of histidine measured by loss from the perfusate. This inhibition was not corrected by adding pyridoxinc, pyridoxal, or pyridoxal phosphate to the perfusate. Intravenous administration of pyridoxal or pyridoxal phosphate partially alleviated the inhibition. Injection of pyridoxine completely corrected it. Uptake of glycine and glutamine was also inhibited by deoxypyridoxine and corrected by pyridoxine. However, uptake of valine and asparagine did not seem to be sensitive to deoxypyridoxine or pyridoxine. Valine uptake was inhibited by 2,4-dinitrophenol, suggesting that it is an active transport process.

Munck[351] summarized previous reports suggesting three systems for amino acid transport in the intestinal tract: one for proline and hydroxyproline; one for lysine, ornithine, arginine, and possibly cystine; and one for neutral amino acids and histidine. He noted that the last two systems had been shown to be $B_6$ dependent but the first system had not been studied. Munck used hydroxyproline, arginine, and threonine to examine these systems in everted segments of rat intestine in vitro. No effect on amino acid transport was observed when the rats were injected with 0.5 mg deoxypyridoxine at 60 and 30 min. before sacrifice. This conflicts with the results of both Jacobs et al.[220] and Braunstein et al.[66] One obvious difference is that Munck[351] was using different amino acids. In addition, and perhaps more important was the fact that Munck[351] excised the intestinal segments whereas Jacobs et al.[220] and Braunstein et al.[66] left the circulation intact. Both of these reports demonstrated that the effects of deoxypyridoxine are quite short-lived. It may well be that cutting the circulation to the intestine reduced the effectiveness of the injected deoxypyridoxine.

Munck[351] attempted to compensate for this in a second series of experiments by including 0.13 m*M* deoxypyridoxine in the incubation medium. Again no inhibition was noted. However, this was a much lower concentration than the 10 m*M* used by Fridhandler and Quastel[147] to produce inhibition in vitro. Transport of threonine and arginine, but not hydroxyproline, tended to be reduced when the segments were incubated in solutions containing 5 m*M* deoxypyridoxine. Transport of all three amino acids was reduced when the sacs were preincubated with 5 m*M* deoxypyridoxine for 30 min before adding the amino acids. Thus, the results of Munck[351] do not seriously conflict with those of Braunstein et al.[66] Jacobs et al.,[220] and Fridhandler and Quastel.[147]

Tasaki and Yokota[497] reported that absorption of leucine and methionine from chicken intestine was inhibited by deoxypyridoxine more than absorption of glycine. Dose and route of administration were not specified in the abstract.

Rosen et al.[424] reported that deoxypyridoxine (100 mg/kg) injected into $B_6$-deficient rats 4 hr prior to sacrifice almost doubled the liver uptake of an unspecified mixture of 20 labeled amino acids injected 2 hr prior to sacrifice compared with $B_6$-deficient animals not receiving deoxypyridoxine. $B_6$ deficiency alone produced about 50% increase in the radioactivity of liver protein compared with normal animals. Deoxypyridoxine caused no further change. $B_6$ deficiency caused a slight increase in the radioactivity of thymus protein. Injection of deoxypyridoxine had no further effect on the radioactivity of amino acids on proteins in the thymus.

Easton et al.[128] fed adult rats a $B_6$-deficient diet containing 10 mg deoxypyridoxine per kilogram diet for up to 18 days. Serum tyrosine showed little change in adult rats receiving only the $B_6$-deficient diet. In the deoxypyridoxine groups, serum tryosine was increased about 3-fold in young rats and adult males, 10-fold in adult females, and 30-fold in lactating females. Serum phenylalanine was slightly but significantly ($p < 0.05$) increased in lactating females receiving deoxypyridoxine. Urine metabolites of tyrosine were examined semiquantitatively using paper chromatography. $B_6$ deficiency with or without deoxypyridoxine seemed to increase the excretion of tyrosine and *p*-hydroxyphenyl-acetic acid and to decrease the excretion of vanillylmandelic acid, homovanilic acid, and homogentisic acid. There was no detectable change in the excretion of *p*-hydroxyphenylacitic acid or *p*-hydroxyphenyl pyruvic acid. The excretion of cystathionine seemed to increase in $B_6$ deficiency and to increase further with addition of deoxypyridoxine. The authors suggested that deoxypyridoxine inhibited tyrosine transaminase but that does not explain the marked effect of sex and lactation. The authors discussed the effects of deoxypyridoxine on urinary excretion of tryptophan metabolites. However, except for cystathionase, their data show no differences between excretion in the group receiving only the $B_6$-deficient diet and the group receiving the deficient diet plus deoxypyridoxine. Therefore, except for cystathionase, it does not seem appropriate to attribute the changes to deoxypyridoxine.

Rasmussen[403] examined the effects of 500 mg deoxypyridoxine per kilogram diet in adult male rats for 17 days. Neither $B_6$ deficiency nor the antagonists affected plasma concentrations of arginine, histidine, isoleucine, leucine, lysine, methionine, phenylalanine, valine, cystine, aspartic acid, and glutamic acid. $B_6$ deficiency tended to lower plasma threonine and serine. Adding deoxypyridoxine accentuated this effect.

Dietary $B_6$ deficiency tended to produce a slight decrease in plasma alanine, tyrosine, and taurine. Adding deoxypyridoxine to the deficient diet, rather than causing a further decrease, caused an increase which in the case of alanine was highly significant ($p < 0.01$). Isonicotinic acid hydrazide tended to produce a similar effect.

$B_6$ deficiency alone tended to reduce the plasma concentrations of essential amino acids but had little effect on the nonessential amino acids. Combining deoxypyridoxine with normal $B_6$ tended to lower total concentrations of both essential and nonessential amino acids. Removing the $B_6$ further lowered the essential amino acids but tended to raise the concentrations of the nonessential amino acids, although they still did not return to normal concentrations.

Maynert and Kaji[303] reported that 4′-deoxypyridoxine 5′-phosphate produced 34 and 56% reductions in brain concentrations of gamma-aminobutyric acid in mice at doses of 250 and 500 mg/kg intraperitoneally.

Snodgrass and Iversen[467] found that four hours after a single intraperitoneal injection of deoxypyridoxine (30 or 60 mg/kg) the uptake of gamma-aminobutyric acid by rat brain slices was reduced by about 13 and 24%, respectively. Preincubation of brain slices with 1 m*M* deoxypyridoxine in vitro inhibited uptake of gamma-aminobutyric acid about 63% ($p < 0.01$). Although no statistical analysis was presented for the in vivo data, the authors concluded that deoxpyridoxine inhibited uptake of gamma-aminobutyric acid both in vivo and in vitro and that uptake of gamma-aminobutyric acid might be a pyridoxal phosphate-dependent process.

Ebadi and Klangkalya[129] have recently challenged that conclusion. They failed to detect any decrease in the uptake of gamma-aminobutyric acid by synaptosomes isolated from either the cerebral hemispheres or the hippocampus when incubated with 0.1 or 1.0 m*M* deoxypyridoxine. Neither did these concentrations have any effect on the uptake of gamma-aminobutyric acid by nuclei-free homogenates of the cerebral hemispheres or cerebral cortex. Intraperitoneal injection of deoxypyridoxine (200 or 1000 mg/kg) had no statistically significant effect on subsequent uptake studies. However, there was a tendency toward decreased uptake 2 hr after treatment (control 0.97 ± 0.05; 200 mg/kg 0.82 ± 0.06 after 2 hr and 1.15 ± 0.02 after 12 hr; 1000 mg/kg 0.89 ± 0.05 nmol/mg protein per 4 min).

There are several differences in the protocols of the two studies which could complicate comparisons. In the in vivo studies, Ebadi and Klangkalya[129] sacrificed their animals 2 hr after treatment compared with four hours for Snodgrass and Iversen.[467] Ebadi et al.[130] showed that pyridoxal kinase activity in rabbit brain was not markedly affected two hours after deoxypyridoxine treatment but was almost doubled by 4 hr. Therefore, the differences in the delay between treatment and sampling might alter the results. Even so, both groups reported some decrease in the uptake of gamma-aminobutyric acid following deoxypyridoxine treatment. As was discussed earlier (Chapter 2) the fact that Snodgrass and Iversen are the only workers to report deaths with doses as low as 100 mg/kg and Ebadi and Klangkalya are the only ones who failed to observe convulsions at doses above 500 mg/kg suggests to us that there may have been errors in preparing the injections for these dosages. Therefore, we question the validity of the data which show little effect from the 1000 mg/kg dose. In the case of the in vitro studies, Snodgrass and Iversen used a 25-fold higher concentration of gamma-aminobutyric acid than was used by Ebadi and Klangkalya. This could mean that Snodgrass and Iversen were examining a saturated system while Ebadi and Klangkalya may have been working at less than maximum velocity. Considering all of these differences, the relationship between deoxypyridoxine and gamma-aminobutyric acid uptake needs further evaluation.

Rats receiving normal $B_6$ show a marked increase in taurine excretion at about 55 days of age.[313] This increase did not occur when a $B_6$-deficient diet was used but did occur when the $B_6$-deficient diet contained 1 g deoxypyridoxine per kilogram. It is unclear whether deoxypyridoxine stimulated taurine excretion or permitted the normal increase to occur. The authors suggested that the increased excretion might refect increased pyridoxal phosphate concentrations in the liver reported by Johnston et al.[229]

Isonicotinic acid hydrazide caused similar increases. The authors noted that cortisone and adrenocorticotrophic hormones can increase taurine excretion and that there was evidence that isonicotinic acid hydrazide might increase secretion and/or decrease inactivation of corticosteroids. Evidence since that time raises the possibility that the deoxypyridoxine effect might also be mediated through the adrenals. The authors suggested that the increased excretion observed in normal animals might be associated with sex hormones. It is possible that deoxypyridoxine and isonicotinic acid hydrazide might also interact with sex hormones.

### F. Summary

All of these reports were limited to very specific aspects of amino acid or protein metabolism and most were primarily interested in the effects of $B_6$ deificiency rather than the mechanism of deoxypyridoxine action. The result is that no firm conclusions about deoxypyridoxine action are warranted. However, some points are worthy of note.

Two different laboratories[36,494] have reported increased blood urea and there are no conflicting reports.

Of all the individual amino acids examined, glycine shows the most consistent evidence of specific interaction with vitamin $B_6$ and deoxypyridoxine. For example, increasing the glycine intake of $B_6$-deficient animals aggravated the deficiency symptoms[163] while two laboratories[163,191] agreed that increasing the glycine intake of animals receiving deoxypyridoxine minimized the clinical effects. Whereas a $B_6$ deficiency significantly increased plasma glycine concentrations, no increase was noted when deoxypyridoxine was administered with either normal or deficient $B_6$ intake.[403] Deoxypyridoxine caused no decrease in glycine concentrations of the liver or muscle when administered with normal $B_6$ intake.[403] In fact, one report noted increased glycine in the cerebral cortex.[301] However, a simple dietary $B_6$ deficiency tended to increase liver, kidney, and muscle glycine while addition of deoxypyridoxine produced decreases.[403,428] One group suggested that the decrease might be due to increased oxidation.[428] Deoxypyridoxine inhibited glycine uptake by tumor cells[87] and rat intestine.[66] The conversion of glycine to serine and serine to methionine was inhibited by deoxypyridoxine.[361]

The effect of deoxypyridoxine on the intestinal uptake of amino acids has several interesting aspects. First, the effect is very rapid and very short-lived. Single injections of deoxypyridoxine from 0 to 90 min prior to uptake studies produced marked effects as did including deoxypyridoxine in the solution perfusing the lumen. Administration for 5 days had little effect if the last injection was five hours prior to perfusion.[221] Adding $B_6$ compounds to the lumen of the intestinal segments did not offset the effects of deoxypyridoxine even if pyridoxal was added before deoxypyridoxine. Intravenous administration of pyridoxal or pyridoxal phosphate did tend to offset the inhibition. As with $B_6$ transport, some workers concluded that deoxypyridoxine did not interfere with the movement of amino acids from the lumen into the cells but did inhibit the movement from the cells into the blood. Various workers reported the intestinal uptake of alanine, methionine tyrosine, lysine, tryptophan, histidine, glycine, and glutamic acid to be reduced by deoxypyridoxine. Uptake of alanine and asparagine seemed to be unaffected.

All three groups who examined plasma tyrosine reported increases with deoxypyridoxine administration.[47,128,403] Considering all of the evidence, it appears that deoxypyridoxine interferes in the metabolism of a few specific amino acids but does not cause the gross disturbances of protein metabolism which would be expected if there was general inhibition of a broad spectrum of $B_6$ enzymes.

## III. Enzymes

### A. Aminotransferases

Transamination is one of the most thoroughly studied aspects of vitamin $B_6$ action. In vitro studies demonstrating that deoxypyridoxine phosphate could inhibit pig heart transaminase were one of the key factors supporting the concept that 4′-deoxypyridoxine acted through conversion to deoxypyridoxine-5′-phosphate which in turn inhibited enzymes requiring pyridoxal phosphate as a cofactor. Meister et al.[306] conducted these studies using pig heart aspartate aminotransferase. They found that over 90% of the protein-bound pyridoxal phosphate could be removed without loss of enzyme activity thus demonstrating that the cofactor was bound to other sites in addition to the active site. Preincubation of the apoenzyme with 5 μg deoxypyridoxine phosphate for 30 min caused 70% inhibition even when 5 μg pyridoxal phosphate was added after the initial incubation. Twenty micrograms deoxypyridoxine phosphate was required to get the same effect when 5 μg pyridoxal phosphate was added simultaneously with the deoxypyridoxine phosphate. No inhibition was obtained when pyridoxal phosphate alone was incubated with the enzyme prior to the addition of deoxypyridoxine phosphate. Pyridoxine phosphate showed action similar to deoxypyridoxine phosphate. Concentrations required for 50% inhibition were $1.12 \times 10^{-5}$ *M* and $1.50 \times 10^{-5}$ *M* for deoxypyridoxine phosphate and pyridoxine phosphate, respectively. The authors noted that the mechanism of inhibition showed by deoxypyridoxine in these studies may not be the only mechanism which occurs in intact organisms. This work was reviewed by Meister.[304]

Evangelopoulos and Sizer[136] found that 4′-deoxypyridoxine-5′-phosphate was so tightly bound to aspartate apoaminotransferase from pig heart that dialysis for four days was required to regain appreciable amounts of analog. The absorption maximum at 315 nm pH 7.0 was not altered by interaction with the enzyme.

Obviously, both pyridoxal phosphate and deoxypyridoxine phosphate interact strongly with the enzyme. The affinity of deoxypyridoxine phosphate for the enzyme is slightly less than that of pyridoxal phosphate. However, once either compound is bound, it is not readily displaced. In contrast to the tight binding of 4′-deoxypyridoxine-5′-phosphate, the binding of 5′-deoxypyridoxal to aspartate aminotransferase from pig heart was so poor (keq = $5 \times 10^{-4}$ at pH 4.6) that its inhibitory effect was not tested.[145] Since the phosphates of pyridoxal and 4′-deoxypyridoxine bind well, but 5′-deoxypyridoxal does not, the aldehyde group is not as important as the phosphate group for enzyme cofactor interaction in this case. This importance of the phosphate group is supported by the inability of 4′-deoxypyridoxine to affect aspartate aminotransferase from pig heart even at a concentration of 900 μg/mℓ.[522] While 5-deoxypyridoxal does not interact well with the apoenzyme, it can participate in the nonenzymatic transmination of glutamic acid.[316] and in the conversion of serine to pyruvic acid.[316,317] In contrast to the results with aspartate aminotransferase, 5′-deoxypyridoxal was active as a cofactor with pyridoxamine pyruvate aminotransferase. 3- or 4′-Deoxypyridoxines could not serve as cofactors.[466]

Bennett et al.[45] found that diisopropyl-4′-deoxypyridoxine-5′-phosphonate (6.9 g/ℓ) did not inhibit a cell-free beef heart transaminase (not further identified) in vitro. The authors concluded the compound did not displace pyridoxal phosphate from the enzyme. It did not compete with pyridoxal phosphate for the apoenzyme. It did not have pyridoxal phosphate activity in a short term incubation but did show a trace of activity when incubated with the enzyme for an hour. Hruban and Wissler[205] stated without giving numerical data that the activity of beta-2-thienylalanine phenylpyruvate transaminase in the liver and kidney was reduced in rats receiving a $B_6$-deficient diet plus 20 mg deoxypyridoxine per day orally.

While the in vitro results are quite straightforward, the in vivo situation is much more complex. Shwartzman and Hift[453] found that deoxypyridoxine (50 μg s.c. three times per week) in hamsters receiving normal $B_6$ did not affect weight gain or aspartate aminotransferase activity in the heart. Administration of the same dosage to $B_6$-deficient animals increased weight loss and other clinical signs of deficiency but the slight decrease in aminotransferase activity compared with $B_6$ deficiency alone was not statistically significant. The in vitro addition of adenosine triphosphate plus 40 or 80 mg of deoxypyridoxine increased transaminase activity by 50 or 85%, respectively. Pyridoxal phosphate or pyridoxal plus adenosine triphosphate caused a maximum increase of 27%. Two mg pyridoxal phosphate caused only a 21% increase. The authors concluded the increases caused by deoxypyridoxine were not significant. We feel they were rather remarkable.

Caldwell and McHenry[77] noted that in $B_6$-deficient rats 100 μg deoxypyridoxine hydrochloride/day had no effect on liver aspartate aminotransferase but appeared to markedly ($p < 0.01$) increase liver alanine aminotransferase.

Meister et al.[305] fed adult rats a $B_6$-deficient diet and in addition, administered an intraperitoneal injection of 0.4 mg deoxypyridoxine six times per week. Liver alanine aminotransferase was markedly reduced in the deficient animals. However, the fact that deficiency symptoms did not develop until 73 to 91 days of treatment suggests that these effects were primarily those of a dietary deficiency and that the dose of deoxypyridoxine had little influence. The authors noted that dialyzing the homogenate did not affect the aminotransferase. The activity of glutamine phenylpyruvic aminotransferase was not reduced by the $B_6$ deficiency. Neither was the transamination of pyruvic acid diminished when glutamine rather than glutamic acid was used as the substrate.

Dietrich and Shapiro[116] found treatment of mice with a $B_6$-deficient diet produced a 38% decline in liver aspartate aminotransferase activity while a normal diet plus 150 mg deoxypyridoxine per kilogram per day intraperitoneally produced a 50% decrease.

Shapiro et al.[452] observed that intraperitoneal injection of deoxypyridoxine (150 to 175 mg/kg) in mice receiving a normal $B_6$ intake produced a progressive decline in the aspartate aminotransferase activity of testes, lung, liver, kidney, heart, and 755 mammary adenocarcinoma over 16 to 21 days. In some instances the activity reached a minimum after 12 days and then increased slightly. However, this effect was not consistent. The percentage of inhibition tended to be highest in tissues with normally low levels of aminotransferase activity. For example, heart was inhibited only about 20% whereas lung and testes were inhibited about 80%.

Dietrich and Shapiro[117] demonstrated that deoxypyridoxine (100 mg/kg/day i.p.) in combination with a normal $B_6$ diet produced about a 50% reduction in aspartate aminotransferase activity in 10 to 16 days in mouse liver and 755 mammary tumor tissue homogenates. The occurrence of marked reduction of aminotransferase activity in this series of studies demonstrates that the lack of weight loss or dermatitis in animals receiving deoxypyridoxine in conjunction with an adequate $B_6$ intake does not guarantee that no biochemical changes are occurring. However, the decrease in aminotransferase activity was avoided when pyridoxine (300 mg/kg body weight i.p. was injected 30 min prior to each of two injections of deoxypyridoxine (40 mg/kg body weight twice daily). We suspect that with such a massive pyridoxine load, the timing of the pyridoxine injection was not a crucial factor.

Hartsook et al.[189] noted a decrease in liver aspartate aminotransferase activity in rats treated with a $B_6$-deficient diet containing 5 mg deoxypyridoxine per kilogram compared with animals receiving only the deficient diet.

Talageri[494] treated male Wistar rats with a $B_6$-deficient diet containing 20 mg deoxypyridoxine hydrochloride per kilogram. Aspartate and alanine aminotransferase activity in the blood of the deoxypyridoxine group was lower than in controls but higher than in the group receiving only the deficient diet. In the liver, aspartate aminotransferase activity was decreased in the group with dietary deficiency alone but activity in the deoxypyridoxine group was slightly higher than in the *ad libitum* controls and even higher in the pair-fed controls. In the case of liver alanine aminotransferase, the deficient group again showed a decrease and the pair-fed controls an increase in activity compared with the *ad libitum* controls. However, activity in the deoxypyridoxine group was 40% lower than in the group with a dietary deficiency. The authors noted that Awapara[15a] has postulated that alanine and aspartate aminotransferases are probably associated with different portions of the cell or different types of protein.

Koivusalo et al.[241] treated chick embryos with 10 to 30 μmol oxypyridoxine. Liver alanine aminotransferase activity was increased while aspartate aminotransferase decreased. There were no statistically significant changes in lactate dehydrogenase, a ketoglutaric acid, or pyruvic acid concentrations in the liver. They noted it was surprising that deoxypyridoxine should act similarly to hydrazides and commented that further work is needed to explain the differing effects on the two transaminases.

Vaughan and Winders[526] reported that liver aspartate aminotransferase and alanine aminotransferase were reduced 24% and 37%, respectively, in rats receiving a $B_6$-deficient diet containing 200 mg deoxypyridoxine per kilogram compared with the $B_6$-deficient diet alone. There were no differences in enzyme activity if deoxypyridoxine administration was terminated 2 weeks before testing.

Wada and Morino[533] noted that *Escherichia coli* grown in the presence of deoxypyridoxine contained 3- to 4-fold more pyridoxamine transaminase than normal. The activity of aspartate aminotransferase was not changed much, but the concentration of apoenzyme was increased.

Iwanowska[217] found no change in the activity of aspartate or alanine aminotransferases in the blood of rats treated with up to 400 μg deoxypyridoxine per day with or without adequate $B_6$.

Rosen et al.[424] examined the effects of deoxypyridoxine (100 mg/kg body weight) on liver tyrosine aminotransferase, amino acid uptake, and synthesis of nucleic acids and protein in rats. In $B_6$ deficiency, tyrosine aminotransferase activity was reduced to 48% of the control value and addition of deoxypyridoxine caused a further reduction to about 17% of the control value. Addition of pyridoxal phosphate to the assay increased activity by 63, 137, and 388% in the control, $B_6$-deficient, and $B_6$-deficient plus deoxypyridoxine groups. However, even the addition of pyridoxal phosphate failed to equalize the tyrosine transaminase activity in the three groups suggesting that the groups differed in the amount of apoenzyme as well as the amount of cofactor present.

In vitro studies indicated that both deoxypyridoxine and deoxypyridoxine phosphate inhibited tyrosine transaminase. The results were reported only for $10^{-3}$ *M* deoxypyridoxine combined with pyridoxal phosphate concentrations ranging from $10^{-4}$ *M* to $5 \times 10^{-5}$ *M*. The inhibitory effect remained around 15% in spite of the tenfold variation in pyridoxal phosphate concentration. In contrast to the minor effect of deoxypyridoxine, $10^{-4}$ *M* deoxypyridoxine phosphate caused 61% inhibition in the presence of $10^{-4}$ *M* pyridoxal phosphate. Reducing the pyridoxal phosphate concentration tenfold increased the inhibition only to 66%. The authors did not provide any values with no added pyridoxal phosphate. The fact that a tenfold variation in cofactor concentration had little effect on activity suggests that the enzyme may have been fairly well saturated with pyridoxal phosphate even without additional cofactor.

Bayoumi et al.[30] used deoxypyridoxine (1 mg/m*l* drinking water) in conjunction with a normal $B_6$ intake. Over ten weeks growth was reduced about 25% compared with normal $B_6$ rats. Brain weight was not affected after 29 days but did tend to be reduced by 48 days. The activities of brain glutamic acid decarboxylase and gamma-aminobutyric acid aminotransferase were significantly reduced by dietary $B_6$ deficiency. However, addition of pyridoxal phosphate to the assay eliminated the difference, suggesting the amount of apoenzyme was unchanged. Enzyme activities were less affected by the deoxypyridoxine treatment. Comparing activities before and after adding pyridoxal phosphate suggests that there was less apoenzyme but that it had a normal degree of cofactor saturation. The concentration of gamma-aminobutyric acid tended to be reduced in both treatments. No convulsions were observed in any animals.

Kawakita et al.[234] reported that deoxypyridoxine phosphate was a competitive inhibitor ($K_i = 0.84 \times 10^{-7}$ *M*) against pyridoxal phosphate with the aspartate aminotransferase from baker's yeast and with the tyrosine aminotransferase from rat liver ($K_i = 2.2 \times 10^{-7}$ *M*).

Compared with activity in animals receiving a normal diet, aspartate aminotransferase activity in the erythrocytes of five adult male monkeys receiving a $B_6$-deficient diet plus deoxypyridoxine (5 mg i.m. on alternate days for 14 months) was reduced to 26% ($p < 0.001$) of control values.[260] Endogenous activity in the liver was slightly but not significantly reduced. However, when pyridoxal phosphate was added to the assay mixture, the activity in the treated animals was significantly ($p < 0.02$) lower than in the controls. Alanine aminotransferase in the liver was reduced to about 50% ($p < 0.01$) of control values.

Much of the apparent conflict in the effect of deoxypyridoxine on aminotransferases can be resolved by examining Table 11. When compared with controls receiving normal $B_6$, deoxypyridoxine generally causes decreased aminotransferase activity. However, when compared with $B_6$-deficient animals, deoxypyridoxine often has little effect on aminotransferase activity.

Furbish et al.[150] studied the interaction between apoaspartate aminotransferase and various $B_6$ analogs including 5′-deoxypyridoxal. An equimolar mixture of apoenzyme and 5′-deoxypyridoxal ($6 \times 10^{-5}$ *M*) produced only a slight change in the spectrum (mainly increased absorbance around 430 nm) suggesting weak binding. However, despite this limited binding, there was rapid reaction with glutamate producing complete turnover to deoxypyridoxamine.

Several reports show evidence of interactions between hormones and aminotransferases. In rats receiving a normal $B_6$ intake, testosterone had no effect.[117] However, when administered in combination with deoxypyridoxine (100 mg/kg/day i.p.) the aminotransferase activity in mouse liver and 755 mammary tumor tissue homogenates fell more rapidly than with deoxypyridoxine alone, although the final level was similar both with and without testosterone. Similar results were obtained in kidney, heart, and lung. Alanine aminotransferase in liver and tumor tissue homogenates showed a similar response. Testosterone also sped up the effect of deoxypyridoxine on dihydroxyphenylalanine decarboxylase. The combination had no effect on liver cysteine desulfhydrase. Another indication of possible interaction with sex hormones is the observation that compared with rats receiving a normal $B_6$ intake, plasma aspartate aminotransferase activities in rats receiving a $B_6$-deficient diet containing deoxypyridoxine (50 mg/kg diet) declined 35 to 58% except in females receiving corn oil where the decline was only 10 to 15%.[395]

Milholland et al.[328] examined the effect of deoxypyridoxine on liver tyrosine aminotransferase and its induction by cortisol in rats. Feeding a $B_6$-deficient diet decreased liver tyrosine and alanine aminotransferase and brain glutamic decarboxylase activity. Brain alanine aminotransferase was unchanged. Treatment with the $B_6$-deficient diet plus deoxypyridoxine (100 mg/kg probably intraperitoneally) caused an additional 40% reduction in liver tyrosine aminotransferase and a 38% reduction in brain alanine

Table 11
SUMMARY OF EFFECTS OF DEOXYPYRIDOXINE ON AMINOTRANSFERASES

| Species | Organ | $B_6$[a] | Deoxypyridoxine | Activity[b] | $B_6$ intake of controls[a] | Ref. |
|---|---|---|---|---|---|---|
| | | | **Aspartate Aminotransferase** | | | |
| Rat | Blood | − | 20 mg/kg diet | + | − | 494 |
| Rat | Plasma | − | 50 mg/kg diet | − | + | 395 |
| Rat | Blood | − | 400 μg/day | 0 | − | 217 |
| Rat | Blood | + | 400 μg/day | 0 | + | 217 |
| Rat | Liver | − | 100 μg/day | 0 | − | 77 |
| Rat | Liver | − | 5 mg/kg diet | − | − | 189 |
| Rat | Liver | − | 20 mg/kg diet | + | − | 494 |
| Mouse | Liver | + | 150 mg/kg | − | + | 116 |
| Mouse | Liver | + | 100 mg/kg | − | + | 117 |
| Mouse | Testes | + | 150 mg/kg | − | + | 452 |
| Chick embryo | Liver | | 10 to 30 μmol | − | | 241 |
| Hamster | Heart | + | 50 μg 3 ×/week | 0 | − | 453 |
| Pig | Heart | | 0.9 mg/m*l* in vitro | 0 | | 522 |
| Monkey | Erythrocytes | − | 5 mg/kg on alternate days | − | + | 260 |
| Monkey | Liver | − | 5 mg/kg on alternate days | 0 | + | 260 |
| | | | **Alanine Aminotransferase** | | | |
| Rat | Blood | − | 20 mg/kg diet | − | − | 494 |
| Rat | Blood | − | 400 μg/day | 0 | − | 217 |
| Rat | Blood | + | 400 μg/day | 0 | + | 217 |
| Rat | Liver | − | 100 μg/day | + | − | 77 |
| Rat | Liver | − | 0.4 mg 6×/week | − | + | 305 |
| Rat | Liver | − | 20 mg/kg diet | − | − | 494 |
| Rat | Liver | − | 100 mg/kg | + | − | 328 |
| Mouse | Liver | + | 100 mg/kg | − | + | 117 |
| Chick embryo | Liver | | 10—30 μmol | + | | 241 |
| Monkey | Liver | − | 5 mg/kg on alternate days | − | + | 260 |
| | | | **Thienylalanine Phenylpyruvate Aminotransferase** | | | |
| Rat | Liver | − | 20 mg/day | − | + | 205 |
| Rat | Kidney | − | 20 mg/day | − | + | 205 |
| | | | **Tyrosine Aminotransferase** | | | |
| Rat | Liver | − | 100 mg/kg | − | − | 424 |
| Rat | Liver | − | 100 mg/kg | − | − | 328 |

[a] + = normal B; − = B-deficient diet
[b] Activity relative to controls: + = increase, 0 = no change, − = decrease.

aminotransferase activity. Liver alanine aminotransferase activity was intermediate between the $B_6$-deficient and control values. The authors suggested that the conformation of the enzyme in the region of the cofactor may determine whether inhibition occurs. Intracellular location and turnover time may also influence the response to an antagonist. Deoxypyridoxine treatment (100 to 500 mg/kg body weight i.p.) 30 min before a cortisol injection reduced the induction of liver tyrosine aminotransferase 4 hr later. The same dosage of deoxypyridoxine had little effect on the activity of the enzyme in the absence of cortisol. Perfusion with 1 n*M* deoxypyridoxine also reduced aminotransferase activity, indicating that the mechanism involved a direct effect on the liver.

Cortisol induction in perfused liver was also blocked by deoxypyridoxine. When the liver homogenates were reassayed after 24 hr storage at 4°C, the tyrosine aminotransferase activity in the inhibited livers had increased almost to the level of the induced controls. The authors interpreted this as suggesting that deoxypyridoxine altered the conformation of the enzyme rather than inhibiting the synthesis. Addition of deoxypyridoxine (200 mg/kg diet) to diets containing 0, 18, and 50% protein for 2 weeks produced inhibition of liver tyrosine aminotransferase. Combining the deoxypyridoxine treatment with 75% protein produced an increase in liver tyrosine aminotransferase, possibly as a result of stress-induced adrenal action. In adrenalectomized rats receiving either 18 or 50% protein, 200 mg deoxypyridoxine per kilogram diet had little effect on tyrosine aminotransferase activity. When an 18% protein, $B_6$-deficient diet was supplemented with 50, 100, or 200 mg deoxypyridoxine per kilogram of diet, the 50 mg treatment produced the greatest tyrosine aminotransferase inhibition in intact rats, suggesting that the higher amounts stimulated an adrenal reaction which partially offset the inhibitory effects of deoxypyridoxine.

Smejkal and Smejkalova[462] initiated studies to determine whether the inhibition of aminotransferases by thyroid hormones is mediated by $B_6$. Aspartate aminotransferase activity in leukocytes from healthy volunteers was studied. The enzyme inhibition produced by $1.25 \times 10^{-6}$ mol of thyroxine was almost completely overcome by an equal concentration of pyridoxine or pyridoxal phosphate. Similar concentrations of pyridoxamine or deoxypyridoxine restored the activity to 85% or 72% of the original value. The authors suggested that the positive effect of deoxypyridoxine might be due to pyridoxine contamination. However, their results would be comparable to about 15% contamination. This might be possible since the author admitted being able to detect pyridoxine contamination chromatographically.

Gilmer et al.[167] studied the interaction between 5′-deoxypyridoxal and pyridoxamine-pyruvate transaminase (E.C. 2.6.1.30). The 5′-deoxy compound was selected in order to eliminate intramolecular hemiacetal formation. They noted that the reaction of 5′-deoxypyridoxal with enzyme shows two stages: a rapid reaction with active-site lysine and a slower reaction with lysine at other sites. The latter reaction is negligible at pH 7.15. Under the conditions used, about 35 to 45 nonactive-site lysine residues of the 60 total residues reacted with 5′-deoxypyridoxal. The enzyme was stabilized and activated by potassium ion. Gilmer and Kirsch[166] presented additional data on the interaction between 5′-deoxypyridoxal and the enzyme. Since no other $B_6$ forms were included, no conclusions can be drawn about the effects of modifications in the $B_6$ molecule on these interactions.

### B. Decarboxylases

#### *1. Nonenzymatic Decarboxylation*

Several studies using 5′-deoxypyridoxal to gain insight into the possible mechanism of pyridoxal-catalzyed decarboxylations were mentioned in Chapter 1.[500-503]

#### *2. Tyrosine Decarbosylase*

A series of in vitro studies demonstrated that 4′-deoxypyridoxine-5′-phosphate, but not deoxypyridoxine itself, would inhibit tyrosine decarboxylase. Beiler and Martin[41] showed that deoxypyridoxine at concentrations up to 300 μg/mℓ did not inhibit tyrosine decarboxylase from *Streptococcus faecalis* R. However, deoxypyridoxine-5′-phosphate at a concentration of about 7 μg/mℓ produced 20 to 35% inhibition.

Umbreit and Waddell[522] observed that the tyrosine decarboxylase from dried cells of *Streptococcus faecalis* was inhibited only by deoxypyridoxine-5′-phosphate and only if the apoenzyme had not previously combined with pyridoxal phosphate. Deoxypyridoxine at concentrations up to 900 μg/mℓ did not inhibit the decarboxylase. Deoxypyridoxine did inhibit the conversion of pyridoxal to pyridoxal phosphate particularly in less active preparations. The use of whole cell preparations does introduce some extraneous factors such as entry into the cell.

Hurwitz[210] reported that 4′-deoxypyridoxine; 4,5-dihydroxymethyl-2-methylpyridoxine; and 3-amino-4,5-dihydroxymethyl-2-methylpyridine had no effect on tyrosine decarboxylase before phosphorylation but did inhibit after phosphorylation. The following compounds had no effect on the decarboxylase: 2-ethyl-3-amino-4-ethoxymethyl-5-aminomethylpyridine; 2-methyl-3-hydroxy-4-hydroxymethylpyridine; 3,4-dihydroxymethyl-2,6-dimethylpyridine; 3-amino-3-methyl-4,5,6-trihydroxymethylpyridine; 5-hydroxymethyl-2,4,6-trimethylpyridine; and 3-hydroxy-5,6-dihydroxymethyl-2-methylpyridine. As was observed with the aminotransferases, neither 3- nor 4′-deoxypyridoxine produced significant inhibition if the enzyme was incubated with pyridoxal phosphate prior to the addition of the analog.

In further studies, Hurwitz[211] found that pyridoxine phosphate was about four times more effective than deoxypyridoxine phosphate as an inhibitor of tyrosine decarboxylase from *Streptococcus faecalis.* At a concentration of about $7.5 \times 10^{-8}$ *M* pyridoxal phosphate, a 20-fold excess of pyridoxine phosphate was needed for 50% inhibition compared with a 94-fold excess for deoxypyridoxine phosphate. After reviewing the literature describing the biological activities of various analogs, he concluded that any generalization concerning the mode of action of antagonists was premature.

Long and Morrison[279] cited unpublished evidence that deoxypyridoxine-3-phosphate was about half as active as the 5′-phosphate in inhibiting tyrosine decarboxylase. We did not encounter any in vivo studies of this enzyme.

### *3. Dihydroxyphenylalanine Decarboxylase*

Martin and Beiler[298] found that deoxypyridoxine phosphate did not exert an inhibitory effect on dihydroxyphenylalanine decarboxylase from rat kidney in vitro. The concentrations used were not specified but probably included 300 μg/mℓ. However, Dietrich and Shapiro[116] found that treatment of mice with a $B_6$-deficient diet produced a 15% decline in liver dihydroxyphenylalanine decarboxylase activity while a normal diet plus 150 mg deoxypyridoxine per kilogram per day intraperitoneally produced about a 20% decline. The effect of deoxypyridoxine on dihydroxyphenylalanine decarboxylase was enhanced by testosterone.[117]

In a later study relating changes in brain amines to seizure activity, Sawaya et al.[441] found that 60 min after intraperitoneal administration of 4′-deoxypyridoxine (250 mg/kg) to mice apparently receiving a normal $B_6$ intake, dihydroxyphenylalanine decarboxylase activity was reduced to 50% of control values. Failure to note this as a statistically significant decrease was probably a typographical error since the smaller decreases produced by thiosemicarbazide and methyldithiocarbazinate were designated as statistically significant. Addition of pyridoxal phosphate to the assay mixture restored activity to 97% of control values. Despite the drop in enzyme activity, dopamine concentrations in the brain were unaffected. After 60 min. norepinephrine concentrations decreased to 76% ($p < 0.05$) of control values and homovanillic acid increased to 126% ($p < 0.05$). The authors suggested that the increased homovanillic acid resulted from increased turnover of dopamine.

### *4. Aspartate-4-Decarboxylase*

Meister et al.[307] studied the enzymatic decarboxylation of aspartic acid to alanine. The enzyme was isolated from *Clostridium wechii* (ATCC 6784). Pyridoxal phosphate activated the reaction to about the same extent as pyruvate. Adding both had no greater effect than either alone. Adding three times as much deoxypyridoxine did not reduce the activation produced by pyridoxal phosphate.

Novogrodsky and Meister[367] found that deoxypyridoxine-5′-phosphate plus an alpha keto acid helped to protect aspartate decarboxylase against inactivation. While the mechanism is not clear, the authors suggested that the analog helped to stabilize the enzyme and slow the release of pyridoxamine phosphate.

Tate and Meister[498] found that 4′-deoxypyridoxine-5′-phosphate stimulated the conversion of the 6S apoenzyme form of L-aspartate-decarboxylase from *Alcaligenes faecalis* to the 19S holoenzyme but the enzyme showed no decarboxylase, desulfinase, or transaminase activity. The results suggested that Schiff base formation is not an essential step in assembly of the enzyme. Here again, no in vivo data were available.

*5. Glutamate Decarboxylase*

Fonda[145] reported an equilibrium constant of $9.0 \times 10^{-4}$ *M* at pH 4.6 for the binding of 5′-deoxypyridoxal to glutamate apodecarboxylase from *Escherichia coli.* This compares with a value of about $10^{-7}$ for pyridoxal phosphate.

Mekhanik et al.[308,309] found that after simultaneous incubation of glutamate decarboxylase from *Escherichia coli* with $5 \times 10^{-6}$ *M* pyridoxal phosphate and $3.3 \times 10^{-4}$ *M* analog, pyridoxine phosphate produced 76% inhibition ($K_i$ $2 \times 10^{-5}$ *M*), 4′-deoxypyridoxine phosphate 46% (no $K_i$ value), 5′-deoxypyridoxal 19% ($K_i$ $4 \times 10^{-4}$), 3-deoxypyridoxal phosphate 35% ($K_i$ $1 \times 10^{-4}$), and pyridoxamine phosphate 10% ($K_i$ $2 \times 10^{-3}$). Based on data from these and other analogs, the authors concluded that glutamate decarboxylase is more susceptible to alterations in cofactor structure than arginine decarboxylase or aspartate aminotransferase. They also noted that while the 2-methyl and 5′-phosphate groups are not essential for nonenzymic decarboxylation, they are essential for proper interaction with enzymes. 3-Deoxypyridoxal failed to form in aldimine bond with the enzyme as evidenced by the failure of treatment with sodium borohydride to permanently inactivate the enzyme.

The possible role of gamma-aminobutyric acid in seizure activity has stimulated a number of studies of this enzyme in vivo. Several laboratories found that $B_6$ deficiency with or without deoxypyridoxine will reduce the degree of cofactor saturation of glutamate decarboxylase in the brain. However, the concentration of apoenzyme did not appear to be affected. Roberts et al.[414] used an oral dose of 5 mg deoxypyridoxine per day plus a $B_6$-deficient diet. After 10 days, the deoxypyridoxine was raised to 7.5 mg for 6 days. This reduced the degree of saturation of the glutamate decarboxylase in the brain to about 60% of the value obtained with normal $B_6$, but did not appear to reduce the concentration of apoenzyme. The authors concluded that the treatment did not facilitate depletion of glutamate decarboxylase activity in the brain.

Pol et al.[395] also reported that compared with rats receiving normal $B_6$, animals receiving a $B_6$-deficient diet containing deoxypyridoxine (50 mg/kg) showed no change in the concentration of apoglutamate decarboxylase or gamma-aminobutyric acid in the brain. No data were reported for endogenous activity without added pyridoxal phosphate.

Milholland et al.[328] noted that rats receiving a $B_6$-deficient diet showed about a 20% reduction in the activity of glutamic decarboxylase in brain. Incorporation of 100 mg deoxypyridoxine per kilogram in the deficient diet produced an additional 11% reduction.

However, the effects of combining deoxypyridoxine with a normal $B_6$ diet are less clear. Rosen, Milholland, and Nichol[423] reported that intraperitoneal injections, of 3-, 4′-, or 5′-deoxypyridoxine (100 mg/kg) for up to 7 days in conjunction with a diet containing 1 mg pyridoxine per kilogram had no significant effect on glutamic acid decarboxylase in rat brain either before or after addition of pyridoxal phosphate to the assay. Similar treatment with omega-methylpyridoxine did produce a 57% decrease in endogenous activity but only a 15% decrease after pyridoxal phosphate was added. However, only 3-deoxypyridoxine produced convulsions under these conditions. When 3-deoxypyridoxine was adminstered at 150 mg/kg in conjunction with a commercial stock diet, no convulsions were observed but a 36% decrease in endogenous decarboxylase activity was noted. A 200 mg/kg dose of omega-methylpyridoxine produced a 65% decrease in activity without producing convulsions. Since this is the only report in which 4′-deoxypyridoxine failed to reduce at least the endogenous activity of brain glutamate decarboxylase, the validity of some of these data may be questionable. However, their conclusion that the decarboxylase effect was not correlated with convulsive activity seems valid and agrees with the conclusion of Maynert and Kaji[303] that hydrazide convulsions were not directly related to the total concentration of gamma-aminobutyric acid in whole brain. Nozaki[368] reported that injection of 4′-deoxypyridoxine inhibited glutamic decarboxylase in the brain of mice. Under in vitro conditions only 4′-deoxypyridoxine-5′-phosphate was effective.

Bayoumi et al.[30] found a decrease in the cofactor saturation of glutamic acid decarboxylase in the brains of rats receiving a normal $B_6$ diet plus 1 mg deoxypyridoxine per millilter drinking water.

Horton and Meldrum[203] reported that in mice glutamic acid decarboxylase activity, measured without adding pyridoxal phosphate, declined sharply during the first 30 min after deoxypyridoxine administration (1.18 mmol/kg i.p.) then more slowly reaching a minimum about 90 min after injection which correlated with observation of seizures. Activity was restored to normal by in vitro addition of pyridoxal phosphate. In vitro addition of deoxypyridoxine (0.02 to 0.04 *M*) caused 46 to 100% inhibition of glutamic acid decarboxylase activity in brain homogenates. Although deoxypyridoxine is not usually an effective inhibitor with purified enzymes in vitro, it would be likely that crude tissue homogenates would retain the ability to phosphorylate the molecule, thus producing an active inhibitor. Sawaya et al.[441] found that 60 min after intraperitoneal administration of 4′-deoxypyridoxine (250 mg/kg) to mice apparently receiving a normal $B_6$ intake, the activity of glutamic acid decarboxylase in whole brain homogenates was reduced to 33% of control values ($p < 0.01$). Addition of pyridoxal phosphate to the incubation mixture restored the activity to 88% of the control value but this was still a statistically significant ($p < 0.05$) difference.

No brain amine concentrations were significantly altered 20 min after deoxypyridoxine administration. Dopamine was still unaffected after 60 min. However, by that time gamma-aminobutyric acid concentrations had declined to 51% of control values ($p < 0.01$). Norepinephrine decreased to 76% ($p < 0.05$). After 60 min, 5-hydroxytryptamine had decreased to 79% of control values but the change was not statistically significant. The concentration of 5-hydroxyindoleacetic acid was increased to 196% ($p < 0.01$) of control values 60 min after treatment with deoxypyridoxine. The authors suggested that the increase in 5-hydroxyindoleacetic acid probably reflected increased release of 5-hydroxytryptamine although they raised the possibility of effects on reuptake or storage of 5-hydroxytryptamine. The authors reaffirmed their earlier conclusion that 4-deoxypyridoxine produces seizures by inhibiting the synthesis of gamma-aminobutyric acid.

### 6. *Phosphatidylserine Decarboxylase*

Suda[484] found that deoxypyridoxine inhibited phosphatidylserine decarboxylase from the nuclear fraction of mouse liver preparations.

### 7. *Arginine Decarboxylase*

Blethen et al.[57] found that neither pyridoxal nor 5′-deoxypyridoxal activated purified arginine apodecarboxylase from *Escherichia coli*, but both interfered about equally with the reactivation of the enzyme by pyridoxal phosphate. However, kinetic studies indicated that pyridoxal was a competitive inhibitor while 5′-deoxypyridoxal showed a mixed mechanism.

## C. Cysteine Desulfhydrase

Cysteine desulfhydrase is an enzyme which is very susceptible to a dietary $B_6$ deficiency but shows little response to deoxypyridoxine. Meister et al.[305] fed adult rats a $B_6$-deficient diet and administered 0.4 mg deoxypyridoxine intraperitoneally six times per week. Deficiency symptoms did not develop for 73 to 91 days suggesting that the dose of deoxypyridoxine may have been too low to have much effect. Liver cysteine desulfhydrase activity was markedly reduced in the deficient animals. Addition of pyridoxal phosphate to the incubation mixture restored most of the activity, indicating that the amount of apoenzyme was not greatly decreased. Dialysis tests suggested that alanine aminotransferase has a much greater affinity than cysteine desulfhydrase for pyridoxal phosphate.

Dietrich and Shapiro[116] reported a 50% decrease in liver cysteine desulfhydrase activity in mice receiving a $B_6$-deficient diet but no decrease in mice receiving 150 mg deoxypyridoxine per kilogram per day intraperitoneally. However, they apparently overlooked the possibility that the 30% increase in activity shown in their data might represent significant stimulation. They cite the decreased activity reported by Meister et al.[305] as support for discounting the possibility that deoxypyridoxine might activate cysteine desulfhydrase. However, we feel that the results of Meister et al.[305] are more typical of dietary $B_6$ deficiency than of deoxypyridoxine action.

Dietrich and Shapiro[117] reported that while testosterone facilitated the effects of deoxypyridoxine on aminotransferase and decarboxylase activity, the combination had no effect on cysteine desulfhydrase.

Deitrich and Borries[115] reported that deoxypyridoxine phosphate at a final concentration of 25 $\mu g/m\ell$ neither inhibited cysteine desulfhydrase nor prevented activation by 2.5 $\mu g$ pyridoxal phosphate per milliliter. Therefore, it appears that pyridoxal phosphate interacts with cysteine desulfhydrase differently than with tyrosine decarboxylase or aminotransferases.

Hartsook et al.[189] noted a slight tendency toward decreased cycsteine desulfhydrase activity in the livers of rats receiving a $B_6$-deficient diet containing 5 mg deoxypyridoxine per kilogram compared with animals receiving only the $B_6$-deficient diet.

## D. Cystathionine Enzymes

Treatment of five adult male monkeys with a $B_6$-deficient diet plus deoxypyridoxine (5 mg/kg i.m. on alternate days) for 14 months reduced liver cystathionase activity to 64% ($p < 0.05$) and cystathionne-beta-synthase to 44% ($p < 0.01$) of the activity in control animals receiving normal $B_6$ without deoxypyridoxine.[260]

Efremova et al.[132] studied in inhibition of cystathionine beta syntheses from the livers of rats and chickens by various $B_6$ analogs after simultaneous incubation of the apoenzyme with the analog ($10^{-4}$ *M*) and pyridoxal phosphate ($2 \times 10^{-3}$ *M*). The article refers to the chicken enzyme as serine sulfhydrase (E.C. 4.2.1.22) and the rat enzyme as cystathionine-beta-synthase (E.C. 4.2.1.21). However, the 1972 revision of the IUB recommendations combined these entries under one heading, cystathionine-beta-synthase (E.C. 4.2.1.22). Pyridoxine phosphate was neither an activator nor an inhibitor of the enzymes. 4′-Deoxypyridoxine-5′-phosphate was not an activator but apparently was not tested as an inhibitor. 3-Deoxypyridoxal-phosphate produced 33% inhibition in the chicken enzyme and 20% inhibition in the rat enzymes. As was the case with glutamate decarboxylase there was no evidence that the 3-deoxypyridoxal formed an aldimine bond with the enzyme. 5′-Deoxypyridoxal produced 15 and 30% inhibition of the rat and chicken enzymes, respectively.

Goryachenkova et al.[171] compared cystathionine-beta-synthase (formerly serine sulfhydrase E.C. 4.2.1.22) and cystathionine-gamma-lyase (E.C. 4.4.1.1, formerly gamma-cystathionase 4.2.1.15). For cystathionine-beta-synthase, inhibition was determined after preincubating the apoenzyme simultaneously with the analog ($10^{-4}$ *M*) and pyridoxal phosphate ($10^{-5}$ *M*). For cystathionine-gamma-lyase the concentrations were $2.5 \times 10^{-5}$ *M* and $10^{-6}$ *M* for the analog and cofactor, respectively. 3-Deoxypyridoxal phosphate did not affect the lyase but produced 33% ($K_i = 4.5 \times 10^{-4}$ *M*) inhibition of the synthase. Again there was no evidence of formation of an aldimine linkage. 5′-Deoxypyridoxal inhibited both the lyase (24% $K_i = 9.1 \times 10^{-5}$) and the synthase (30%, $K_i = 1.2 \times 10^{-3}$). 4′-Deoxypyridoxine phosphate was not tested.

**E. Miscellaneous Enzymes**

Similar to the results obtained with aminotransferase,[306] 4′-deoxypyridoxine and deoxypyridoxine phosphate were effective inhibitors of alanine racemase from *Streptococcus faecalis* only if preincubated with the enzyme before pyridoxal phosphate was added.[36a] Deoxypyridoxine (1.25 m*M*) produced 43% inhibition when preincubated with the enzyme prior to the addition of pyridoxal phosphate. Deoxypyridoxine phosphate was even more effective with a concentration of only 0.25 m*M*, producing 58% inhibition. 5′-Deoxypyridoxal (0.94 m*M*) was more effective in competing with pyridoxal phosphate, producing 81% inhibition when preincubated alone, 46% when incubated with pyridoxal phosphate, and 29% when the enzyme was preincubated with pyridoxal phosphate alone. The concentration of pyridoxal phosphate used in these experiments with both 4′- and 5′-deoxy compounds was 0.125 m*M*. Therefore, just a 2-fold excess of deoxypyridoxine phosphate was able to produce 58% inhibition. Thus while 5′-deoxypyridoxal was more effective than 4′-deoxypyridoxine phosphate in direct competition with pyridoxal phosphate, deoxypyridoxine phosphate was more effective when preincubated with the enzyme alone. Neither 4′-deoxypyridoxine nor 4′-deoxypyridoxine phosphate inhibited the growth of *Streptococcus faecalis* when used at inhibitor:pyridoxamine ratios of $2.4 \times 10^5$ and $4.8 \times 10^4$, respectively. This lack of effect was predicted on the premise that $B_6$ would be present simultaneously with the inhibitor. As expected, 5′-deoxypyridoxal did inhibit growth. In view of the greater effectiveness of 5′-deoxypyridoxal in competing with pyridoxal phosphte, the authors suggested that 5′-deoxypyridoxal might be a more effective antagonist than 4′-deoxypyridoxine in animals. 5′-Deoxypyridoxal is more toxic (see Table 3) than 4′-deoxypyridoxine, but we have found no data elucidating the mechanism of its toxicity in vivo. Since the conversion of pyridoxine to pyridoxal normally involves primarily the phosphorylated forms, the loss of the 5′-hydroxy group may have prevented 5′-deoxypyridoxine from being converted efficiently to the aldehyde. Therefore, use of 5′-deoxypyridoxal itself is probably the only way to effectively determine whether it is a more potent antagonist than 4′-deoxypyridoxine.

Matano[300] reported that intravenous injection of normal mice with 300 mg deoxypyridoxine per kilogram produced convulsions accompanied by a marked decrease of cholinesterase activity in the cortex.

Kumar and Mahadevan[261] observed that deoxypyridoxine (concentration unspecified) did not inhibit 3-indoleacetaldoxime hydrolyase alone but did inhibit activation by added pyridoxal-5′-phosphate.

Vogel et al.[532] included deoxypyridoxine in their comparison between aromatic and nonaromatic compounds as enzyme inhibitors. Deoxypyridoxine hydrochloride, pyridoxine hydrochloride, and pyridoxamine hydrochloride all caused 50% inhibition of carboxypeptidase B at about $7 \times 10^{-5}$ *M* and catalase at $1 \times 10^{-4}$ *M*. Pyridoxal phosphate was slightly more effective. None of these compounds inhibited carboxypeptidase A or peroxidase. Deoxypyridoxine (1.5 mg/day per mouse) did not affect the response of uterine peroxidase or liver catalase to estrogen. Pressey[397] studied the effect of pyridoxine analogs on invertases from potato, *Neurospora*, and yeast. Pyridoxine and deoxypyridoxine were generally comparable in their inhibitory effect which the author postulates results from the pyridine ring analogous to the effect of aniline, another aromatic base.

Feldman et al.[138] found that potentially active glycogen phosphorylase monomers could be covalently bound to sepharose and then activated by noncovalent interaction with soluble subunits. If the soluble subunits carried inactive pyridoxal phosphate analogs such as 5-deoxypyridoxal, the activated dimer showed only about half the activity obtained from dimers containing pyridoxal phosphate.

Pocker[394] found that while Compound *A* produced 30% reactivation of rabbit muscle glycogen phosphorylase, Compound *B* produced less than 1% reactivation.

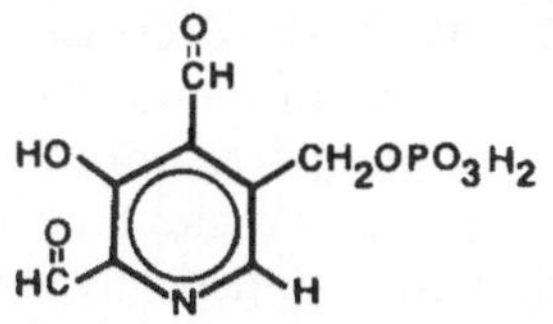


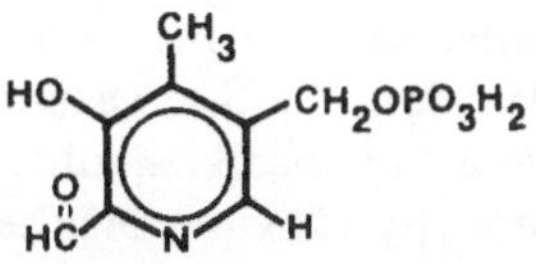


Rosen et al.[424] reported that deoxypyridoxine in vivo (100 mg/kg diet) had no effect on the activity of serine hydroxymethyltransferase in liver, spleen, or thymus. Deoxypyridoxine ($10^{-3}$ *M*) also had no effect on this enzyme in vitro. In a study of the metabolism of the malarial parasite, *Plasmodium lophurae*, Platzer et al.[393] noted unpublished results that serine hydroxymethyltransferase from the parasite and from duckling erythrocytes had a similar affinity for deoxypyridoxine. Neither $B_6$ deficiency nor deoxypyridoxine treatment altered serine dehydratase activity in the liver.[424]

The data of Milholland et al.[328] showed a slight increase in the activity of serine dehydratase in the liver of rats receiving a $B_6$-deficient diet containing 100 mg deoxypyridoxine per kilogram ($1.8 \pm 0.29$ μmol/mg protein per hour) compared with animals receiving only the $B_6$-deficient diet ($1.3 \pm 0.16$) or a normal $B_6$ diet ($1.4 \pm 0.25$).

5′-Deoxypyridxal at unspecified concentrations showed no shift in its absorption maximum at 388 nm in the presence of serine dehydratase (E.C. 4.2.1.14) from *Escherichia coli*.[124] Neither did it inhibit nor enhance the activity of the enzyme. Thus there was no evidence of interaction.

## IV. HORMONES

While the data are not sufficient to warrant any broad generalizations, there are several suggestions of interactions between deoxypyridoxine and hormones.

### A. Adrenals

The possible interaction between deoxypyridoxine and the adrenal glands has been studied through organ weights, histological studies, and functional studies. An increase in the weight of the adrenal glands relative to body size is commonly observed. Mushett et al.[355] reported increased adrenal size in $B_6$-deficient dogs receiving 5 mg deoxypyridoxine per kilogram per day and in normal monkeys receiving 100 mg/kg/day. Gregoire[173] found no change in the absolute weight of the adrenal glands in rats receiving a $B_6$-deficient diet plus 25 to 250 μg deoxypyridoxine per day. However, because of the lower body weight in the treated group, the relative weights were higher than in the normal controls. A similar increase in the relative weight of the adrenals was also noted in pair-fed controls which received only a limited amount of normal $B_6$ diet. Beare et al.[34] found that relative adrenal weights were increased in rats receiving a $B_6$-deficient diet plus deoxypyridoxine (100 μg/day) compared with animals receiving the same treatment plus pyridoxine. Administration of growth hormone further increased the adrenal hypertrophy in the animals not receiving pyridoxine.[40] Olsen and Martinadale[371] confirmed this relative increase in $B_6$-deficient rats. $B_6$-deficient rats treated with deoxypyridoxine for a portion of the experiment showed no further increase.

While it seems that most of the relative increase can probably be attributed to losses in body weight, the situation is complicated somewhat by suggestions of a sex difference. Hsu and Kawin[208] noted that a $B_6$ deficiency increased the relative weight of the adrenals 37% in male rats but only 10% in females. Yet, the average body weight of the females had declined almost as much (25.7%) as in the males (28.5%). Combining the $B_6$-deficient diet with deoxypyridoxine (probably 0.75 mg i.p. three times per week) produced a 74% increase in relative adrenal weight in males and a 23% increase in females compared with controls receiving adequate $B_6$. Again the decreases in body weight compared with the normal controls were quite similar in males (43.0%) and females (38.5%). The increases in relative adrenal weight compared with normal $B_6$ controls were statistically significant ($p < 0.01$) in the $B_6$-deficient males both with and without deoxypyridoxine. Neither group of females showed a statistically significant change. Part of the difference between the sexes stems from the higher relative weight of the adrenals in the female controls (30 ± 1 mg/100 g) compared with the males (19 ± 2). It appears that the absolute weight was relatively unaffected by the treatments in the males while it declined about 25% in the females treated with deoxypyridoxine.

Histological studies have produced varying results probably due at least in part to the differences in the selection of parameters for study. Sudan IV staining of the adrenal glands of $B_6$-deficient dogs receiving 5 mg deoxypyridoxine per kilogram per day or normal monkeys receiving 100 mg deoxpyridoxine per kilogram per day revealed a depletion in lipoid content, particularly in the *zona fasciculata* and *zona reticularis.*[355] Weir and Heinle[542] noted mild adrenal cortical hyperplasia as evidenced by a measurable increase in the width of the cortex in mice receiving a normal $B_6$ diet containing 0.05 to 0.15% deoxypyridoxine compared with normal animals. In three other reports[192,370,456] no histological changes were noted in the adrenals of rats receiving a $B_6$-deficient diet plus deoxypyridoxine. However, Hawkins and Evans[192] used a dose of only 100 μg/day. Olsen and Martindale[370] administered deoxypyridoxine only for a portion of the experimental period, and Silva[456] did not specify the dose. Therefore, the lack of histological effect may have reflected reduced exposure to deoxypyridoxine.

There have been several in vivo studies of adrenal activity. A study[347] of the interaction between the adrenals and the lymphoid system will be discussed in Chapter 5. Stoerk[476] noted that treatment of pyridoxine-deficient animals with cortisone leads to early death even in nontumor-bearing animals. The author suggested that glucocorticoids aggravate the $B_6$ deficiency symptoms while androgenic steroids ameliorate them and prolong survival. Beare et al.[34] observed that adrenalectomy prevented the dermatitis produced in rats by administration of a $B_6$-deficient diet plus an oral dose of 100 μg deoxypyridoxine per day. Adrenocortical extract did not prevent dermatitis. Riggs and Walker[412] used a $B_6$-deficient diet alone and with 4 mg deoxypyridoxine hydrochloride per kilogram diet. However, in their data and discussion they do not separate the two groups. $B_6$ deficiency was associated with an increased concentration of alpha-aminoisobutyric acid in the liver. This did not occur if the animals were adrenalectomized. This suggested that the enlargement of the adrenals occasionally reported in $B_6$ deficiency may represent true hypertrophy. Milholland et al.[328] found that a single intraperitoneal dose of deoxypyridoxine (100 to 500 mg/kg) interfered with the induction of tyrosine aminotransferase by cortisol in the liver of rats receiving normal $B_6$.

The possibility of interaction between $B_6$ metabolism and adrenal function is enhanced by recent work with glucocorticoid receptors. Cake et al.[76] have noted that pyridoxal phosphate can interfere with the binding of glucocorticoid receptors to nuclei in vitro and proposed that pyridoxal phosphate may be an endogenous inhibitor of glucocorticoid action in liver by blocking the receptor sites. Cidlowski and Thanassi[88] extended this work to extracting glucocorticoid receptors from thymus nuclei from adrenalectomized rats. They found that pyridoxine, pyridoxamine phosphate, and 5-deoxypyridoxal had little effect. Pyridoxal was slightly active possibly due to conversion to pyridoxal phosphate. The high pyridoxal kinase activity of the adrenals (Table 7) also suggests that vitamin $B_6$ has an important role in adrenal function. However, Swell et al.[490] found that dietary $B_6$ deficiency in rats did not affect the total steroid production of adrenal slices in vitro. Neither did it alter the fatty acid content in adrenal lipids.

### B. Thyroid

Gregoire[173] reported that a relative hypertrophy of the thyroid gland was observed in deficient animals and signs of inactivity were more apparent in the deoxypyridoxine groups and their pair-fed controls than in the animals receiving the $B_6$-deficient diet *ad libitum*. No specific data were given.

### C. Testes

Olsen and Martindale[370] found no histological changes in the testes of $B_6$-deficient rats.

Boone et al.[63] found no changes in the weight of the testes in rats receiving a $B_6$-deficient diet plus 100 $\mu$g deoxypyridoxine per day intraperitoneally.

Hsu and Kawin[208] reported an increase in the relative weight of the testes in $B_6$-deficient rats which was further increased by treatment with deoxypyridoxine (probably 0.75 mg i.p. three times per week). It appears that this increase was due primarily to the decreased body weight of the animals and that the absolute weight of the testes actually decreased.

Pol et al.[395] compared the effect of various lipids in rats receiving normal $B_6$ with animals receiving a diet containing 1.75 mg pyridoxine and 50 mg deoxypyridoxine per kilogram diet. They state, "Contrary to results of similar experiments on this subject reported in the literature, the addition of deoxypyridoxine to the diet did not depress the weight of the testes." However, they give no references and none of the earlier articles mentioned above noted a significant decrease in the weight of the testes. Actually, the data of Pol et al.[395] do show a slight decrease (5 to 6.5%) in absolute testicular weight in two of their three groups receiving deoxypyridoxine. The third group showed a 1% increase in absolute weight. All three deoxypyridoxine groups showed an increase in relative weight ranging from 23 to 40% of control values. It appeared that clinical signs of deficiency were more severe in males than in females. This impression was supported by the fact that all four deaths in the deoxypyridoxine groups occurred in males. This contrasts with the data of Easton et al.[128] who found lactating rats and adult females to show a greater change than adult males in serum tyrosine after receiving a $B_6$-deficient diet containing 10 mg deoxypyridoxine per kilogram.

Dietrich and Shapiro[117] found that testosterone had no effect on the activity of aspartate aminotransferase or alanine aminotransferase in mouse liver or muscle. Neither was there any effect on aspartate aminotransferase activity in kidney, heart, or lung. Liver dihydroxyphenylalanine decarboxylase activity was also unchanged. Intraperitoneal administration of deoxypyridoxine (100 mg/kg) for 16 days to mice receiving a stock diet produced about a 50% decline in liver aspartate aminotransferase, a 35% decline in alanine aminotransferase, and a 20% decline in dihydroxyphenylalanine decarboxylase. These seemed to be the maximum effects for the conditions employed. Extending the treatments did not cause further change. Combining testosterone with deoxypyridoxine produced this maximal inhibition in a shorter time than when deoxypyridoxine was used alone. Similar results were observed with aspartate aminotransferase activity in kidney, heart, and lung. Cysteine desulfhydrase was not affected by deoxypyridoxine or testosterone. The data showed no detectable difference between the responses of males and females to the treatments. The authors cited unpublished evidence that testosterone lowered the $B_6$ content of normal tissues (liver, lung, and stomach) from female mice and in tumors from male animals. Based on their usage in another paper[452] "normal tissues" do not necessarily come from tumor-free animals but are tissues which did not appear to be directly involved with the tumor itself.

### D. Estrogen

Paul et al.[385] noted that deoxypyridoxine (1.5 mg/day per mouse) significantly decreased uterine growth following estrogen treatment. Part of this effect may have been due to general poor health. Deoxypyridoxine did not alter the response of uterine peroxidase, DNA-phosphorus, RNA-phosphorus, or liver catalase to estrogen.

### E. Summary

There seems to be no doubt that deoxypyridoxine interacts with both adrenal and sex hormones either directly or indirectly. One key question is whether the increased relative weight of the endocrine organs produced by deoxypyridoxine represents true hypertrophy. Since the absolute weights differ little from those of controls and since there seem to be no marked histological changes it is our impression that the growth of these organs proceeds almost normally while overall weight gain is reduced.

## V. WATER

Beare et al.[33] examined the effect of pyridoxine deficiency on growth hormone action in rats receiving 100 μg deoxypyridoxine per day, orally. Controls were supplemented with 50 μg pyridoxine hydrochloride per day in addition to the deoxypyridoxine. The authors stated that the treatments did not affect body composition. While no statistical analysis was provided, the water content of both the liver and carcass was lowest in the controls receiving growth hormone and highest in the deficient animals treated with growth hormone. A tendency toward increased water content of the carcass was also noted in the animals not supplemented with pyridoxine in another study.[34]

Ross and Pike[425,427] found no change in the water content of the liver of rats treated with a $B_6$-deficient diet containing 5 mg deoxypyridoxine per kilogram for 7 to 30 days prior to mating. Pike and Brown[391] increased the deoxypyridoxine content of the diet to 40 mg/kg. Treatment of some animals was started on the day of mating and of others 6 to 26 days prior to mating. In nonpregnant animals, the deoxypyridoxine group showed an increase in serum protein concentrations and hematocrit suggesting hemoconcentration. Returning to a normal diet after 21 days of treatment with deoxypyridoxine produced a fall in serum protein and in hemoglobin and hematocrit. The authors interpreted this as indicating restoration of normal water balance. Brown and

Pike[70] reduced the deoxypyridoxine content of the diet to 20 mg/kg and again concluded that the 6-day depletion period may produce hemoconcentration. Brown and Pike[69] used thiocyanate to examine water balance more directly. Nonpregnant rats receiving a $B_6$-deficient diet containing 20 mg deoxypyridoxine per kilogram showed little change in thiocyanate space although body weight declined. Therefore, the authors suggested that the weight loss might represent loss of fat. Pregnant animals receiving the antagonist showed less increase in thiocyanate space during gestation than pregnant controls receiving a normal $B_6$ intake. Body specific gravity was not correlated with thiocyanate space.

Dietrich and Shapiro[117] cited unpublished data showing that the water content and size of unspecified organs (probably liver, kidney, heart, and lung) were unaffected by deoxypyridoxine and/or testosterone.

Remaggi[410] examined the effect of daily intraperitoneal injections of 6 μg deoxypyridoxine per gram body weight on the consumption of bitter, sweet, salty, and acidic solutions by rats receiving a normal $B_6$ intake. Controls were injected with 3 μg pyridoxal and 3 μg pyridoxamine per gram body weight in place of deoxypyridoxine. The two groups showed no difference in their total consumption of salty, bitter, or acidic solutions over the five day experimental period. However, the groups receiving deoxypyridoxine had wider fluctuations than the controls in daily intake especially with the bitter solution, leading Remaggi to suggest that deoxypyridoxine affected water balance.

## VI. MINERALS

Deoxypyridoxine seems to stimulate sodium retention and potassium loss. Christensen et al.[87] found that in vitro incubation of Ehrlich mouse ascites tumor cells with deoxypyridoxine caused loss of potassium, chloride, and water with a gain of sodium.

Hartsook et al.[189] fed rats (43 to 70 g body weight) a $B_6$-deficient diet for six weeks and then added 5 μg deoxypyridoxine or 10 μg pyridoxine hydrochloride per gram diet. Metabolic studies were conducted after 7 days on the treatment. However, the deoxypyridoxine group died before the collection period. In the next experiment, the animals were maintained on a normal diet for 4 weeks during which time they attained weights of 129 to 150 g. Then they were placed on the treatments for 18 days followed by 7 days of urine and feces collections. Compared with dietary deficiency alone, addition of deoxypyridoxine to the deficient diet decreased weight gain, liver aspartate aminotransferase, calcium retention, and sodium retention. Loss of magnesium and potassium was increased. There was a slight tendency toward decreased liver cysteine desulfhydrase. Only the change in potassium excretion was statistically significant.

Hartsook and Hershberger[190] used 5 μg deoxypyridoxine per gram $B_6$-deficient diet over 34 to 56 days in rats with an initial weight of 110 to 120 g. Deoxypyridoxine significantly ($p < 0.05$) reduced weight gains. $B_6$ deficiency significantly ($p < 0.05$) increased the potassium content of the liver and kidneys. Deoxypyridoxine lowered it to normal in these tissues and decreased ($p < 0.05$) it below normal in the brain. The treatments did not significantly alter the potassium content of muscle, heart, blood plasma, or blood cells.

In $B_6$ deficiency, the sodium concentration was lowered ($p < 0.05$) in the liver and increased in the blood cells. However, addition of deoxypyridoxine raised sodium concentrations significantly above those found in the liver, kidneys, and blood plasma of both the normal and $B_6$-deficient animals. The sodium content in the blood cells averaged higher than in the deficient animals but the increase was not statistically significant. Neither the deficiency nor deoxypyridoxine significantly altered the sodium content of the brain, muscle, or heart. Inanition was not a major factor in these studies.

In a later experiment,[191] the deoxypyridoxine content was increased to 10 mg/kg diet for 28 days. Again $B_6$ deficiency alone tended to increase the potassium content of the blood, brain, liver, and muscle. Only in the case of muscle was the increase statistically significant ($p < 0.01$), In contrast to the earlier study where deoxypyridoxine tended to lower potassium, the authors concluded in this study deoxypyridoxine tended to aggravate the increase. However, in no case were potassium concentrations in the deoxypyridoxine groups significantly different from the groups receiving only the $B_6$-deficient diet. $B_6$ deficiency alone caused no statistically significant changes in the sodium concentrations of the brain, liver, muscle, or blood. The deoxypyridoxine groups did have a significant ($p < 0.05$) increase in blood sodium compared with the groups receiving only the $B_6$-deficient diet.

Hsu et al.[207] injected $B_6$-deficient female rats with deoxypyridoxine (0.75 mg i.p. 3 times per week). Deoxypyridoxine reduced growth and significantly ($p < 0.01$) lowered serum potassium compared with the normal or low $B_6$ diets. $B_6$ deficiency significantly ($p < 0.01$) increased serum sodium and treatment with deoxypyridoxine had no further effect.

Levine et al.[273] found no change in the mechanical activity or the sodium and potassium content of the atria of rabbits treated with 4 mg deoxypyridoxine per day for 10 days. In vitro 0.001 *M* pyridoxal caused a loss of potassium and a gain of sodium. Deoxypyridoxine (0.002 *M*) did not alter this effect.

Matano[300] reported a marked decrease of cholinesterase activity and an increase of water content of the cerebral cortex in young mice fed a $B_6$-deficient diet for 30 days. There was little change in the sodium or potassium content of the cortex. Injection of deoxypyridoxine (0.3 mg/g body weight intravenously) in normal mice produced convulsions with a marked decrease of both cholinesterase activity and potassium content, and an increase of sodium content in the cortex.

Chan and Kare[81] recently reported observing increased intake of a sodium chloride solution by rats with a dietary $B_6$ deficiency. In an earlier study, Remaggi[410] failed to note such a difference between rats receiving normal $B_6$ plus a daily intraperitoneal injection of 6 μg deoxypyridoxine per gram body weight compared with controls receiving an additional 3 μg pyridoxal and 3 μg pyridoxamine per gram body weight per day.

Aikawa et al.[8] studied the relationship between $B_6$ and magnesium metabolism in adult male rabbits receiving a stock diet of unspecified $B_6$ content. Deoxypyridoxine was administered intravenously at 40 mg/kg body weight per day. After the third injection of deoxypyridoxine, the animals became hyperirritable. About 2 hr after the fifth and sixth injections, generalized convulsions lasting several minutes appeared in an unspecified number of animals. Deoxypyridoxine significantly reduced $^{28}Mg$ uptake by lung, kidney, and bone, and increased serum magnesium but did not alter the uptake by muscle, skin, appendix, liver, and heart. The authors mentioned unaltered urinary excretion but provided no numerical data. They concluded that pyridoxine may influence the distribution of magnesium between intra- and extracellular spaces.

While the above reports (summarized in Table 12) suggest an interaction between vitamin $B_6$ and mineral metabolism, further work is needed to clarify the nature of that interaction. Alterations in electrolyte balance could be mediated through the adrenal glands.

## VII. NUCLEIC ACIDS

Cerecedo et al.[80] fed drinking water containing 300 μg deoxypyridoxine per mℓ to normal and $B_6$-deficient C3H mice bearing a lymphosarcoma. In animals receiving deoxypyridoxine with a normal $B_6$ intake, the DNA content of the liver, lung, kidney,

### Table 12
### EFFECT OF DEOXYPYRIDOXINE ON SODIUM AND POTASSIUM METABOLISM

| $B_6$[a] | Deoxypyridoxine dosages | Species | Tissue | Na | K | Ref. |
|---|---|---|---|---|---|---|
| – | 5 mg/kg diet | Rat | Brain | 0[b] | – | 190 |
| – | 10 mg/kg | Rat | Brain | 0 | 0 | 191 |
| – | 5 mg/kg diet | Rat | Liver | + | | 190 |
| – | 10 mg/kg | Rat | Liver | 0 | 0 | 191 |
| – | 5 mg/kg diet | Rat | Muscle | 0 | 0 | 190 |
| – | 10 mg/kg | Rat | Muscle | 0 | 0 | 191 |
| – | 5 mg/kg diet | Rat | Plasma | + | 0 | 190 |
| – | 10 mg/kg | Rat | Plasma | + | | 191 |
| – | 5 mg/kg diet | Rat | Heart | 0 | 0 | 190 |
| – | 5 mg/kg diet | Rat | Kidney | + | | 190 |
| – | 5 mg/kg diet | Rat | Retention | – | – | 189 |
| – | 0.75 mg 3×/week | Rat | Serum | | – | 207 |
| ?[c] | 4 to 30 m*M* | Mouse | Ascites tumor | + | – | 87 |
| + | 300 mg/kg | Mouse | Brain | + | – | 300 |
| ?[c] | 4 mg/day | Rabbit | Heart | 0 | 0 | 273 |

[a] – = $B_6$-deficient diet; + = normal $B_6$ intake.
[b] – = Decrease, 0 = No change, + = Increase.
[c] ? = Authors did not specify the $B_6$ content of the diet.

and spleen increased during tumor growth reaching a maximum after about 17 days. RNA followed a similar trend in the liver, lung, and kidney but remained fairly constant in the spleen. In animals receiving deoxypyridoxine with a $B_6$-deficient diet, the increases in DNA in the liver, lung, and kidney, and RNA in the liver and lung were reduced. In the kidney, RNA actually decreased slightly. In the spleen, the DNA thymine, cytosine, guanine, and adenine concentrations showed a marked decline in the animals receiving deoxypyridoxine plus a $B_6$-deficient diet. The authors stated that RNA and uracil both remained fairly constant in the spleen. Since it was unqualified, presumably that statement was meant to apply to the groups receiving both normal and deficient $B_6$ intakes. The authors concluded from these results that deoxypyridoxine may inhibit nucleic acid synthesis. However, both groups received deoxypyridoxine. The only difference between them was the $B_6$ intake. Therefore, while the results do suggest that $B_6$ deficiency does interfere with nucleic acid synthesis, the experimental design does not separate the effects of deoxypyridoxine from those of a dietary deficiency.

LaPage and Greenlees[272] reported that incubation of Ehrlich ascites cells for 60 min in 12 m*ℓ* of solution containing 1 to 2 mg deoxpyridoxine had little effect on incorporation of glycine into adenine, but did show a 14 to 20% reduction in incorporation into guanine. Increasing the dosage to 3 mg/12 m*ℓ* reduced incorporation into adenine by 20% and into guanine by 28%. With 6C3HED cells, incorporation into adenine was slightly reduced but incorporation into guanine was slightly increased.

Heidelberger and Keller[193] injected (i.p.) tumor-bearing (Flexner-Jobling carcinoma) rats with 200 mg deoxypyridoxine per kilogram body weight at 27 and 3 hr before sacrifice. They observed a 22% reduction in protein synthesis in the tumor in vivo, a 25% reduction in the incorporation of formate into adenine in the spleen, and a 25% increase in the incorporation of formate into guanine in the tumor.

Clark and Stoerk[89] used a $B_6$-deficient diet plus 2 mg deoxypyridoxine HCl per day by an unspecified route for 10 days in rats. Compared with normal animals, there was a 25% loss in body weight, a 60% loss in thymus weight, and a 40% loss in lymph node weight. Uptake of $^{32}P$ was reduced in thymus by about 40% and lymph node by about 60%. Incorporation of $^{32}P$ into DNA was reduced in thymus by about 50% but increased in lymph node by 100%. The authors do not discuss these effects. The authors cited references that 40 to 60% of the thymus DNA is renewed in 24 hr. Hydrocortisone also caused thymus atrophy.

Mickelson and Flippin[320] reported that intraperitoneal administration of 150 mg deoxypyridoxine per kilogram per day for 9 days in mice with and without sarcoma 180 implants receiving a normal $B_6$ diet tended to increase liver RNA while decreasing DNA. The DNA effect was even more pronounced in tumor tissues.

In attempting to elucidate the effect of $B_6$ deficiency on the immune systems, Axelrod and Trakatellis[23] found that incorporation of labeled valine and leucine into proteins was decreased. The amount of cells and DNA per milligram spleen tissue was also decreased. Incorporation of serine, adenine, thymidine, and deoxyuridine into DNA was diminished in $B_6$ deficiency. These changes were reversed by administration of pyridoxine 24 hr prior to sacrifice. $B_6$-deficient animals possessed fewer polysomes per unit of liver or spleen than controls. RNA synthesis and particularly mRNA synthesis was decreased. Therefore, the authors concluded that the effects of $B_6$ deficiency resulted from its ability to catalyze the biosynthesis of nucleic acids which are required for both cellular proliferation and synthesis of specific immune proteins.

Paul et al.[385] found no effect of deoxypyridoxine (1.5 mg/day) on the DNA-phosphorus and RNA-phosphorus of the uterus in mice treated with estrogen. The lack of effect could be due to the low dosage combined with an apparently normal $B_6$ intake.

McDonald et al.[291] examined the effects of various carcinostatic compounds on DNA synthesis in Ehrlich ascites carcinoma in CF-1 mice. A single dose of deoxyridoxine (112 mg/kg i.p.) produced 90% inhibition of DNA synthesis and tumor growth after 24 hr. However, both returned to normal after 96 hr. Estimation of DNA synthesis was based on uptake of tritiated thymidine. Therefore, it is conceivable that the effect of deoxypyridoxine might have been on transport rather than directly on DNA synthesis.

Rosen et al.[424] reported that neither $B_6$ deficiency nor deoxypyridoxine treatment (100 mg/kg diet) for 1 week affected liver nucleic acid metabolism as measured by the concentrations of RNA and DNA and thc rates of incorporation of labeled orotic acid and thymidine. However, in the thymus, deoxypyridoxine caused a 40% reduction in the concentration of RNA and DNA and decreased incorporation of label into RNA. Neither labeling of DNA nor the activity of serine hydroxymethyltransferase were affected. Therefore, the authors concluded that ". . . although impaired synthesis of nucleic acids may underlie the regression of lymphoid tissues produced by deoxypyridoxine, this effect is not due to inhibition of serine hydroxymethylase."

In summary, it appears that nucleic acid metabolism of the thymus is particularly susceptible to alteration by deoxypyridoxine. This observation is in agreement with the marked thymus atrophy associated with deoxypyridoxine action. The reports also suggest that at higher dosages, nucleic acid metabolism in other tissues is also affected. This difference might result simply from slower turnover rates in other tissues rather than differences in susceptibility to deoxpyridoxine. In this regard, it might be appropriate to note that the skin, which is a very rapidly dividing tissue, is one of the tissues most susceptible to deoxypyridoxine action.

A key question is whether deoxypyridoxine interference with nucleic acid synthesis is direct or indirect. The strong interaction between deoxypyridoxine and glycine metabolism suggests one possible indirect mechanism.

# VIII. LIPIDS

The most visible effects of deoxypyridoxine treatment are dermatitis and rapid weight loss. Both of these symptoms could readily be explained by interference in lipid metabolism. However, compared with the interest in $B_6$ and amino acid metabolism, there has been rather limited study of the effect of deoxypyridoxine on lipid metabolism.

Beare et al.[32] examined carcass composition in rats exposed to a $B_6$-deficient diet combined with 100 μg deoxypyridoxine per day. Total crude fatty acids in the carcass were decreased about from 12.8% in controls to 6.5% in deficient animals. The decrease in crude fatty acids persisted even when the diet was supplemented with 5 or 20% corn oil. The authors noted that this conflicted with the previous work in which corn oil did counteract the decreased fatty acid content caused by $B_6$ deficiency.[79] They suggested that the difference might be due to deoxypyridoxine. Also, since deficiencies of several B vitamins produced similar effects on carcass fat content, the authors suggested a defect in energy metabolism may be involved.

In another study, Beare et al.[34] reported that compared with animals receiving the $B_6$-deficient diet plus 100 μg deoxypyridoxine plus 50 μg pyridoxine, animals not receiving the pyridoxine showed no change in the crude fatty acid content of the liver but a 50% decrease in the fatty acid content of the carcass.

Talageri[494] also concluded that treatment of male rats with a $B_6$-deficient diet containing 20 mg deoxypyridoxine per kilogram produced no appreciable effect on the moisture or fat content of the liver. However, they apparently were looking only for increases typical of liver damage and did not run a statistical analysis. Actually, their data suggests that the increased liver moisture and decreased lipid content observed in the deoxypyridoxine group was probably statistically significant. It is particularly interesting to note that liver lipid increased from 6.25 ± 0.38% in the ad libitum controls to 6.98 ± 0.70% in the animals receiving only the deficient diet, and to 6.48 ± 0.43% in the pair-fed controls, but decreased to 5.27 ± 0.27% in animals receiving the deficient diet plus deoxypyridoxine. These results do not necessarily conflict with those of Beare et al.[32] Talageri was reporting total lipid content and was comparing groups with and without deoxypyridoxine. Beare et al.[32] discussed only crude fatty acids and all groups received deoxypyridoxine. More detailed studies of the effects of deoxypyridoxine on lipid metabolism have been conducted by other workers. Holman[200] reported that feeding mice a $B_6$-deficient diet plus deoxypyridoxine (150 mg/kg diet) and essential fatty acids caused deposition of less nonlinoleate polyunsaturated fatty acids. They are not clear whether the comparison is with a normal diet or a $B_6$-deficient diet without deoxypyridoxine. In their figure, it appears that compared with dietary deficiency alone, addition of deoxypyridoxine affects mainly the linoleic acid. Increasing the dosage to 450 mg/kg diet markedly reduced the linoleic acid content of the mice with little effect on the other polyenoic acids compared with mice receiving the $B_6$-deficient diet alone. Combining 30 mg pyridoxine per kilogram with 450 mg deoxypyridoxine per kilogram restored the fatty acid composition to normal. The content of linoleic acid and other polyenoic acids was higher in this group than in the group receiving only the $B_6$-deficient diet. This work was reviewed later.[201]

Peluffo et al.[386] used a $B_6$-deficient diet plus 450 mg deoxypyridoxine hydrochloride per kilogram diet. $B_6$ deficiency alone had little effect on the conversion of linoleic to arachidonic acid. $B_6$ deficiency plus deoxypyridoxine approximately doubled the concentration of eicosatrienoic acid in the heart and liver, indicating that deoxypyridoxine did not inhibit the conversion of oleic to eicosatrienoic acid in rats deficient in essential fatty acids. However, conversion of linoleic to arachidonic acid did seem to be slightly impaired.

Table 13
EFFECT OF DEOXYPYRIDOXINE ON INCORPORATION OF ACETATE INTO CHOLESTEROL AND PHOSPHOLIPIDS[292]

| | Cholesterol (% of controls) | Phosphatides (% of controls) |
|---|---|---|
| Liver | 301 | 148 |
| Aorta | 113 | 102 |
| Adrenal | 140 | 110 |

Pol et al.[395] included deoxypyridoxine in their studies of the relationship between vitamins A, E, and $B_6$ and lipid metabolism in rats. Deoxypyridoxine was incorporated in the pyridoxine-deficient (1.75 mg/kg) diets at 50 mg/kg. All changes refer to comparisons between animals receiving a normal $B_6$ intake and those receiving a $B_6$-deficient diet plus deoxypyridoxine. All diets were fed ad libitum. The authors noted that the results of the histological examinations would be reported elsewhere, but we have not located them. The deoxypyridoxine groups showed reduced food intake and weight gains, particularly with corn oil as the source of fat. Growth in the groups receiving the corn oil diet was also reduced by its low vitamin A content. The animals did develop the dermatitis characteristic of deoxypyridoxine after about 16 weeks. $B_6$ deficiency did reduce liver stores of vitamin E, particularly in females. In animals receiving adequate vitamin A, $B_6$ deficiency plus deoxypyridoxine decreased liver vitamin A while increasing plasma vitamin A. Plasma lipids tended to increase in the $B_6$-deficient groups receiving butter or margarine but decreased in the corn oil groups. $B_6$ deficiency plus deoxypyridoxine reduced the oleate and linoleate concentrations in liver lipids and increased the arachidonate. Females showed similar changes compared with males. The results did not confirm previous reports of decreased liver pantothenic acid in pyridoxine-deficient rats. While the authors frequently refer to changes as being caused by deoxypyridoxine, the design does not permit a distinction between the effects of deoxypyridoxine and dietary $B_6$ deficiency.

Several workers have examined other lipids. Gambassi[156] treated rats for 20 days with 300 μg deoxypyridoxine per 100 g body weight. Neither the $B_6$ content of the diet nor the route of administration is specified. No data are presented. The author stated that increased synthesis of cholesterol and phosphatides was noted in the liver, adrenals, and slightly in the aorta while there was simultaneously a decrease in hepatic cholesterol and phospholipids. The author commented on the paradox that although the results suggest increased catabolism of lipids, similar conditions in monkeys led to atherosclerotic lesions.

Maggi et al.[293] found changes in the specific activity of lipids in rats treated with deoxypyridoxine (600 μg/day i.p.) and $^{14}C$-acetate probably in conjunction with a normal $B_6$ intake (Table 13). Despite the increased incorporation of acetate into lipid, the absolute amount of cholesterol and phospholipid declined in the livers of the deoxypyridoxine group. This suggested increased catabolism. The authors emphasized that deoxypyridoxine does not necessarily accurately reproduce a dietary $B_6$ deficiency.

Maggi et al.[292] used adult female rats on the McCollum diet treated for 20 days with 0.6 mg deoxypyridoxine per rat per day intraperitoneally. Compared with 3 controls, the myocardium of the treated rats contained almost twice as much ($p < 0.01$) cholesterol. The phospholipid content tended to be decreased and the glycogen content tended to increase. However, apparently either the table or the text is misprinted since the text states that cholesterol was unchanged but the phospholipids were increased.

**Table 14**
**EFFECTS OF $B_6$ DEFICIENCY AND DEOXYPYRIDOXINE ON THE CONCENTRATIONS OF CHOLESTEROL AND TRIGLYCERIDES IN PLASMA[90a,93]**

| Pyridoxine[a] | Deoxypyridoxine[b] | Cholesterol (mg/dℓ) | Triglycerides (mg/dℓ) |
|---|---|---|---|
| 0 | 0 | 96 ± 24 | 133 ± 42 |
| 0 | + | 94 ± 20 | 59 ± 17 |
| + | 0 | 94 ± 18 | 199 ± 118 |
| + | + | 128 ± 14 | 303 ± 49 |

[a] 22 mg/kg diet.
[b] 1 g/kg diet.

In 5 adult male monkeys treatment for 14 months with a $B_6$-deficient diet plus deoxypyridoxine (5 mg/kg i.m. on alternate days) produced slight but not significant decreases in serum cholesterol and triglycerides.[260] The total lipid content of the arterial wall was also slightly reduced. The percentages of cholesterol and phospholipid in the arterial wall were both increased ($p < 0.05$) by about 30% in the deficient group compared with controls receiving normal $B_6$. The concentration of triglycerides in the arterial wall was reduced ($p < 0.02$).

Here again, there is evidence of a deoxypyridoxine effect, but the wide variations in experimental design prevent any definite conclusions. The complexity of the situation is illustrated by some data from our laboratory[93] (Table 14). $B_6$ deficiency with or without deoxypyridoxine had little effect on plasma cholesterol. Adding deoxypyridoxine to the $B_6$-deficient diet significantly ($p < 0.05$) reduced plasma triglycerides. Subsequent pair-feeding studies suggested that this decrease was due primarily to the reduced food intake in the deoxypyridoxine group. When deoxypyridoxine was added to the normal $B_6$ diet, both cholesterol and triglyceride concentrations were significantly ($p < 0.05$) increased. In views of these interactions, care must be used in attributing such changes to deoxypyridoxine alone.

## IX. CARBOHYDRATES AND ENERGY METABOLISM

There appears to be a complex interaction between deoxypyridoxine and carbohydrates. Beaton et al.[36] found little change in blood glucose in rats treated with a $B_6$-deficient diet plus 100 μg deoxypyridoxine per day. Beaton et al.[40] noted increased blood sugar in deficient animals after lysine administration. In further studies, Beare et al.[33] showed that rats receiving a $B_6$-deficient diet plus an oral dose of 100 μg deoxypyridoxine per day and 50 μg pyridoxine hydrochloride per day showed increased blood sugar 1 hr after an alanine load. In three out of four comparisons, groups not receiving the pyridoxine supplement had higher blood sugar than the controls at 0 time, showed a decrease after 1 hr, and a peak after 3 hr.

Beare et al.[34] again reported no consistent change in blood sugar in rats receiving the $B_6$-deficient diet plus deoxypyridoxine comared with animals receiving the same treatment plus pyridoxine. Administration of a large dose (12 dog units per rat per day) of adrenocortical extract for 5 days to animals receiving the deficient diet plus deoxypyridoxine plus pyridoxine produced a 29% increase in blood sugar. In animals not receiving the pyridoxine supplement, the hormone treatment produced a 64% increase in blood sugar.

Fridhandler and Quastel[147] reported that deoxypyridoxine inhibited the uptake of glucose and the conversion of fructose into glucose by isolated guinea pig intestine. The authors concluded that deoxypyridoxine may interfere with an aspect of active transport that is common to both amino acids and sugars.

In a study of the effects of vitamin $B_6$ on taste, Remaggi[410] found that rats receiving a complete natural diet plus a daily intraperitoneal injection of 6 μg deoxypyridoxine per gram body weight per day consumed about 50% more sucrose solution than controls receiving the same diet plus injections of 3 μg pyridoxal and 3 μg pyridoxamine per gram body weight. Remaggi postulated that this reflected increased reliance on substances which could be metabolized without pyridoxine. In a more recent study, Chan and Kare[81] failed to note any increased intake of sweet solution by rats with a simple dietary $B_6$ deficiency but they used the nonnutritive sweetener saccharin rather than sucrose.

Hruban et al.[206] reported that deoxypyridoxine (20 mg/kg) in conjunction with a $B_6$-deficient diet decreased the relative and absolute amounts of liver carbohydrate in tumor-bearing (water soluble components) by almost 50% compared with tumor-bearing controls receiving normal $B_6$.

Koide and Furuta[240] reported that large doses of pyridoxal phosphate suppressed the oxidation of carbohydrates in vivo and in vitro but stimulated the oxidation of fat only in vivo. Pyridoxamine phosphate and deoxypyridoxine stimulated carbohydrate oxidation and suppressed the oxidation of fat in vivo but showed no direct effect on oxidation of carbohydrate and fat in vitro. 2-Deoxyglucose prevented the stimulation of carbohydrate oxidation.

Olsen and Martindale[371] found that $B_6$ deficiency reduced the oxygen uptake of liver and kidney preparations in vitro. Deoxypyridoxine treatment had no additional effect. The authors concluded that the oxygen uptake of deficient tissues could be restored to normal by adding pyridoxal in vitro. Their data for the kidney support this conclusion, but addition of pyridoxal in the liver corrected only about 50% of the decrease. Deoxypyridoxine slightly reduced oxygen uptake administered in conjunction with a normal $B_6$ intake.

Guha et al.[176] reported that deoxypyridoxine (0.001 *M*) produced a 60% increase in ammonia formation by rat brain homogenates without affecting oxygen uptake.

Pal[381] observed that pyridoxal phosphate at concentrations of 1 m*M* and 2 m*M* caused 20% and 32% inhibition of the swelling of rat liver mitochondria, respectively. Deoxypyridoxine caused 47% and 75% inhibition at 1 m*M* and 2 m*M*, respectively. Deoxypyridoxine (1 m*M*) plus pyridoxal (1 m*M*) produced only 22% inhibition. However, deoxypyridoxine (1 m*M*) plus pyridoxal phosphate (1 m*M*) produced 100% inhibition. Deoxypyridoxine and 2,4 dinitrophenol both caused a several-fold release of adenosine triphosphatase activity. When deoxypyridoxine was combined with pyridoxal phosphate there was no release of adenosine triphosphatase.

Szmelcman and Guggenheim[491] observed that administering 0.5 g starch per 100 g body weight along with a tryptophan load prevented deoxypyridoxine from reducing the concentration of tryptophan in the portal vein.

Horton and Meldrum[204] have examined the relationship between body temperature, brain glucose, and convulsions in mice apparently receiving a normal $B_6$ intake plus a single intraperitoneal injection of 250 mg deoxypyridoxine per kilogram. In mice kept at room temperature, deoxypyridoxine produced a 6° fall ($p < 0.01$) in body temperature within 30 min compared with saline-injected controls. This decrease was more rapid than that produced by allylglycine, methionine sulfoximine, or isonicotinic acid hydrazide. At the same time, brain glucose rose from $1.45 \pm 0.11$ μmol/g in the controls to $2.47 \pm 0.15$ in the deoxypyridoxine group ($p < 0.01$) and brain glycogen from $1.46 \pm 0.11$ to $2.05 \pm 0.25$ μmol/g ($p < 0.05$). Blood glucose was not significantly altered. Brain glucose and glycogen were significantly ($p < 0.05$) elevated even when body temperature was maintained by placing the animals in an incubator. Under similar conditions, isonicotinic acid hydrazide (150 mg/kg) caused a significant ($p < 0.05$) decrease in brain glucose and glycogen. While hypothermia has some role in causing the increased glucose and glycogen levels, it was not the sole factor since some increase occurred even when body temperature was maintained. Neither could the increases be due to altered blood glucose concentration since that showed little change. The authors speculated about possible alterations in the glucose transport system, utilization of carbohydrate, or inhibition of glutamic decarboxylase.

Chapter 5

# IMMUNE SYSTEM

## I. INTRODUCTION

The immune system is so sensitive to $B_6$ deficiency that Axelrod[16] suggested that antibody formation might be a better indicator of nutritional status than growth. Deoxypyridoxine appears to be especially antagonistic to this aspect of $B_6$ action. Studies of this area can be divided into three broad categories: gross examination of lymphoid tissues; specific tests of antibody response; and tissue transplantation.

## II. LYMPHOID TISSUE

The close dependence of the thymus gland on vitamin $B_6$ was noted by Stoerk and Zucker[479] as early as 1944. Even in rats receiving a 1 μg/day pyridoxine supplement which permitted growth of about 1.5 g/day, there was marked thymus atrophy. In further studies,[473] $B_6$-deficient mice were injected with 0.2 mg deoxypyridoxine twice daily for 3 days. In the injected mice, the thymus weighed 50% less than in the noninjected animals and showed pycnosis, fragmentation of lymphocytes, and reduced cortical tissue.

Stoerk, Eisen, and John[478] found that thiamin deficiency caused thymus atrophy without reducing antibody formation, while $B_6$ deficiency decreased both thymus size and antibody production. Histological examination of thymi from thiamin-deficient rats showed well-preserved lymphocytes and distinct corticomedullary differentiation. In $B_6$-deficient animals, the thymi were totally depleted of lymphocytes, the cortex had disappeared, and the glands consisted almost exclusively of epithelial cells and stroma. In a further study in mice, 0.03 mg deoxypyridoxine per day produced weight loss and thymus atrophy not seen with the deficient diet alone.[477] Dosages up to 50-fold higher showed no additional effect. The changes were prevented by 0.3 mg pyridoxine per day even in the presence of 0.9 mg deoxypyridoxine. Deoxypyridoxine caused thymic atrophy even in adrenalectomized rats. Effects on the thymus were not relieved by terminating deoxypyridoxine administration. In older rats, the deficient diet alone did not produce weight loss or dermatitis but did cause thymus atrophy.

Mushett et al.[355] conducted similar studies with chicks, puppies, and monkeys. Administration of 2 mg deoxypyridoxine per kilogram diet combined with a slightly suboptimal $B_6$ intake produced changes in spleen weight in chicks similar to a total $B_6$ deficiency. The most prominent feature was hypoplasia or failure of development of lymphoid elements. Pyridoxine-deficient pups showed a leukopenia with lymphocytosis and neutropenia. Treatment with deoxypyridoxine caused leukopenia with lymphopenia and polynucleosis.

Monkeys maintained on a natural stock diet which were injected with 100 mg deoxypyridoxine per kilogram body weight per day in two divided doses developed microcytic anemia, leucopenia, and lymphopenia. Monkeys given 50 mg/kg daily for 144 days showed only transient blood changes.

At autopsy, the most striking change seen in both dogs and monkeys was atrophy of the spleen, thymus, and lymph nodes. The femoral bone marrow of the monkeys was pale yellow and gelatinous in consistency. The authors described the pathology as follows:

> Histological sections of the spleen of the monkeys injected with 100 mg per kg revealed an extensive loss of lymphoid elements. No germinal centers were present in any of the sections, and remnants of Malpighian

follicles were represented by an occasional small accumulation of lymphocytes. Very few lymphocytes remained in the red pulp. As a result of the profound atrophy of the organ, there was an apparent increase in trabecular and reticular tissue. The capsule frequently appeared to be thickened. Many of the central arterioles were very tortuous, due presumably to the shrinking of the spleen as a whole.

Lymph nodes taken from many sites in the body also showed a striking diminution in lymphoid elements, particularly in the follicles. The reticular tissue appeared to be increased in amount, and the demarcation between cortical and medullary sinuses, and erythrophagocytosis provided a prominent feature.

The blood-forming constituents and fat of the femoral marrow were strikingly reduced. Both erythroid and myeloid elements were affected. Many congested sinusoids were present. Between the small and widely scattered loci of active hematopoietic cells, a pale, eosinophilic matrix was present, which in some areas appeared relatively homogeneous and in others appeared to be made up of large swollen cells with indistinct cell boundaries.

In the pups treated with deoxypyridoxine, the lymphoid tissue also showed histological evidence of atrophy. The degree of change was comparable to that present in the monkeys given repeated doses of 100 mg per kg.

The thymus showed, in addition to lymphoid hypoplasia, many areas of necrosis. Demarcation between cortex and medulla was frequently indistinct. Hassall's corpuscles were decreased in number and showed morphological abnormalities. Whereas the marrow of the pups on the pyridoxine-deficient diet alone was hyperplastic, the marrow of the pups which were maintained on the pyridoxine-deficient diet and treated with desorypyridoxine was relatively normal.

Sudan IV stains of the adrenal glands of dogs and monkeys treated with desoxypyridoxine showed a depletion of lipoid content, particularly in the *zona fasciculata* and *zona reticularis.* The adrenal glands of such animals were usually larger than normal. Since lymphoid tissue mass as well as the number of circulating lymphocytes is said to be under the control of adrenal cortical hormones, the question arises as to whether the lymphoid atrophy in desoxypyridoxine-treated animals resulted from a stimulation of the adrenal glands or from a direct action of the vitamin analogue on the lymphoid tissue. . . .

In addition to the changes in tissues and in the blood already described as effects of deoxypyridoxine treatment, other signs of pyridoxine deficiency were seen in animals given this drug. These include dryness of the hair, scaliness of the skin, tongue lesions, hyperirritability, and epileptiform convulsions. *

Ling et al.[277] noted generalized atrophy of lymphoid tissues in rats that died after receiving a $B_6$-deficient diet containing 20 mg deoxypyridoxine per kilogram. Mortality from pericardial hemorrhage during heart puncture was greater for the normal rats than deficient animals.

Gregoire[173] also studied the effects of deoxypyridoxine on lymphoid tissue in the rat. The animals received a $B_6$-deficient diet plus 250 μg deoxypyridoxine per day for 8 days. The $B_6$-deficient diet was continued for another 12 to 16 days at which time the animals were sacrificed. $B_6$ deficiency caused a decrease in the weight of the thymus, lymph nodes, and spleen. In relationship to body weight, the thymus showed atrophy, whereas the lower weights of the lymph nodes and spleen were appropriate for the lower body weight of the deficient animals. Deoxypyridoxine in conjunction with the $B_6$-deficient diet caused a further loss in thymus weight but no additional weight loss in the lymph nodes or spleen.

Histological examination of the thymus in $B_6$-deficient rats revealed tissues ranging from almost normal to severe atrophy with extensive depletion of the lymphoid cortical elements and diffuse caryoclasis. No major changes were observed in the lymph nodes or spleen even in animals with severe thymus atrophy.

Histologically, the thymus from $B_6$-deficient animals treated with deoxypyridoxine showed the dark medulla and light cortex characteristic of inversion. In the cortical area, the epithelial reticulum was clearly seen. Small, dark lymphocytes were rare. Again, there were no consistently related changes in the lymph nodes or spleen. Attempts to observe which portions of the lymphoid tissue first responded to pyridoxine supplementation after deoxypyridoxine treatment were unsuccessful.

Weir, Heinle, and Welch[544] found that leukocyte changes were erratic in mice receiving a $B_6$-deficient diet plus 1 mg deoxypyridoxine per day. A dose of 4 mg/day in

* From Mushett, C., Stebbins, R., and Barton, M., *Trans. N.Y. Acad. Sci.*, 9, 294, 1947. With permission.

combination with a natural diet produced a complete reversal of the ratio of granulocytes to lymphocytes which could be reversed by supplementation with 6 to 12 mg pyridoxine. In preliminary studies, smaller amounts of $B_6$ failed to correct the leukocyte pattern. No dermatologic or neurologic symptoms of $B_6$ deficiency were noted; neither was growth impaired. Even after 32 days with an abnormal leukocyte pattern, no histological changes were evident in the thymus glands, lymph nodes, spleen, or adrenals. Two animals showed some liver necrosis. The authors suggested that the granulocytosis was due to metabolic blocks leading to accumulation of substances used to produce granulocytes.

Weir and Heinle[542] added 0.05, 0.1, and 0.15% deoxypyridoxine to a Purina® Dog Chow diet and fed it *ad libitum* to groups of 5 young adult Swiss mice. Hematologic changes occurred in all groups within 5 days. There were no differences between the groups. After 32 days, the 12 surviving mice showed significant weight loss. Six animals with moderately severe dermatitis were sacrificed and showed histological changes in the spleen. There was slight enlargement of the spleens with extensive infiltration of the pulp by neutrophils and large, immature mononuclear cells with occasional mitoses. Megakaryocytes were slightly increased in number and erythropoietic elements were abundant. Atrophy of the splenic follicles did not occur. Changes in the liver were not striking, only a few small areas of infiltration with neutrophils and mononuclear cells being present. Atrophy of lymphoid tissue might have been expected but none occurred. In addition to the intact splenic follicles, the thymus and lymph nodes were of normal size and histologic structure.

Citing these results, Hawkins and Evans[192] later commented that lymphoid atrophy had been observed in all species tested except the mouse. They apparently overlooked the work of Stoerk[473] in mice. In any case, the comments of both Weir and Heinle, and Hawkins and Evans reemphasize the need to consider the experimental conditions in interpreting the results. In the previous studies, thymic atrophy was produced in conjunction with either $B_6$-deficient diets or very high doses (e.g. 100 mg/kg body weight) of deoxypyridoxine. Weir and Heinle used a natural diet in conjunction with a moderate dose of deoxypyridoxine. We interpret their results as indicating that the treatment was adequate to produce some changes in leukocytes but not quite strong enough to produce histological evidence of atrophy. Therefore, we see no reason to postulate any unique resistance to lymphoid atrophy in the mouse.

Mueller, Weir, and Heinle[347] noted that granulocytosis, lymphopenia, and myeloid metaplasia in the spleen had been produced either by treatment with cortisone or by deoxypyridoxine. Therefore, the possible role of the adrenals in deoxypyridoxine action was studied. One group of mice received a normal $B_6$ ration plus 0.1% deoxypyridoxine hydrochloride. A second received the diet plus cortisone (1 mg s.c. per day). The third group received both deoxypyridoxine and cortisone. Either deoxypyrodoxine or cortisone caused a sevenfold increase in the granulocyte count and a 50% decrease in the lymphocyte count within 10 days. Combining both treatments produced an even greater effect. There was no difference in the histologic characteristics between the three groups. Adrenalectomy alone produced no hematologic changes and did not prevent the changes produced by deoxypyridoxine administration. Neither did administration of $B_6$ vitamers prevent the hematologic effects of cortisone. Weir and Heinle[543] provided a general summary of their work.

Hawkins and Evans[192] examined the effects of deoxypyridoxine (100 μg/day orally) on white cells and lymphoid tissue in rats. Deoxypyridoxine produced statistically significant ($p < 0.01$) leukopenia with lymphopenia and increased neutrophils. Histological examination revealed atrophy of the lymphoid elements of the thymus with loss of corticomedullary differentiations. However, the weights were significantly higher than in normal or $B_6$-deficient controls apparently due to accumulation of fat. The spleen was also heavier. No histological changes were noted in the spleen or adrenals. Addition of 40 μg pyridoxine per day to the deoxypyridoxine treatment prevented the changes in organ weights and restored body weight to that of $B_6$ deficiency alone. It tended to ameliorate the change in the white cells.

Two 4-month-old dogs were given 6 mg deoxypyridoxine per day in conjunction with a $B_6$-deficient diet for 30 days and 20 mg/day for 19 days. Growth was inhibited. Mild anemia and marked lymphopenia appeared by 50 days. The authors noted that in all species reported to date, deoxypyridoxine causes lymphopenia usually associated with lymphoid atrophy. The authors further suggested that the shorter life span of lymphocytes may make them more susceptible than erythrocytes to the effect of deoxypyridoxine.

Beare et al.[34] found that treatment with a $B_6$-deficient diet, plus 100 μg deoxypyridoxine per day increased adrenal and liver weights and decreased thymus weight in relationship to body size compared with rats receiving the same treatment plus pyridoxine. Administration of growth hormone to rats receiving a $B_6$-deficient diet plus an oral dose of 100 μg deoxypyridoxine per day tended to increase the atrophy of the thymus and the hypertrophy of the adrenals.[33]

Olsen and Martindale[370] found no histological changes in the spleen of $B_6$-deficient rats. No hematological changes were noted. Since deoxypyridoxine (12.5 mg/kg diet) was administered only at the beginning and end of the experiment, it is uncertain how deoxypyridoxine influenced the results.

Boone et al.[63] reported that compared with $B_6$-supplemented animals or animals receiving only the $B_6$-deficient diet, $B_6$-deficient rats receiving 100 μg deoxypyridoxine per day intraperitoneally showed thymic atrophy. Spleen weight was not affected.

Bianchi Donnasibilla and Maggiora[51] studied the effect of deoxypyridoxine (400 μg/day i.m.) on blood and bone marrow in rats receiving a natural diet. There was a marked decline in the leukocyte count and a reversal of the lymphocyte:granulocyte ratio. The bone marrow showed significant increases in hemocytoblasts, eosinophilic and eosinophilic myelocytes, promyelocytes, and a decrease in lymphocytes. There were also slight decreases in neutrophilic myelocytes and metamyelocytes. The authors noted that the blood lymphocyte count seemed to stabilize after 40 days even though the number of lymphocytes in the bone marrow continued to fall slightly. Cytochemical tests using Sudan black and the Hotchkiss reaction failed to detect any differences from the control group.

Anglesio et al.[14] reported that intraperitoneal injection of 75 mg deoxypyridoxine per kg per day for 5 days in 100 to 240 g rats apparently receiving a normal $B_6$ diet caused no significant change in body weight, peripheral leukocyte count, intestinal mitotic count, or tumor tissue. It did seem to decrease the cellularity of the bone marrow and increase the fat. We suspect that the lack of effect is due primarily to the brevity of the treatment although the deoxypyridoxine:pyridoxine ratio may also have been a factor.

Doctor[122] administered deoxypyridoxine (100 mg/kg/day i.p.) to adult male rats receiving a commercial stock diet. Growth was reduced about 20% over a 4-week period. During the same period, the leukocyte count fell 40%. Deoxypyridoxine at 20 mg/kg/day had little effect on growth or leukocytes over 12 weeks. However, when combined with aminopterin there was a synergistic effect on leukocytes but little evidence of other toxic effects such as loss of appetite, diarrhea, and hemorrhage. Increasing the dosage of deoxypyridoxine to 40 mg deoxypyridoxine/kg/day produced a 36% drop in leukocyte count within 2 weeks. Again, the addition of aminopterin produced a synergistic effect. The differential count showed a 45% decrease in lymphocytes and an 86% increase in granulocytes in the rats receiving 40 mg/kg/day deoxypyridoxine. Histological examination of sternal smears and decalcified sternum revealed no change

**Table 1**
**EFFECT OF DEOXYPYRIDOXINE (20 mg/kg BODY WEIGHT) ON LEUKOCYTES AND ERYTHROCYTES IN RATS[206]**

| Treatment | Cells/mm³ | | | Hematocrit |
|---|---|---|---|---|
| | Leukocytes | Granulocytes | Lymphocytes | (%) |
| $B_6$ deficiency | 12088 ± 865 | 1715 ± 130 | 9886 ± 784 | 49.96 ± 0.48 |
| $B_6$ deficiency plus deoxypyridoxine | 5744 ± 312 | 3417 ± 282 | 1991 ± 140 | 51.74 ± 0.36 |

in cellularity of the marrow in the deoxypyridoxine group. Marrow from animals receiving deoxypyridoxine plus aminopterin was moderately hypocellular.

Deoxypyridoxine caused a 7% decrease in liver weight, and marked decrease in thymus weight in rats receiving a $B_6$-deficient diet plus 20 mg deoxypyridoxine per day orally.[205] The weights of the spleen and kidneys were unchanged. Dermatological evidence of $B_6$ deficiency was not observed during the 8-day course of the experiment. Deoxypyridoxine produced a moderate depression of splenic small cell immune reaction, and of bone marrow meylopoiesis. The authors also noted slight depression of splenic white pulp, and marked depression of pyroninophilic cells in the red pulp.

Hruban et al.[206] observed that deoxypyridoxine (20 mg/kg body weight per day) combined with a $B_6$-deficient diet for 8 days altered the total and differential leukocyte count but not the hematocrit count in rats (Table 1). There was no marked change in body weight, or the weights of the liver, spleen, and kidney during this short time. The authors cited a reference that brief restriction of movement can cause relative lymphocytosis in rats.

Hsu and Kawin[208] noted little change in the spleen in $B_6$-deficient males but a statistically significant ($p < 0.01$) increase in the relative weight in females. Compared with normal $B_6$ intake, the relative weight of the thymus decreased about 30% in $B_6$-deficient animals and about 70% in animals receiving the deficient diet plus deoxypyridoxine (0.75 mg deoxypyridoxine hydrochloride i.p. 3 times per week).

Rosen and Milholland[422] and Rosen et al.[421] found that in dogs, $B_6$ deficiency alone caused a slight increase in white cell count but no change in ratio of neutrophils to lymphocytes. Feeding a normal $B_6$ diet containing 500 mg deoxypyridoxine per kilogram for 14 days produced granulocytosis and lymphopenia with no depression in total cell count. In other experiments, as little as 10 mg deoxypyridoxine per kilogram diet caused a 65% reduction in leukocytes within 7 days due to lymphopenia. Similar effects were obtained when deoxypyridoxine was administered by stomach tube or subcutaneous injection. Intraperitoneal administration was less effective.

In rats treated for 7 days with a $B_6$-deficient diet containing 50 mg deoxypyridoxine per kilogram, circulating lymphocytes doubled within 2 days after restoration of a normal diet and attained normal levels within 2 weeks. If deoxypyridoxine was withdrawn but the $B_6$-deficient diet continued, the leukocyte cell count returned to normal but the granulocyte:lymphocyte ratio did not. Therefore, the authors suggested that induction of lymphopenia might be a sensitive test for $B_6$ antagonists.

The authors noted that neither $B_6$ deficiency nor deoxypyridoxine alone produced the rapid, severe leukopenia produced by the combination. Also, the acute clinical symptoms were associated with essentially normal hepatic concentrations of pyridoxine and pyridoxal phosphate and with elevated transaminase activity. They admitted a lack of explanation for these apparently paradoxical effects.

Osetrova[373] found that in rats with skin homografts, 375 μg deoxypyrodixine per day for 10 days reduced the weight of lymph node, thymus, and spleen, suggesting suppression of the immune response to the transplants. Neither the route of administration nor the $B_6$ content of the diet was specified in the abstract.

Van Bijsterveld[524] examined the possible effect of a $B_6$-deficient diet plus deoxypyridoxine (50 mg/kg diet) in guinea pigs on phagocytosis. Deoxypyridoxine concentrations of 5 to 120 μg/mℓ did not affect cell vitality or phagocytosis in vitro. Carefully conrolled procedures were needed to detect statistically significant differences. The best parameter was the average number of ingested bacteria *(Moraxella nonliquefaciens)* per polymorphonuclear leukocyte. The animals receiving the deficient diet plus deoxypyridoxine did show reduced phagocytic capacity.

Robson and Schwarz[416] again confirmed that compared with rats receiving a commercial stock diet, deoxypyridoxine (0.1 mg/mℓ drinking water) combined with a $B_6$-deficient diet reduced the total leukocyte count and increased neutrophils while decreasing lymphocytes. The concentration of cells in thoracic duct lymph was reduced about 90% in the treated animals. Mixed lymphocyte response as measured by thymidine uptake was reduced 55%. The normal lymphocyte transfer reaction was also reduced. Uridine incorporation into small lymphocytes was decreased.

Deficiencies in the immune system were also seen in the progeny of $B_6$-deficient dams.[417] A $B_6$-deficient diet plus drinking water containing 0.1 mg deoxypyridoxine hydrochloride per milliliter was administered from day 4 to 21 of pregnancy. Normal $B_6$ intake was then restored and deoxypyridoxine was removed. When tested at 3 months of age, the progeny showed no differences in the number of lymphocytes and neutrophils in the peripheral blood when compared with progeny of normal $B_6$ controls. There was a tendency toward smaller thymus size in the progeny of deficient dams. The concentration of cells in thoracic duct lymph was slightly (16%) reduced. Mixed leukocyte reactions to pokeweed mitogen and normal lymphocyte transfer reactions were also reduced in the progeny from the deficient dams. However, uridine uptake was not affected. Thus, the prenatal treatment produced long-term impairment of the immune system. Unfortunately, the design did not permit isolation of the effects of deoxypyridoxine from those of dietary $B_6$ deficiency.

Most recently, Willis-Carr and St. Pierre[552] used a $B_6$ deficient diet plus 0.1 mg deoxypyridoxine per milliliter drinking water. The treatment caused such a severe depletion of lymphocytes in the thymus that the corticomedullary boundary was not clearly defined. These changes were corrected if a normal $B_6$ intake was restored. Thymic epithelial cells showed no abnormalities under the electron microscope. Lymphocytes from treated rats showed impaired response to the mixed leukocyte reaction and to the thymic lymphocyte mitogens, concanavalin A and phytohemaglutinin. Thymic epithelial monolayers from treated animals had a reduced capacity to convert immunoincompetent precursors to functional T lymphocytes. However, the precursors from treated animals responded normally to thymic epithelial cells from normal animals. Therefore, the authors concluded that the main effect of the treatment was to impair the ability of thymic epithelial cells to activate precursor cells.

We feel there are two reasons for further study of the effects of deoxypyridoxine on the lymphoid system. First, if deoxypyridoxine is acting solely as a $B_6$ antagonist, identifying the reactions responsible for these very selective effects should help to pinpoint the role of $B_6$ in specific metabolic processes in vivo. Second, these observations may be due to effects of deoxypyridoxine independent of its $B_6$ antagonism. If so, it would be useful to be aware of these additional mechanisms.

# III. ANTIBODY RESPONSE

The first direct test of the effect of deoxypyridoxine on the immune response was the report of Bodian,[60] who observed occurrence of paralytic poliomyelitis in an uninoculated rhesus monkey (a very rare occurrency). For 17 days previous to this, the animal had been receiving subcutaneous injections of 100 mg deoxypyridoxine per kilogram body weight (twice daily during the week, once on weekends) along with a normal $B_6$ diet. The treatment produced no symptoms other than lack of weight gain. Because this suggested that deoxypyridoxine prevented normal immune function, further tests were conducted with 11 monkeys injected with 127 mg deoxypyridoxine per day for 17 days. Six monkeys received an equal amount of pyridoxine in addition to the deoxypyridoxine. The animals were innoculated with virus orally, a procedure which is normally ineffectual in transmitting the disease. Two of the 11 animals receiving deoxypyridoxine developed polio. The authors noted that the infections in the animals seemed to progress more rapidly than normal. The authors suggested a possible defect in the immune system or conditions favoring increased invasiveness of the virus.

Stoerk[477] suggested that possible mechanisms for the effect of $B_6$ deficiency on the immune response might be excessive destruction of the antigen; decreased ability of the antibody-forming apparatus to take up antigens; a reduced production of antibody protein; and an excessive destruction of properly formed antibody. $B_6$-deficient rats receiving deoxypyridoxine produced lower levls of circulating antibodies to sheep cells than rats receiving only the deficient diet. Electrophoresis detected no changes in the protein fractions; neither was there any detectable difference in the rats of disappearance of antigen or antibody from serum. Based on amount of phagocytized material in histocytes, Stoerk concluded the reticuloendothelial system seemed to be functioning normally. Therefore, Stoerk concluded that the immune deficiency probably resulted from impaired antibody formation. He also noted that in many cases, animals receiving deoxypyridoxine died with diarrhea and emaciation in spite of regression of the tumors. When such mice were autopsied, they showed gross and microscopic evidence of Tyzzer's disease. This condition is due to *Bacillus pyriformis* and susceptibility appears to be genetically restricted.[520] The fact that Tyzzer's disease had not been previously observed in this strain of mice and was not observed in those receiving only the $B_6$-deficient diet is further evidence that deoxypyridoxine interferes with the immune process.

Forse and Webster[146] noted that dogs receiving a $B_6$-deficient diet plus deoxypyridoxine (5 mg/kg/day orally) were unusually susceptible to infection.

Bianchi Donnasibilla and Cortesi examined the effect of $B_6$ deficiency and deoxypyridoxine on the immune system in rats[49] and rabbits[50] In one experiment, each rat received 15 μg pyridoxine per day orally and some received 150 μg deoxypyridoxine in addition. These two groups showed no difference in their response to a vaccine from *Salmonella typhi*. In another study in rats, daily injections of 4 μg pyridoxine and 40 μg deoxypyridoxine were used, and in a third the effect of a dietary deficiency alone was examined. The authors cited unpublished work as their basis for selecting a 1:10 ratio between pyridoxine and deoxypyridoxine. Both dietary deficiency and deoxypyridoxine failed to alter the concentration of serum complement but did increase the hemolysin titer. Deoxypyridoxine did not affect the response to the typhus vaccine under the conditions of this study.

Since the above results conflicted with other reports that $B_6$ deficiency interfered with the immune system, another study was conducted on rabbits.[50] Again, the authors cited unpublished work that a 1:8 ratio of $B_6$:deoxypyridoxine is best. The rabbits received 3 mg deoxypyridoxine per day apparently combined with a normal $B_6$ intake. The tryptophan load test was used to evaluate $B_6$ status. The antimetabolite treatment did not affect the response to typhoid vaccine, diphtheria toxoid, nor did it alter the concentration of serum complement or hemolysin.

Axelrod[17] demonstrated that a deficiency of any one of several nutritional factors including pyridoxine for 2 to 4 weeks prior to injection of rats with diphtheria toxoid and continuing for 3 weeks after anitgenic stimulation produced an impaired antibody response which prevented a normal anamnestic response to a second toxoid injection even if dietary therapy was instituted at the time of the second injection.

In order to study the effect of a pyridoxine deficiency only during the anamnestic phase, the pyridoxine intake during the primary phase was reduced to 10 $\mu$g/day instead of the usual 50 $\mu$g/day. The reduced intake still supported good growth and antibody production. Following the booster injection, the animals were placed on a $B_6$-deficient diet and treated with 5 to 10 mg deoxypyridoxine per rat per day intraperitoneally. One group was injected daily with 10 mg deoxypyridoxine plus 1 mg pyridoxine. The deoxypyridoxine groups showed a reduced anamnestic response detectable as early as 3 days after the booster injection. The deoxypyridoxine group supplimented with pyridoxine showed a normal anamnestic response. Thus, the data indicate that pyridoxine deficiency can interfere with the initial antibody response and also with the anamnestic response even in cases where proper nutrition permitted development of a normal primary response. The authors noted that deoxypyridoxine treatment induced a continuing weight loss which amounted to about 80 g/rat after 5 weeks. Other symptoms of $B_6$ deficiency appeared within the first week.

Axelrod et al.[20] studied the effects of deoxypyridoxine on immune response in the guinea pig. All animals received normal diets for 4 weeks and then some were placed on a $B_6$-deficient diet with or without 10 mg deoxypyridoxine HCl per 100 g/day intraperitoneally. In the next 4 weeks, the deoxypyridoxine group gained an average of only 28 g compared with 166 g for the group receiving pyridoxine. The only physical symptom noted in the deoxypyridoxine group was an unkempt appearance of the fur. The authors, therefore, concluded that guinea pigs are more resistant than rats to deoxypyridoxine. The data clearly showed that a dietary $B_6$ deficiency either alone or in conjunction with deoxypyridoxine depressed circulating antibody response and skin hypersensitivity reactions to diphtheria toxoid. However, the experimental design does not permit comparison of the dietary deficiency with the deoxypyridoxine group to detect possible difference in the action of the two processes.

Axelrod et al.[24] extended these studies to include delayed hypersensitivity reactions. Again, the experiment was not designed to detect differences between the action of a dietary deficiency and deoxypyridoxine. Either treatment depressed skin reactivity. However, in vitro and passive transfer tests demonstrated that the cells from deficient animals were sensitized. Therefore, the authors suggest that $B_6$ deficiency may not prevent sensitization but may instead impair the ability of the sensitized cells to react.

Trakatellis et al.[507] used the same guinea pigs as Axelrod et al.[24] Deoxypyridoxine was administered intraperitoneally at 10 mg/100 g body weight per day in conjunction with a $B_6$-deficient diet. In animals immunized to Bacillus Calmette-Guerin (BCG), $B_6$ deficiency plus deoxypyridoxine reduced systemic reaction to purified protein-derivative-tuberculin but did not alter anaphylactic activity to bovine serum albumin. Skin reactivity was also depressed but returned to normal after 10 days on a normal diet. Neither inanition alone nor vitamin C deficiency depressed skin reactivity. The design of the study ruled out poor growth of BCG and poor physical conditions as the causes of reduced systemic response. The authors speculated that pyridoxine deficiency interfered with the reaction between sensitive cells and antigen.

Stinebring et al.[472] extended these studies further. The treatment did not alter the systemic response of guinea pigs to endotoxin at either lethal or sublethal doses. Therefore, the authors concluded that endotoxin acts via a different mechanism than protein-derivative-tuberculin.

Axelrod and Trakatellis[22] reviewed their work with $B_6$ deficiency and immune phenomena. While $B_6$ deficiency impaired both the systemic and skin reactivity to a purified protein derivative in sensitive BCG immunized guinea pigs, there was no difference in response to endotoxin. Skin homotransplants survived better in pyridoxine-deficient rats. The number of syngeneic splenic cells needed to induce immune tolerance is decreased if the cells are administered during a period of pyridoxine deficiency. In this case, the deficiency was induced by a $B_6$-deficient diet coupled with 100 μg deoxypyridoxine per 100 g body weight per day intraperitoneally. The experimental design did not permit evaluation of the effect of deoxypyridoxine itself. Pyridoxine deficiency also facilitated the repression of immune response by splenic microsomal RNA. The interference with nucleic acid metabolism and protein synthesis observed by these workers was discussed in Chapter 4. Another review was published in 1971.[18]

Gandullia and Cordone[157] studied antibody formation in growing rats receiving 0.5 mg deoxypyridoxine per day intramuscularly for 50 days starting at 7 days of age. The $B_6$ content of the diet is not given but survival of the animals and amount of growth observed suggest that it contained a normal amount of $B_6$. Growth was reduced about 50%. A decrease in erythrocytes and leukocytes was noted. In the final week of treatment, convulsions were noted.

Human albumin was administered intraperitoneally after 50 days. Antibody formation was determined after 18 and 25 days during which deoxypyridoxine treatment was continued. None of the 10 animals failed to develop a detectable antibody titer. The authors concluded that compared with antimitotic agents and steroid hormones, the side effects of deoxypyridoxine are minimal and that it shows great promise as an antimmune agent.

Cordoba et al.[102] showed that deoxypyridoxine was not a very effective inhibitor of precipitation of antibodies to pyridoxal phosphate. Deoxypyridoxine phosphate was not tested. Therefore, the conclusion that the lack of effect reflected the lack of phosphate was not fully confirmed.

Gershoff et al.[164] found that a dietary $B_6$ deficiency reduced antibody formation. Supplimenting the $B_6$-deficient diet with 5 or 10 mg deoxypyridoxine per kilogram diet almost completely eliminated antibody formation. Growth was reduced about 50% by the dietary deficiency alone and about 95% by the combination with deoxypyridoxine. Glycine supplementation tended to partially offset the inhibition. In the one study where it was used, serine supplementation was more effective than glycine. Restricted feeding alone had no effect on antibody production. The study included germ-free rats but not in the $B_6$ or deoxypyridoxine treatments. The authors suggested one action of deoxypyridoxine was to block the availability of serine and glycine.

Davis[110] used a $B_6$-deficient diet plus drinking water containing 0.1 mg/mℓ deoxypyridoxine for the last 15 to 17 days of gestation after which the Sprague-Dawley rats were returned to a normal $B_6$ intake with no deoxypyridoxine. At 6 weeks of age, the animals were immunized with killed BCG bacilli in Freund's adjuvant. Two weeks later they were injected with 250 tuberculin units of purified protein derivatives. After 24 hr, the diameter of induration was significantly less than any other group. No group received the $B_6$-deficient diet alone. Adding pyridoxine to the deoxypyridoxine treatment, either in an equimolar amount or at 1 mg/mℓ water, permitted normal sensitization. Decreased sensitivity was also observed in similarly treated Lewis rats immunized with killed *Mycobacterium tuberculosis.* In contrast to the reduced hypersensitivity test, the $B_6$-deficient deoxypyridoxine group showed no reduction in the development of circulating antibodies to sheep erythrocytes or *Escherichia coli* compared with controls receiving normal $B_6$. Histological examination of the thymus, spleen, and lymph nodes in rats exhibiting decreased development or delayed hypersensitivity revealed no abnormalities. A variety of control groups excluded dietary factors other than $B_6$ and maternal protein calorie malnutrition as factors. Unfortunately, the design did not separate the effects of a dietary $B_6$ deficiency from those of deoxypyridoxine. It should be noted that the reduced capacity for delayed hypersensitivity was seen in apparently healthy animals who had received a normal $B_6$ intake since birth. However, the mortality of this group seemed quite high. Of 429 pups, 207 died or were eaten the first day. An additional 94 pups died during the first 6 weeks leaving only 128 (30%). Eight developed a runting syndrome. This is another case where some of the effects of $B_6$ deficiency and deoxypyridoxine during gestation continued to be shown long after normal $B_6$ was restored.

Two studies have been conducted in humans. Wayne et al.[538] observed no marked decrease in antibody response to typhoid vaccine and A and/or B blood group substance in 4 chronically ill subjects who received 300 mg deoxypyridoxine plus about 2 mg vitamin $B_6$ per day for 9 weeks.

Hodges et al.[199] reported that $B_6$ deficiency alone did not reduce response to typhoid and tetanus immunization. Addition of deoxypyridoxine (100 mg three times daily for 6 weeks) markedly inhibited the immune response. In a further 7-week study of 5 men using the same protocol except that the oral dose of 100 mg deoxypyridoxine was increased to four times daily, the immune response to typhoid and tetanus but not polio was impaired. The authors suggested that these deficiency states may be useful tools for studying the immune process with possible applications to organ transplants and autoimmune disorders.

## IV. TISSUE TRANSPLANTATION

Probably the most extensive study of the effect of $B_6$ deficiency on tissue transplantation has come from the laboratory of A. E. Axelrod. Through careful experimental design, these workers have demonstrated that the effects to be described are not due to the inanition which often accompanies vitamin deficiencies. Fisher et al.[141] examined the effect of deoxypyridoxine on skin homograft survival in a study involving a total of 1529 rats. Deoxypyridoxine was administered intraperitoneally at 5 mg/kg body weight per day for 10 days following the graft in conjunction with a $B_6$-deficient diet. Then the $B_6$-deficient diet alone was continued for 5 more days. Then the animals received 10 μg pyridoxine per day orally for 3 to 5 weeks and were restored to a normal diet. $B_6$ deficiency with or without deoxypyridoxine improved graft survival. There were differences in response to deoxypyridoxine between Long-Evans and Wistar rats. Grafts from Wistar rats to Wistar or Long-Evans rats were 80% successful after 30 days. Grafts from Long-Evans donors showed poor survival whether the recipients were Long-Evans or Wistar rats. The authors concluded that the technique might have potential clinical application. Axelrod et al.[19] also found that $B_6$ deficiency either alone or combined with intraperitoneal injection of 250 or 500 μg deoxypyridoxine per day increased the survival of skin grafts in rats.

Fisher and Schewe[143] found that a $B_6$-deficient diet combined with deoxypyridoxine (7.5 mg/kg/day i.p.) permitted survival of 27% of the homografts for at least 40 days compared with none for the normal $B_6$ controls. When normal $B_6$ was restored to these animals, a second graft was rejected while most of the original grafts survived. These and other experiments led to the conclusions that survival of the grafts was not due to persistency of the deficiency after reestablishing normal $B_6$ intake; a state of acquired tolerance; complete inability of the graft to respond to antibodies; or incorporation of the graft into host tissue. Therefore, the most likely explanation seemed to be that with the passage of time, grafts become less antigenic or less responsive to antibody.

Axelrod[21] extended attempts to induce immune tolerance with tissue extracts. They assumed that mitochondria were free of ribonucleic acid and cholinesterase. Cholinesterase was used as a microsomal marker. Animals received a $B_6$-deficient diet plus 25 μg deoxypyridoxine per 100 g body weight per day intraperitoneally for 3 weeks. Growth ceased and dermatitis appeared. After the second week, splenic ribonucleic acid was injected. After the third week the animals returned to a normal diet. After 1 week on the normal diet, the animals received a skin graft. Ribonucleic, acid from the whole homogenate or from the soluble or cytoplasmic fractions had no effect on graft survival. However, ribonucleic acid from the microsomal fraction markedly prolonged graft survival. Dosage seemed to be a factor since the highest doses of ribonucliec acid were less effective than lower dosages. The authors assumed that injection of ribonucleic acid led to formation of donor antigens in the host. In further studies of this phenomenon, the splenic microsomal extracts were effective in prolonging skin grafts only if they were from animals from the same genetic strain and were injected into animals which were receiving the $B_6$-deficient diet plus deoxypyridoxine rather than normal $B_6$. Since no animals received only the $B_6$-deficient diet, the importance of the deoxypyridoxine could not be differentiated from a simple dietary deficiency.[281]

Axelrod and Trakatellis[23] injected (i.p.) mice receiving a $B_6$-deficient diet with 100 μg deoxypyridoxine per 100 g body weight per day for 3 weeks. At the end of the third week, the animals returned to the $B_6$-adequate control diet. After 12 more days, the animals received a skin graft. Graft survival was not improved in animals that received no spleen cells or spleen cells from a strain other than the skin donor. However, survival markedly improved in $B_6$-depleted animals which received spleen cells from the same strain as the skin donor. The experiment did not include a group with only dietary deficiency. Rejection of successful grafts could be induced by injecting the animals with sensitized spleen cells obtained from animals in a stage of active graft rejection. The authors cited unpublished observations that pyridoxine deficiency limits the availability of one-carbon units required for biosynthesis of DNA and thereby depletes the number of splenic cells. They also noted observing an accompanying decrease in ribonucleic acid synthesis with presumed involvement of messenger ribonucleic acid.

Trakatellis and Axelrod[505] extended their work to test the effect of deoxypyridoxine on the sex-linked histocompatibility system in C57B1/6J mice. Graft recipients were placed on a normal diet for 0 to 14 days and then received a $B_6$-deficient diet supplemented with 100 μg deoxypyridoxine per 100 g body weight per day (i.p.) from 15 to 35 days. A single injection of splenic cells was administered on the 28th day. The normal diet was restored on day 36 and grafts were implanted on day 47. Doses of 1, 2.5, and $5 \times 10^7$ splenic cells combined with the deoxypyridoxine treatment produced immune tolerances of 84, 91, and 100%, respectively. The same doses in normal $B_6$ animals produced at most 46% immune tolerance. The authors stated that this technique warrants clinical attention.

Other workers have also found that deoxypyridoxine improved graft survival. Parkes[383] found that injecting mice with 2.5 mg deoxypyridoxine per day prolonged the survival of ovarian homografts. Increasing the dosage to 5 mg/day did not further improve survival. The effect could be offset by pyridoxine administration. Survival of heterografts involving rat ovarian tissue in mice was also improved.

Osetrova and Eingorn[374] noted that administration of deoxypyridoxine (375 mg/day) to $B_6$-deficient rats for 8 days after transplantation increased the survival of skin homografts threefold. Tubazid (10 mg) increased survival twofold.

Dobblestein et al.[120] found that $B_6$ deficiency enhanced skin transplant survival. However, whereas Trakatellis and Axelrod[505] were able to obtain 90% tolerance, Dobblestein et al. obtained only 19%. Even so, since deoxypyridoxine seemed to have a fairly specific immunosuppressive action which was readily reversible, the authors recommended further studies. In such studies, rats were treated for 3 weeks with a $B_6$-deficient diet plus 120 to 150 mg deoxypyridoxine per day in the drinking water.[121] After renal transplantation, the deoxypyridoxine was continued for 7 days and the $B_6$-deficient diet for 14 days. The controls received the $B_6$-deficient diet plus 0.5 to 0.75 mg vitamin $B_6$ per day in the drinking water. Over the 3-week period before transplantation, the average weight of the treated rats dropped 72 g from 391.8 to 319.8 g while the weight of the controls increased from 349.3 to 390.1 g. In the control group, 23 out of 25 rats died between 7 and 13 days after transplantation. In the treated group, only 16 out of 38 died within that time. Not included in these figures were 5 control and 16 treated animals which died within 3 days after transplantation and were considered to result from failure to survive the surgery rather than from transplant rejection. The authors state that the increased surgical losses in the experimental group were due to the poor general condition of the animals and that the difference in surgical losses between the two groups was not statistically significant. Even if they were included, the results would still show only 1 animal out of 30 in the control group surviving for more than 35 days compared with 17 out of 54 in the experimental group. Out of these 17, 11 survived for more than 100 days and only 2 died in less than 60 days. Therefore, the percentage of transplants reaching their criterion for tolerance (survival for more than 35 days) was much higher in the deoxypyridoxine group even when surgical losses were included.

Histological signs of rejection were divided into two classes: cellular (referring to cellular infiltration of the tissue), and humoral (referring to necrosis of the endothelium and mesangium). The two groups showed no difference in cellular response, but the humoral response was reduced in the deoxypyridoxine group. Therefore, the authors concluded that cellular infiltration had less influence than humoral-vascular factors on the functional integrity of the organ. The authors speculated that once deoxypyridoxine or some other immunosuppressive treatment carried the animals through the initial postoperative phase, autochthonous factors might arise to provide further protection, possibly even over a normal life span.

The poorest results were obtained by Bento and Silva[46] who treated rabbits receiving a nutritionally inadequate diet with deoxypyridoxine (200 μg/100 g body weight per day) administered in four doses as follows: 1/3 orally, 1/6 intravenously, 1/6 intramuscularly, and 1/3 orally. After 20 to 35 days, autografts or homografts were made and the treatment continued for 14 to 16 days after which they were returned to a normal stock diet. Supplementation of the experimental diet with 300 mg pyridoxine hydrochloride per day per rabbit did not prevent loss of weight, suggesting the diet was also deficient in other nutrients. None of the treatments affected the progress of autografts. Supplementation with pyridoxine did not accelerate the rejection of homografts compared with the experimental diet alone. The grafts in both groups were judged dead after an average of 11 or 12 days. Treatment with deoxypyridoxine for 20 days prior to grafting did not seem to affect graft survival in the 2 animals available for observation. However, pretreatment for 30 to 35 days increased graft survival in the 5 surviving rabbits to an average of 21 days. There were no consistent changes in the electrophoretic patterns of the serum during the course of the experiment. Since these are the only data on rabbits, it is not clear whether the poor results are due to the experimental conditions or to a species difference. Similarly, in the only study in humans, Fisher[142] found no increase in homograft survival in men receiving 250 mg deoxypyridoxine four times per day in conjunction with apparently normal $B_6$ intake. Deficiency symptoms such as stomatitis and other skin problems were noted. Autografts did survive indicating that failure of the homografts was not due solely to the poor skin condition associated with the treatment. We have shown that rabbits and deoxpypyridoxine differently than rats. However, since guinea pigs which exhibit the same metabolic pathway as rabbits and humans, were susceptible to inhibition of antibody formation, the difference in deoxypyridoxine metabolism does not necessarily prevent all effects on the immune system.

## V. SUMMARY

The possible use of deoxypyridoxine as an immunosuppresive agent is attractive because deoxypyridoxine is relatively nontoxic, it seems to produce quite dramatic inhibition of the immune system, and its effects can be easily reversed by pyridoxine. The major impediment to this application is the apparent necessity for a simultaneous dietary $B_6$ deficiency. Better understanding of the metabolic relationships between vitamin $B_6$ and deoxypyridoxine might suggest ways to overcome this problem.

Chapter 6

# CANCER

## I. INTRODUCTION

Examination of deoxypyridoxine as a possible antitumor agent started soon after the identification of deoxypyridoxine as a $B_6$ antagonist and has continued ever since. The following discussion will start with animal studies and conclude with clinical trials in humans.

## II. ANIMALS

As would be expected, the results of deoxypyridoxine treatment are highly dependent on the dosage and availability of vitamin $B_6$. Therefore, deoxypyridoxine has not shown much promise when it was included in broad screening programs which made no allowance for the fact that deoxypyridoxine is usually most effective when the $B_6$ supply is restricted. For example, in a study of 28 potential antitumor agents, deoxypyridoxine failed to show activity in any of the 15 tumor systems and in 44 of the 59 nontumor systems.[161] However, Hitchings and Elion,[198] who conducted the test with *Lactobacillus casei,* specifically noted that deoxypyridoxine would not have been expected to show activity under the conditions employed in their assay. Clarke[90] commented that the lower activity against mouse sarcoma 180 in this study compared with earlier results could have been due to different nutritional status in the host animals or to changes in the characteristics of the tumor itself. Another summary[2] also showed only limited success for deoxypyridoxine. At doses up to 225 mg/kg there was some activity against sarcoma 180 but not against carcinoma 755. No activity against L1210 was observed at dosages up to 50 mg/kg.

DiPaolo and Moore[118] found deoxypyridoxine at concentrations of 0.01 to 1 mg/m*ℓ* inactive against ascites cells in vitro. The culture medium was not $B_6$ deficient.

Smith et al.[463] found that deoxypyridoxine showed marginally useful cytotoxic activity in their system using Eagle's KB strain of human carcinoma cells.

Korytnyk et al.[250] found it necessary to reduce the amount of serum in order to obtain inhibition from $B_6$ analogs. They concluded that the $B_6$ content of the serum was the reason. Deoxypyridoxine produced 50% inhibition of cultures of sarcoma 180 and mammary adenocarcinoma (TA3) at concentrations of 0.1 and 1 $\mu M$, respectively. While the authors concluded that the S-180 cells were more sensitive to deoxypyridoxine, it also appears that the TA3 cultures contained higher concentrations of pyridoxine. Thus, it would not be surprising that more antagonist was required to counteract it. The 4-vinyl analog was six- to tenfold more effective than 4′-deoxypyridoxine but had previously been found to be a potent convulsing agent in mice.[249]

Korytnyk and Potti[254,255] found that $6.8 \times 10^{-8}$ $M$ deoxypyridoxine produced 50% inhibition of the growth of mouse mammary adenocarcinoma (TA3) in Eagle's medium containing $1 \times 10^{-7}$ $M$ pyridoxal. It is very unusual for deoxypyridoxine to produce such a large effect at a deoxypyridoxine:$B_6$ ratio of less than 1.0. Increasing the pyridoxal content to $10^{-5}$ $M$ reversed the inhibition. The *N*-oxide with an $ID_{50}$ of $5 \times 10^{-5}$ $M$ was less inhibitory.[248]

While the great variety of protocols used for in vivo studies makes a casual encounter with the literature very frustrating, the summary in Table 1 reveals that the results are not as contradictory as they seem at first glance. Deoxypyridoxine reduced tumor growth in 9 out of 11 cases where it was used in conjunction with a $B_6$-deficient diet. It was effective in only 4 out of 17 cases where it was used with a normal $B_6$ intake.

### Table 1
### EFFECT OF 4′-DEOXYPYRIDOXINE ON TUMOR GROWTH

| Tumor | $B_6$ intake[a] | Deoxypyridoxine dose | $B_6$ intake of controls[a] | Effect[b] | Ref. |
|---|---|---|---|---|---|
| | | Rat | | | |
| Lymphosarcoma | – | 0.25 mg/day | – | – | 173 |
| | – | 2 mg/day | – | + | 476 |
| | – | 20 mg/kg | + | + | 205 |
| | – | 20 mg/kg | – | + | 422 |
| Walker carcinoma | – | 20 mg/kg | – | + | 422 |
| | – | 25 mg/kg | | – | 158 |
| Reticulo sarcoma | + | 0.35 mg/day | + | – | 266 |
| Fibrosarcoma | + | 5 mg/day | + | – | 278 |
| Walker carcinoma | + | 25 mg/kg | + | + | 158 |
| | | Mouse | | | |
| Lymphosarcoma | – | 0.9 mg/day | – | + | 474, 476 |
| Sarcoma 180 | – | 18 mg/kg | – | + | 67 |
| | – | 43 mg/kg | – | + | 435 |
| | – | 75 mg/kg | – | + | 460 |
| | – | 100 mg/kg | – | + | 324 |
| Leukemia | + | 0.5 mg/day | + | + | 512 |
| | + | 3 mg/day | + | – | 544 |
| | + | 150 mg/kg | + | – | 483 |
| Sarcoma 180 | + | 18 mg/kg | + | – | 67 |
| | + | 85 mg/kg | + | – | 459 |
| | + | 100 mg/kg | + | – | 324 |
| | + | 125 mg/kg | + | ± | 486 |
| | + | 150 mg/kg | + | + | 320 |
| | + | 150 mg/kg | + | – | 460 |
| | + | 150 mg/kg | + | + | 319 |
| Ascites | + | 0.07 mg/day | + | – | 74 |
| Ehrlich ascites | + | 125 mg/kg | + | – | 486 |
| Krebs ascites | + | 125 mg/kg | + | – | 486 |
| Mammary adenocarcinoma | + | 175 mg/kg | + | ± | 450 |

[a] (–) = $B_6$-deficient
(+) = Normal $B_6$

[b] (–) = No effect on tumor growth
(+) = Inhibition of tumor growth
(±) = Possible inhibition

Even when deoxypyridoxine is used in conjunction with a $B_6$ deficiency, the effect seems to be stasis rather than toxic. Tumor growth resumes if $B_6$ is restored. Stoerk[474-476] noted that deoxypyridoxine in doses as low as 0.9 mg/day caused regression of lymphosarcoma implants in pyridoxine-deficient mice. Dietary pyridoxine deficiency alone had no such effect. However, complete realimentation with pyridoxine led to recurrence of the tumors and death of the animals.

Weir et al.[544] treated leukemic mice receiving a commercial stock diet with deoxypyridoxine (1.5 mg s.c. twice daily). Survival decreased from 31.8 days in the controls to 24.6 days in the treated mice. No differences in the histology of the liver, spleen, bone marrow, and lymph nodes were seen between the two groups.

Gregoire[173] found that a $B_6$-deficient diet inhibited the growth of lymphosarcoma by 80% in rats. Adding 250 μg deoxypyridoxine daily for the first 8 days of the 3-week study caused no further inhibition. Larger doses (up to 1.5 mg/day) were used in a second experiment under conditions selected to maximize tumor growth. While the treatment produced 65% inhibition compared with the normal $B_6$ controls, there was no group which received only the $B_6$-deficient diet. Thus, the effects of deoxypyridoxine can not be separated from those of $B_6$ deficiency alone.

Stoerk[476] used adult male rats receiving a $B_6$-deficient diet plus 2 mg deoxypyridoxine per day by stomach tube. After 3 weeks, average tumor (lymphosarcoma) weight on the diet alone was 25 g compared with 61 g in controls. Treatment with cortisone or methyltestosterone further reduced tumor weight.

Loefer[278] injected rats intraperitoneally with eight 5 mg doses of deoxypyridoxine over 14 days in conjunction with a normal $B_6$ diet at which time a fibrosarcoma was implanted. In another experiment, 13 doses of 5 mg/day were used. In a third experiment, 11 injections each of 10 mg deoxypyridoxine before implantation and 4 more after implantation were used. While pyridoxine supplementation increased the number of takes as measured after 13 to 25 days compared with controls, deoxypyridoxine did not significantly reduce the number of takes. Only a few animals showed minor dermatitis. Our demonstration that deoxypyridoxine is rapidly excreted, combined with the author's use of a normal $B_6$ diet and termination of deoxypyridoxine treatment at the time of implantation, makes it unlikely that any effect would be observed.

Cerecedo et al.[80] examined the growth of a lymphosarcoma in normal and $B_6$-deficient C3H mice. Both groups received drinking water containing 300 μg deoxypyridoxine per milliliter. The tumors regressed in mice receiving the $B_6$-deficient diet.

Tedeschi[499] observed an inhibitory action of deoxypyridoxine against sarcoma induced by benzopyrene in rats particularly if the tumor was not well established. Also, regardless of the degree of development, he always observed significant changes in the gross and histological characteristics of the tissue. No specific details were given.

Skipper and Thomson[459] found that injection of 50 to 85 mg deoxypyridoxine per kilogram every other day over 8 days had no effect on sarcoma 180 or leukemia in mice receiving a normal $B_6$ diet. When the deoxypyridoxine treatment was combined with a $B_6$-deficient diet, survival of the mice seemed to be improved but the treatment still failed to meet the authors' criteria for tumor inhibition.

Sugiura and Creech[487] found that intraperitoneal administration of 125 mg deoxypyridoxine per kilogram per day in mice presumably receiving a normal $B_6$ diet had no effect on Ehrlich, Krebs 2, or sarcoma 180 ascites tumors, or on solid sarcoma 180 tumors. There was some inhibition of solid Ehrlich tumors.

Sassenrath et al.[437] reported that deoxypyridoxine injected (probably i.p.) at 112 mg/kg/day for 7 to 10 days ($B_6$ content of diet not specified) produced a statistically significant reduction of Ehrlich C3H-SX ascites tumors in C3H mice, and sarcoma 180 ascites tumor in Swiss mice. However, the inhibition did not reach the 75% threshold which the authors arbitrarily selected as a biologically useful value. Deoxypyridoxine had no significant activity against 6C3H-ED lymphosarcoma ascites tumor in C3H mice.

Skipper and Thomson[460] studied the effect of combining dietary amino acid deficiencies with deoxypyridoxine on the growth of sarcoma 180 in mice. The route of deoxypyridoxine administration is not specified and it is not clear whether dosages are given in terms of kilograms body weight or kilograms diet. Dosages of 20 to 75 mg deoxypyridoxine per kilogram did not produce consistent inhibition of tumor growth on a complete diet. However, when combined with a diet deficient in isoleucine, or glutamic and aspartic acids, inhibition ranged from 43 to 88%.

Rosen and Milholland[422] found that a dose of about 20 mg deoxypyridoxine per kilogram body weight per day in conjunction with a $B_6$-deficient diet produced 80% inhibition of Murphy-Sturm lymphosarcoma and lymphosarcoma P-1798 and 90% inhibition of Walker carcinoma 256.

Ono et al.[372] found that tumor size and levels of histamine and serotonin were markedly decreased when the mice were fed a $B_6$-deficient diet. Except for serotonin, the changes were greater when deoxypyridoxine hydrochloride (15 μg/g/day) was given with the deficient diet. The abstract does not specify whether the dose refers to gram body weight or gram diet. Neither does it specify the route of administration or the tissue in which histamine and serotonin levels were determined.

Rosen et al.[421] found that a $B_6$-deficient diet containing 0.001 to 0.005% deoxypyridoxine caused a significant reduction in tumor size in mice. They concluded that lack of total regression was due to the simultaneous inhibition of host defenses.

Tsuji[512] reported that subcutaneous injection of 0.5 mg deoxypyridoxine hydrochloride per mouse per day increased the survival time of leukemic mice to 16.7 days compared with 14.2 days for controls. The mice were fed a commercial stock diet.

Fujimoto[148] administered 20 mg/kg/day intraperitoneally for 21 days to rats receiving a $B_6$-deficient diet. The deoxypyridoxine treatment reduced both body weight and growth of ascites hepatoma tumor, while $B_6$ deficiency alone had little effect.

Yuspa and Blaustein[565] used a $B_6$-deficient diet plus 0.3 mg deoxypyridoxine per milliliter drinking water for about 21 days in rats. The deoxypyridoxine was stopped at the time of tumor (osteogenic sarcoma 344) transplantation. In young females (77 to 122 g), the treatment significantly reduced tumor growth. In a second study using older (144 to 165 g) females, a similar trend was observed but the differences were not statistically significant.

LaPesa et al.[266] inoculated rats receiving a normal $B_6$ diet with cells from a solid reticulosarcoma (IRE) and then treated the rats intramuscularly with 0.35 mg deoxypyridoxine per rat per day. The treatment did not affect latency, survival, or weight changes of the animals.

Grimaldi et al.[174] used ascites cells in mice receiving a normal $B_6$ diet. Deoxypyridoxine (0.07 mg per mouse per day i.m.) was administered 7 days prior to inoculation and continued to the end of the experiment. In another group, treatment was started after inoculation, and in another deoxypyridoxine was injected intraperitoneally. The survival of the various groups was not significantly affected although the average survival for the controls, intramuscular, and intraperitoneal groups was 11, 12, and 13 days, respectively.

A few workers have tried deoxypyridoxine phosphate with mixed results. Tedeschi[499] reported that deoxypyridoxine phosphate was effective against sarcomas induced by benzopyrene even in the presence of normal $B_6$. Unfortunately, no experimental details were presented. Skipper and Thomson[460] found that administration of 75 mg deoxypyridoxine or its phosphate per kilogram (route unspecified) in conjunction with a $B_6$-deficient diet produced about 75% inhibition of sarcoma 180 growth in mice. The deficient diet alone produced only 7% inhibition. Deoxypyridoxine phosphate was also tried at the lower dose of 18 mg/kg and produced 65% inhibition. With normal $B_6$ and a dose of 150 mg/kg, deoxypyridoxine and its phosphate produced 22% and 13% inhibition, respectively. In discussion following the paper, Skipper mentioned observing a synergistic effect on tumor growth between penicillamine and deoxypyridoxine with a normal $B_6$ diet.

Skipper et al.[461] later noted that deoxypyridoxine was inhibitory to sarcoma 180 at doses that were only 1% of the $LD_{10}$. Such low doses were not possible with most of the chemotherapeutic agents under study at that time. Deoxypyridoxine and deoxypyridoxine-5′-phosphate were equally effective. The antitumor effect could not be accounted for by general inanition.

Sugiura[486] injected mice receiving a normal diet with deoxypyridoxine-5′-phosphate (250 mg/kg in carboxymethylcellulose) intraperitoneally for 7 days after tumor implantation. No effect was observed on sarcoma 180 in solid or ascitic form or on solid Ehrlich carcinoma. A marginal effect was seen on Ehrlich ascites tumors.

Considering that tumor inhibition was obtained consistently only when deoxypyridoxine treatment was combined with a $B_6$-deficient diet and that tumor growth resumed if $B_6$ was restored, it seems unlikely that deoxypyridoxine alone would be a useful tumor treatment. However, it might be possible to take advantage of the synergistic effect between deoxypyridoxine and some other antitumor agents.

In the first of several studies, Shapiro and Gellhorn[450] found that deoxypyridoxine (175 mg/kg body weight per day i.p. for 10 days) significantly inhibited the growth of the 755 mammary adenocarcinoma in mice in experiments in which a total of 40 mice were treated. In two other experiments involving treatment of a total of 30 mice, deoxypyridoxine had no statistically significant effect. A normal $B_6$ diet was used in all experiments In only one of eight experiments was deoxypyridoxine treatment associated with noticeable weight loss. In five of seven experiments, the combination of deoxypyridoxine with 8-azaguanine produced significantly greater inhibition than 8-azaguanine alone. The authors concluded that a deoxypyridoxine dose of 150 mg/kg body weight is probably the threshold range of a combination effect. They also concluded that the decreased tumor growth could not be accounted for by loss of body weight.

Further studies[451,452] showed that when a noncarcinostatic dose (80 mg/kg/day) of deoxypyridoxine was combined with testosterone and 8-azaguanine, the carcinostatic effect was consistently greater than with 8-azaguanine alone in mice bearing 755 mammary adenocarcinomas and receiving a normal $B_6$ intake. The authors suggested that the slight increase in deaths among the deoxypyridoxine groups might be due to determining dosage solely by body weight rather than by $B_6$ status which their data showed to vary widely between animals. The pyridoxine content of tumors from testosterone-treated mice was lower than in controls that received no hormone.

Shapiro et al.[448] combined deoxypyridoxine (40 mg/kg twice daily, 4 hr apart) with azaguanine, testosterone, and 6,7-dimethyl-9-hydroxyethyl-isoalloxazine. Again, the combination was more effective than individual agents and more effective than the triple combination without the isoalloxazine compound.

Shapiro et al.[449] found that 30 mg deoxypyridoxine per kilogram body weight injected twice daily and combined with testosterone, 8-azaguanine, and niacinamide produced more regression of 755 mammary adenocarcinoma in mice than any agent individually. No work with deoxypyridoxine alone was presented.

Hruban and Wissler[205] also found that testosterone enhanced the effects of antitumor agents. Oral administration of 20 mg deoxypyridoxine per kilogram per day in conjunction with a $B_6$-deficient diet produced a 50% decrease in growth of Murphy-Sturm lymphosarcoma. Raising the dose to 30 mg/day yielded a 63% decrease in tumor growth. Adding DL-beta-3-thienylalanine further enhanced the antitumor effect. The authors concluded the effects were additive rather than synergistic. Testosterone further increased the antitumor activity of this combination. The tumors resumed normal growth when the treatment was stopped. Deoxypyridoxine had little or no effect against Jensen tumor, breast carcinoma, and Novikoff hepatoma. In further studies, treatment with 20 mg deoxypyridoxine per kilogram per day produced a 20 to 80%

decrease in the size of Murphy-Sturm lymphosarcoma.[206] The effect was again enhanced by thienylalanine. Johnson and McColl[228] patented the use of deoxypyridoxine and testosterone to potentiate the anitiumor activity of 6-aminonicotinamide.

Several workers have reported an interaction between deoxypyridoxine and isonicotinic acid hydrazide. Garattini and Vegeto[158] used adult rats implanted with Walker carcinoma. Deoxypyridoxine (25 mg/kg i.p.) was administered for about 3 weeks to animals receiving normal and $B_6$-deficient diets. In conjunction with a normal $B_6$ intake, deoxypyridoxine alone or combined with isonicotinic acid hydrazide produced about a 30% reduction in tumor size compared with control animals after 20 days. Even in the $B_6$-deficient animals, deoxypyridoxine had little effect on body weight. Isonicotinic acid hydrazide did produce a marked weight loss. The $B_6$-deficient diet reduced tumor growth about 20%. This amount was not increased by addition of deoxypyridoxine to the diet. Isonicotinic acid hydrazide was a more effective inhibitor and seemed to act synergistically with deoxypyridoxine.

Wheeler et al.[547] did observe a synergistic interaction on the inhibition of *Escherichia coli*, adenocarcinoma 755, and sarcoma 180 when deoxypyridoxine was combined with azaserine. There was no potentiation of anticancer activity with leukemia L1210. Deoxypyridoxine also showed synergistic inhibition against *Escherichia coli* and sarcoma 180 when combined with benzoic acid hydrazide, isonicotinic acid hydrazide, *p*-aminobenzoic acid hydrazide, or 1,5 diaminobiuret. Combinations between deoxypyridoxine and 3-(1,2,4-triazolyl) alanine, DL-ethionine, or sulfanilamide showed synergistic activity against *Escherichia coli* but not against the tumors.

Bain and Williams[27] reported that the pyridoxal phosphate content of Erhlich ascites tumor cells in mice dropped 44% with deoxypyridoxine treatment (75 mg/kg) and 62% when isonicotinic acid hydrazide was combined with the deoxypyridoxine treatment. Only the latter combination inhibited tumor growth.

Mihich and Nichol[324] demonstrated that their deoxypyridoxine contained less than 0.01% pyridoxine. In the presence of normal $B_6$, intraperitoneal administration of 50 and 100 mg/kg/day had little effect on tumor sarcoma 180 growth in mice. Combining the treatment with a $B_6$-deficient diet did cause inhibition of tumor growth but not regression, probably because the antagonist simultaneously interfered with host defenses. Isonicotinic acid hydrazide alone had no effect. However, deoxypyridoxine combined with isonicotinic acid hydrazide did produce inhibition of tumor growth even on a normal $B_6$ diet but again failed to produce the regression seen with the $B_6$ deficiency alone. Inhibition of tumor growth was observed if dietary depletion was started 2 weeks prior to tumor implantation but not if the deficient diet was started on the day of implantation. Inhibition of tumor growth was observed if treatment with a $B_6$-deficient diet containing 10 to 50 mg deoxypyridoxine per kilogram was started on the day of implantation, However, tumors in the deoxypyridoxine group failed to regress like the tumors in mice which had been depleted for 2 weeks prior to implantation. Solid tumors were less responsive to the effects of the drugs or $B_6$ deficiency. $B_6$ deficiency alone had no effect on ascitic S-180 inoculated subcutaneously. However, combining the deficient diet with deoxypyridoxine did produce inhibition. Thiosemicarbazide inhibited tumor growth without affecting tumor regression. The authors concluded that clinical trials in humans are warranted.

Other hydrazine derivatives have also yielded positive results. Mickelson and Flippin[319] found that treatment (150 mg/kg i.p.) with either deoxypyridoxine or DL-8-aspartylhydrazine for 10 days produced 30 to 40% inhibition of the growth of sarcoma 180 in BALB/C mice receiving normal $B_6$. When the two compounds were administered simultaneously, 70 to 80% inhibition was obtained. In 26 out of 41 mice in five experiments, the dual treatment did not increase the loss of body weight over that shown by the control animals suggesting that the treatment was reasonably selective for the tumors.

In a subsequent study,[320] a dose of 150 mg/kg for 9 days produced a 30 to 40% inhibition of sarcoma 180 growth in mice receiving normal $B_6$. When combined with a similar dose of DL-aspartylhydrazine, tumor inhibition temporarily approached 80%. However, after 2 weeks of treatment, growth resumed and survival time was not greatly improved. L-Glutamyl hydrazine or DL-histidine hydrazine had no effect alone but exerted a synergistic action with deoxypyridoxine.

Brockman et al.[67] found that in mice receiving adequate $B_6$, combinations of a hydrazide with deoxypyridoxine phosphate were more inhibitory than those with deoxypyridoxine. The effects of deoxypyridoxine could be offset by pyridoxine, pyridoxamine, pyridoxal, or pyridoxal phosphate. Combining hydrazides with deoxypyridoxine on a $B_6$-deficient diet produced a synergistic effect.

Tarnowski and Stock[496] observed no synergistic effect when unspecified amounts of deoxypyridoxine were added to the azaserine regimen for treating RC mouse mammary carcinomas in conjunction with normal or $B_6$-deficient diets.

Mihich and Nichol[325] reported that intraperitoneal administration of deoxypyridoxine (50 to 100 mg/kg/day) decreased the growth of solid sarcoma 180 tumors in mice receiving either normal or $B_6$-deficient diets. Mortality was also increased. Both effects were enhanced when methylglyoxal-bis ($N^8$-methylthiosemicarbazone) was added to the treatment.

Sano et al.[435] examined the effects of deoxypyridoxine (43 mg deoxypyridoxine per kilogram per day i.p.) on the growth of ascites sarcoma 180 in ICR mice. When administered with a $B_6$-deficient diet, deoxypyridoxine was more effective than 3-aminopyridoxine alone, but when combined with the antitumor polysaccharide EA3, 3-aminopyridoxine was more inhibitory.

Little work has been done with other deoxypyridoxines. Harris et al.[188] found that 4 mg 3-deoxypyridoxine per rat in conjunction with a $B_6$-deficient diet produced "significant arrest" of the growth of lymphosarcoma implants compared with controls receiving only the deficient diet. The authors did not specify how often the 4 mg treatment was administered nor did they clarify their use of the term "significant arrest." No specific data were presented. Struck et al.[483] found no antileukemic activity for a variety of analogs including 4′-deoxyisopyridoxal and possibly 4′-deoxypyridoxine at dosages including 400 mg/kg in mice on a normal diet. Toxicity was observed with 4′-deoxyisopyridoxal at dosages above 75 mg/kg.

## III. HUMANS

Gellhorn and Jones[162] first reported an attempt to treat human cancers with deoxypyridoxine. The study involved three patients with disseminated lymphosarcoma and three with acute leukemia. After initial dosages of 25 mg/kg proved toxic (see Chapter 7), the dosage was reduced to 2.5 mg/kg/day in conjunction with a pyridoxine-deficient diet for 4 to 14 days. In two of the three patients with lymphosarcoma, there was evidence of moderate regression in the size of the lymph nodes. However, Gellhorn and Jones attributed this regression to inanition analagous to the thymic involution seen by Stoerk[473] in rats. Stoerk[477] disputed the similarity and commented that he would not expect tissue from a lymphoid neoplasia to behave like normal thymic tissue.

Weir[541] reported treating five adult patients with acute lymphatic leukemia with up to 1.4 g deoxypyridoxine per day for 80 days while on a normal diet. No toxic effects or dermatitis were observed. Two patients showed improvement in both clinical and hematological symptoms. One showed improved clinical condition with little hematological change. Two terminal cases died shortly after the treatment started. It seems likely that this group of patients included the two mentioned by Weir and Heinle.[453]

Vilter et al.[530] observed no beneficial effect of deoxypyridoxine in chromic myeloid leukemia, acute childhood leukemia, lymphocytic lymphoma, or multiple myeloma. There was some improvement in patients with chronic lymphatic leukemia although not as much as reported by Weir.[541] The number of cases of each type of leukemia examined is not clear, but it appears that the total was less than ten.

Weir and Morningstar[545] tested the anticancer effects of deoxypyridoxine (up to 1400 mg/day) in four adults with terminal acute lymphatic leukemia. The relationship between these patients and those reported in 1952 is not stated, but it seems likely that they are the same cases. Some modification of the disease process apparently occurred in all patients. In one case a brief clinical remission occurred. The authors concluded further clinical trials were justified.

Eagle and Foley[127] summarized previous tests as showing that deoxypyridoxine had antitumor activity in 2 of 22 systems tested. They determined that it caused 50% inhibition of the growth of cells from normal human liver and a human epidermoid carcinoma at concentrations of $10^{-5}$ to $10^{-6}$ g/m$\ell$.

Gailani and Holland[153] fed two patients a $B_6$-deficient diet plus 1 to 2 mg deoxypyridoxine for 10 to 14 days. Both developed stomatitis and dermatitis. No tumor regression was observed. Gailani and Holland[153] reported data on nine patients which presumably are included in later, more detailed reports. Gailani et al.[155] obtained improvement in a patient with carcinoid syndrome associated with a metastatic bronchial adenoma through use of deoxypyridoxine in conjunction with a $B_6$-deficient diet. Deoxypyridoxine (30 mg three times daily) was administered for 28 days. It was discontinued because of anorexia, nausea and vomiting, dermatitis, and paresthesia. Although the diarrhea and facial flushing were eliminated, there was no tumor regression. The data suggest that deoxypyridoxine was particularly effective in reducing 5-hydroxyindoleacetic acid excretion, and reticulocyte and lymphocyte counts.

## IV. SUMMARY

Table 1 summarizes some of the simplest animal studies. Many of the reports included various levels of deoxypyridoxine as well as combinations with other drugs. Except for the very low dose (0.25 mg/day) used by Gregoire, every report in which a $B_6$-deficient diet was used found that deoxypyridoxine reduced tumor growth. The results are much less encouraging with a normal $B_6$ intake. Even with the $B_6$-deficient diet, the doses approach or exceed the 25 mg/kg dose which has been found to be convulsive in humans. Another problem is that deoxypyridoxine appears to be more cytostatic than cytotoxic. Tumor growth resumed if the treatment was interrupted. While the data presented here are not encouraging in relationship to clinical applications, it might be that better understanding of deoxypyridoxine effects in vivo would suggest ways to enhance its anitiumor activity while minimizing other effects. Also, in view of the evidence of synergistic effects, deoxypyridoxine might be a useful supplement to other treatments.

Chapter 7

# CLINICAL STUDIES IN HUMANS

## I. INTRODUCTION

Although deoxypyridoxine has been used many times in man, its use has generally been restricted to tests of its antitumor activity or to induction of a $B_6$ deficiency. Our own recent work[92] appears to be the first attempt at studying the metabolism of deoxypyridoxine itself. The cancer studies have already been presented. In this chapter we will examine the more general clinical effects.

## II. PHYSIOLOGICAL EFFECTS

The most serious consequence of deoxypyridoxine is convulsions which were noted by Gellhorn and Jones[162] who were the first to report using deoxypyridoxine in humans. A 52-year-old woman with reticulum cell sarcoma was treated with a $B_6$-deficient diet combined with an oral dose of deoxypyridoxine every three hours, totalling 25 mg/kg body weight per day. After receiving a total of 1.5 g deoxypyridoxine (presumably about 48 hours after starting the treatment), "she developed persistent nausea and suddenly lost consciousness, became cyanotic and exhibited mass movements of extremities of a convulsive character." Similar treatment in a 4-year-old boy with acute leukemia produced two generalized convulsize seizures during the fifth day of treatment. In both cases subsequent mental confusion was noted even though the dose was reduced to 2.5 mg/kg immediately after the seizures. Four other terminal cancer patients received doses of 2 to 2.5 mg/kg for periods of 4 to 14 days. In one case mental dullness was noted. However, in these latter four cases it seems likely that the various symptoms noted (e.g., anorexia, cachexia, somnolence, cough, edema, low grade fever) were associated with the severity of their illness rather than specifically with the effects of deoxypyridoxine. All six patients died from 2 days to 6 weeks after the deoxypyridoxine treatment.

Mueller and Vilter[346] noted a personal communication from Hoster and Zanes confirming that high doses of deoxypyridoxine will cause convulsions in humans, but no further details were given. After these early reports, subsequent authors used lower dosages and there have been no further reports of seizures during experiments involving the use of deoxypyridoxine in human subjects.

Weir and Heinle[543] reported no detectable toxic symptoms in two adults treated over 80 days with up to 1400 mg deoxypyridoxine per day. Seborrheic dermatitis did appear late in the course of therapy. Weir[541] reported treating five patients, probably including the two mentioned above. Weir and Morningstar[545] provided a more complete description of four patients. The relationships of these patients to the earlier reports was not stated but it seems likely they are the same. Two patients received a dose of 1400 mg/day, one for 24 days and one for 30 days. The other two received doses of 1000 mg/day, one for 45 days and one for 11 days. No convulsions were reported. The fact that these patients received a normal diet while Gellhorn and Jones[162] used a $B_6$-deficient diet may account for the lack of convulsions in these patients. In the three cases which were autopsied there was some degeneration of the adrenal glands but the authors suggested that the changes may have been due to postmortem degeneration. The authors concluded that a daily dose of 500 mg deoxypyridoxine or more will produce a $B_6$ deficiency in humans on a normal diet.

Pfeiffer[388] also commented that attempts to induce seizures in two patients with an oral dose of 1.5 g deoxypyridoxine failed to produce seizure activity or recruitment when they were photically stimulated. Fisher[142] reported no convulsions in men apparently receiving a normal $B_6$ intake plus 250 mg deoxypyridoxine four times per day for about 5 weeks. Skin symptoms were observed.

Hansson[181] examined the effect of deoxypyridoxine on the niacin pathway of tryptophan metabolism in four healthy volunteers. Deoxypyridoxine was administered at doses of 200 to 400 mg/day for 3 to 19 days in conjunction with a normal diet. After 10 to 14 days deficiency symptoms (fatigue, paraesthesia, erythema, desquamation, and slight conjunctivitis) appeared. In two of the three subjects showing deficiency symptoms, electroencephalography showed abnormal slowing of basal activity which persisted in one subject for at least 5 weeks after termination of the experiment. Unfortunately, no electroencephalograms had been taken prior to the start of the experiment. Other symptoms disappeared within 3 to 4 days after termination of the experiment.

Gailani et al.[154] reported mild electroencephalographic changes in three patients during $B_6$ deficiency. In one case, the authors specifically noted no deoxypyridoxine had been given. No comment concerning use of deoxypyridoxine was included for the other two patients. Mueller and Vilter[346] found that single intramuscular injections ranging from 1 to 50 mg of deoxypyridoxine produced no clinical symptoms in an unspecified number of patients receiving either a normal or B-complex-deficient diet. Three patients with various noncarcinogenic chronic illnesses who received a B-complex-deficient diet plus 60 mg deoxypyridoxine per day intramuscularly developed dermatitis after 19 to 21 days. The lesions disappeared within 72 hr after administering 100 mg pyridoxine per day. One patient who was started on 10 mg/day and gradually increased to 80 mg/day developed no dermatitis after 4 months even though for the last 3 months he had been receiving succinyl sulfathiazole to minimize any intestinal pyridoxine production.

Four patients received higher doses. One of these received the B-complex-deficient diet plus 100 mg deoxypyridoxine per day for 2 weeks. Three other patients received the deficient diet plus 100 mg deoxypyridoxine per day for 3 weeks. The dose was then increased to 125 mg/day for 5 days and to 150 mg for two days. All of these subjects developed seborrheic dermatitis ranging from minimal to severe within 10 to 21 days. The authors had not observed similar lesions from the diet alone even after several months. One patient also developed nausea, vomiting, weakness, and dizziness. The authors concluded that the inhibition ratio for deoxypyridoxine:pyridoxine was at least 1:1.

Glazer et al.[168] observed seborrhea-like lesions in all eight patients treated with a B-complex-deficient diet plus 200 mg deoxypyridoxine per day. The lesions disappeared within 1 to 4 days after initiating pyridoxine supplementation (5 to 50 mg/day i.m.).

The appearance of skin lesions has probably been the most consistently documented effect of deoxypyridoxine administration in both man and animals. Because $B_6$ deficiency in rats is associated with dermatitis, it has often been assumed that the appearance of skin lesions following deoxypyridoxine treatment is evidence of a generalized $B_6$ deficiency. We feel that deoxypyridoxine has a very specific effect against some aspect of skin metabolism, perhaps involving lipids and that the appearance of skin lesions following deoxypyridoxine treatment does not necessarily reflect a widespread deficit in the $B_6$ systems in other tissues. In this connection it is interesting to note some of the early studies of deoxypyridoxine and pyridoxine in seborrheic dermatitis.

Vilter and Schreiner[531] treated a spontaneous seborrheic dermatitis lesion in each of two patients with an ointment containing 100 mg deoxypyridoxine per gram and a second lesion with just the ointment base. The treated lesion worsened. Temporary remissions of spontaneous lesions could usually be induced by topical but not by systemic administration of pyridoxine. Wayne et al.[538] also showed that topically but not intramuscularly administered pyridoxine increased the $B_6$ concentration in the skin.

Schreiner et al.[442] reported that dermatitis produced in two patients by 100 mg deoxypyridoxine per day in conjunction with a $B_6$-deficient diet could be cleared within 48 hours by topical application of pyridoxine. In a third patient with spontaneous seborrheic lesions, topical application of deoxypyridoxine increased the severity of the lesion. Administration of 300 mg pyridoxine per day did not affect the lesions but topical application cleared them in 10 days. Further studies of 23 patients with spontaneous lesions confirmed that topical application of pyridoxine was quite effective whereas other routes of administration were not. The authors emphasized the interaction between $B_6$ and lipid metabolism.

The most extensive use of deoxypyridoxine in humans involved 50 patients.[530] This study included the eight patients from their original report.[346] A brief summary of this work is also included in a later review.[529] Dosages ranged from 50 to 500 mg/day for periods of 3 to 52 weeks. Thirty-four of the 50 subjects developed symptoms of $B_6$ deficiency. The authors hypothesized that variable contamination of the deoxypyridoxine with pyridoxine probably accounted for the fact that some patients never developed symptoms. Symptoms appeared most rapidly on a low $B_6$ diet. The earliest appearance of symptoms was 5 days with an average of 22 days. They described the effects of deoxypyridoxine as follows:

> Seborrheic dermatitis, indistinguishable from seborrhea sicca, is the most common lesion. It begins as an area of erythema usually in the nasolabial folds, but sometimes around the eyes, in the eyebrows, and at the angles of the mouth. Itching and burning sensations are associated with the erythema, and within twenty-four hours the involved areas are covered with yellowish oily scales. Frequently the lesion spreads from the area involved initially and has been observed over the entire face, behind the ears, in the scalp, down the neck, over the shoulders, in the perineal region, and over the buttocks. This lesion was the first sign of pyridoxine deficiency in 24 patients, and eventually appeared in 29.
>
> Glossitis indistinguishable from that of acute niacin deficiency occurred in five patients as the initial manifestation of pyridoxine deficiency, and ultimately appeared in fourteen before the experimental period was terminated by the administration of pyridoxine. At first the tongue burns as if scalded by hot coffee. Usually within the next twenty-four hours the papillae at the tip of the tongue are red, or areas of red hypertrophied filiform papillae may be found on the sides or dorsal surface of the tongue. Later the tongue swells sufficiently to blot out the structure of the filiform papillae, which remain as red dots. The fungiform papillae stand out as hypertrophied red knobs.
>
> The oral mucous membrane becomes reddened diffusely or focally. Small ulcerations may develop on this reddened background. This lesion is similar to niacin deficiency stomatitis and often resembles the lesions of mild aminopterin toxicity.
>
> Cheilosis, angular stomatitis, and conjunctivitis similar to the lesions of riboflavin deficiency occurred as the initial clinical manifestations in three patients. Before the test periods were terminated, six patients had conjunctivitis. Keratitis was not observed.
>
> Severe sensory neuritis was the initial manifestation of vitamin $B_6$ lack in one patient. Two others developed this lesion before the termination of the experiment. It began as a sensation of tingling and numbness in the hands and feet. In one patient no further progression was noted, but in two the lesion progressed rapidly in ascending fashion. Within four to five days the hands, arms, legs, and feet were hyperesthetic to pinprick. Afterglow was a prominent sign. Perception of vibration and the position of the toes and fingers was greatly impaired, and the deep tendon reflexes became hypoactive. Plantar reversal signs were not observed. After treatment with pyridoxine had eliminated all physical signs of neuritis, these patients experienced the "hot foot" or "electric foot" syndrome described so frequently by prisoners of war in the Pacific area. Nighttime was the most troublesome period. Subjective sensations of burning and electric shocks, particularly in the feet, forced these patients to kick the bed covers off. Gradually over a period of several months, these symptoms subsided.
>
> Two patients developed erythema of the dorsal surfaces of the arms and legs which spread on to the ventral surfaces. In one of these patients the erythema involved the neck in collar distribution and the perineal area. Pigmentation and scaling followed the erythema, and the lesion assumed the morphologic characteristics of chronic pellagrous dermatitis. One of these patients had two separate trials on desoxypyridoxine. On both occasions this same type of dermatitis appeared.

Two patients developed intertrigo in the perineal and inguinal regions, under the breasts, and in the axillae. These lesions were resistant to the usual types of local therapy until pyridoxine was administered. Then they healed rapidly.

Anorexia and drowsiness were the most common subjective manifestations of the deficiency state. Thirteen patients developed moderate degrees of nausea and vomiting. Eight became lethargic, somnolent, and confused. Six patients who were prone to infections because of pre-existing medical disease developed pneumonia, pyelonephritis, or superficial skin infections. In one of these, a patient with paraplegia who developed pyelonephritis, the therapeutic response was poor with antibiotics active against the offending organism in vitro. These same antibiotics were effective after the deficiency state was relieved by pyridoxine. These observations are reminiscent of similar ones by Stoerk in pyridoxine deficient rats. Weight loss was common after deficiency signs appeared. Some patients lost as much as 15 pounds.*

Although oral nicotinamide had no effect, intravenous administration of nicotinamide adenine dinucleotide cured the glossitis and somatitis observed in two patients. The effect of linoleic acid supplementation on dermatitis was examined in four patients. Improvement was noted in the two subjects receiving 8 g or more per day. Most symptoms disappeared within five days after pyridoxine supplementation.

Price et al.[399] observed no clinical symptoms in one patient who received doses of deoxypyridoxine increasing from 75 to 375 mg/day over a period of 5 weeks apparently in conjunction with a normal diet. Results in a second patient were invalidated by pyridoxine contamination in the deoxypyridoxine.

Wayne et al.[538] did observe typical skin and mucous membrane changes in four chronically ill subjects who received 300 mg deoxypyridoxine per day for 9 weeks apparently in conjunction with a dietary intake of about 2 mg vitamin $B_6$ per day. Mild sensory peripheral neuritis developed in two patients. No convulsions or electroencephalographic changes were observed. Webb[539] cited Schrodt et al.[442a] as observing seborrheic dermatitis and glossitis in two carcinoid patients treated with deoxypyridoxine. Webb presented no further details and we were unable to locate the original article.

Hodges et al.[199] observed the usual skin problems in two prisoners who received a $B_6$-deficient diet plus an oral dose of 100 mg deoxypyridoxine three times daily and 1000 mg omega-methyl pantothenic acid four times daily for 6 weeks. Symptoms disappeared within about two weeks of restoring a normal diet. In a further study five men were placed on the $B_6$-deficient diet plus doses of 100 mg deoxypyridoxine and 1000 mg omega-methyl pantothenic acid four times daily for 7 weeks. Clinical symptoms including dermatitis and gastrointestinal disturbances are tabulated in detail for all subjects. There was little weight loss. These appear to be the same subjects described by Faber et al.[137] Mueller and Iacono[344] observed clinical deficiency symptoms (primarily seborrhea-like dermatitis) in 13 out of 17 chronically ill adults receiving a normal hospital diet plus 300, 400, or 500 mg deoxypyridoxine per day for variable periods (apparently about 7 to 14 weeks).

Elder and Mengel[133] noted the development of optic neuritis in a patient with acute intermittent porphyria treated with a $B_6$-deficient diet plus deoxypyridoxine starting at 100 mg/day, increasing to 400 mg/day over four days and maintained at that level for about 10 days.

Gailani et al.[154] reported that paresthesia and dermatitis developed in seven of the nine cancer patients receiving deoxypyridoxine (0.5 to 1.5 mg/kg body weight). Of the patients receiving the deficient diet alone only one developed paresthesia and two showed skin changes. Six of the seventeen patients showed a moderate rise in serum alkaline phosphatase activity. The authors do not state whether any of these persons were receiving deoxypyridoxine in addition to the $B_6$-deficient diet.

Karasek[231] included deoxypyridoxine in development of a method of screening po-

* From Vilter, R., Mueller, J., Glazer, H., Jarrold, F., Abraham, J., Thompson, C., and Hawkins, V., *J. Lab. Clin. Med.*, 42, 335, 1953. © C. V. Mosby Co., St. Louis.

### Table 1
### EFFECT OF DEOXYPYRIDOXINE ON PYRIDOXIC ACID EXCRETION AND LEUKOCYTES[346]

| Subject | DOP (mg/day) | Pyridoxic acid (mg/24 hr) | Leukocytes total (cells/mm) | Lymphocytes | Neutrophils |
|---|---|---|---|---|---|
| M.B. | | | | | |
| Before | | 6.7 | 5000 | 1200 | 3100 |
| After | 100 | 8.0 | 5000 | 1200 | 2950 |
| H.P. | | | | | |
| Before | | 5.9 | 8400 | 2680 | 4150 |
| After | 60 | 6.6 | 8500 | 935 | 5525 |
| E.S. | | | | | |
| Before | | 5.7 | 10600 | 3600 | 6455 |
| After | 60 | 3.3 | 8600 | 1462 | 6280 |
| H.H. | | | | | |
| Before | | 1.2 | 7350 | 1690 | 4850 |
| After | 60—80 | 3.9 | 8200 | 700 | 6300 |
| L.K. | | | | | |
| Before | | 2.6 | 6750 | 1890 | 4185 |
| After | 10—80 | 2.3 | 5800 | 1045 | 3600 |
| R.W. | | | | | |
| Before | | 7.8 | 5150 | 1750 | 3090 |
| After | 100—150 | 2.6 | 5650 | 1075 | 3500 |
| L.L. | | | | | |
| Before | | 0.7 | 5750 | 1955 | 2930 |
| After | 100—150 | 2.0 | 4650 | 1255 | 2790 |
| J.M. | | | | | |
| Before | | 0.8 | 6200 | 2045 | 3530 |
| After | 100—150 | 3.1 | 5750 | 1265 | 3335 |

tential therapeutic agents for psoriasis by measuring epithelial cell growth of human foreskin in vitro. No effort was made to reduce the $B_6$ content of the medium. It is not clear whether deoxypyridoxine had any effect. The text states that there was no effect. However, the tabulated data recorded the 50% inhibition index in the 1000 μg/mℓ column and not in the > 1000 μg/mℓ column. The methods section states that if no effect was observed at 1000 μg/mℓ, the inhibition index would be recorded as greater than 1000 μg/mℓ. At any rate, little inhibition would be expected without reducing the $B_6$ content of the medium.

## III. HEMATOLOGY

Mueller and Vilter[346] closely examined the hematologic status of eight subjects who received a B-complex-deficient diet plus deoxypyridoxine in doses of 60 to 150 mg/day. Only a patient who was also receiving succinyl sulfathiazole developed anemia, and it responded to folic acid. No hypochromia, siderocytosis, or leukopenia was observed. Seven subjects showed lymphocytopenia and two developed eosinophilia at the height of their rash (Table 1).

Data on these patients plus ten more who received 50 to 500 mg/day for periods of 3 to 52 weeks were reported by Vilter et al.[530] Mild normocytic normochromic anemia developed in seven patients. No bone marrow changes were found in the five patients examined. Reticulocytosis did not occur after pyridoxine administration. The authors concluded that no definite relationship between the anemia and $B_6$ deficiency could be established.

Total white counts did not change significantly. Seventeen of the eighteen subjects did develop lymphopenia and there was a tendency toward polymorphonuclearcytosis in ten patients. Eosinophilia was observed in patients with extensive skin lesions. Both lymphocytes and eosinophils returned to pretreatment values after pyridoxine supplementation.

Weir and Heinle[543] observed decreased leukocyte counts, increased hemoglobin and clinical signs of remission in two patients with acute lymphatic leukemia treated over 80 days with doses of deoxypyridoxine up to 1400 mg/day.

Wayne et al.[538] observed a decrease in total lymphocyte counts but no anemia in four patients who received 300 mg deoxypyridoxine plus about 2 mg pyridoxine per day for 9 weeks.

Hodges et al.[199] noted a slight drop in red count, hemoglobin, and hematocrit in two prisoners fed a $B_6$-deficient diet plus an oral dose of 100 mg deoxypyridoxine three times daily and 1000 mg omega-methyl pantothenic acid four times daily for 6 weeks. Erythrocyte sedimentation rate increased slightly. A polymorphonuclearcytosis developed without a rise in the absolute number of circulating lymphocytes. The eosinopenic response to ACTH was abnormal. In a further study of five men who received the $B_6$-deficient diet plus oral doses of 100 mg deoxypyridoxine and 1000 mg omega-methyl pantothemic acid four times daily for 7 weeks, anemia and polymorphonuclear leukocytosis appeared.

Elder and Mengel[133] reported that the lymphocyte count decreased during treatment of a patient with a $B_6$-deficient diet plus 400 mg deoxypyridoxine per day.

Gailani et al.[154] found that $B_6$ deficiency alone produced a definite increase in the reticulocyte count of four of nine patients receiving only the $B_6$-deficient diet and in five of seven receiving the deficient diet plus deoxypyridoxine (0.5 to 1.5 mg/kg/day for 6 to 46 days. These increases occurred without evidence of bleeding, hemolysis, or definite change in hemoglobin concentration. However, the authors did report a slight to moderate decline in hemoglobin in seven of seventeen subjects and a more marked decline in five others. The presentation of the data does not permit correlation of hemoglobin values and reticulocyte counts for individual subects. Changes in leukocyte counts occurred only in lymphocytes and only during deoxypyridoxine supplementation.

Definite lymphopenia developed in four subjects with nonlymphoid neoplasms and presumably normal bone marrow function. Three out of four subjects with lymphoproliferative disease exhibited a higher peripheral lymphocyte count. In order to clarify whether the increases might have resulted from an accelerated disease process rather than deoxypyridoxine treatment, a patient with stable chronic lymphocytic leukemia was tested. No change was observed during 46 days of deoxypyridoxine supplementation. However, a rash and paresthesia appeared 3 days after discharge and her white count increased about 50% within three weeks and remained at the new plateau during one year of follow-up.

## IV. VITAMIN $B_6$

Mueller and Vilter[346] found the pyridoxic acid excretion in eight patients receiving deoxypyridoxine (60 to 150 mg/day) plus a B-complex-deficient diet too variable to warrant any firm conclusions (Table 1). However, soon after deoxypyridoxine treatment started pyridoxic acid excretion increased, suggesting to the authors that cellular pyridoxine was being displaced. Urinary excretion of thiamine and riboflavin was in the low normal range in eight patients receiving deoxypyridoxine (60 to 150 mg/day) plus a B-complex-deficient diet. However, they presented no control values.

Elder and Mengel[133] reported that pyridoxic acid excretion decreased during treatment of a patient with a $B_6$-deficient diet plus 400 mg deoxypyridoxine per day.

Hansson[181] found no significant decrease in plasma pyridoxal phosphate during treatment of four healthy volunteers receiving a normal diet with 200 to 400 mg/day for 3 to 19 days.

Gailani et al.[154] reported that pyridoxal phosphate concentrations in leukocytes and urinary pyridoxic acid excretion declined during treatment of cancer patients with a $B_6$-deficient alone but increased when a deoxypyridoxine supplement was added to the diet. Although the authors noted that their results paralleled the results of Johnston et al.,[229] they apparently overlooked the fact that Johnston et al.[229] associated the change in leukocyte pyridoxal phosphate with an increased number of polymorphonuclear leukocytes. We[295] later confirmed that the pyridoxal phosphate concentration in human polymorphonuclear leukocytes is about three times higher than in lymphocytes. Since Gailani et al.[154] did observe lymphopenia during deoxypyridoxine treatment but did not attempt to correlate the changes in differential count with the changes in leukocyte pyridoxal phophate, it remains uncertain whether the apparent increase in leukocyte pyridoxal phosphate may have simply reflected the increased percentage of polymorphonuclear leukocytes. The authors suggested that the increased pyridoxic acid excretion might be an artifact caused by a metabolite of deoxypyridoxine. Our later studies[92] showed that interference from 4′-deoxypyridoxine-5-pyridoxic acid is eliminated by comparing the lactonized and delactonized forms as suggested in the Reddy[404] procedure for determining 4-pyridoxic acid. Deoxypyridoxine treatment did not seem to alter the rate of pyridoxal phosphate depletion in liver of three patients compared with four others who received only the $B_6$-deficient diet.[154] In one patient, the pyridoxal phosphate content of muscle declined from an initial value of 4.1 μg/g to 2.6 μg/g over a 30-day period which included about 2 weeks of deoxypyridoxine treatment.

Yamada and Tsuji[557,558] studied the transport of vitamin $B_6$ in human erythrocytes. Potassium fluoride inhibited the uptake of pyridoxal, pyridoxine, pyridoxamine, and pyridoxal phosphate by 0, 20, 15, and 60% respectively. Even under control conditions, uptake of pyridoxal phosphate was much less than uptake of the free forms. The authors felt that uptake of pyridoxine and pyridoxamine was more susceptible to potassium fluoride than the uptake of pyridoxal. However, we feel that the very rapid uptake of pyridoxal and pyridoxal phosphate during the first five minutes suggests that this may represent Schiff-base binding on the cell wall rather than true uptake. Equimolar amounts of deoxypyridoxine inhibited the uptake of pyridoxine and pyridoxamine but not of pyridoxal or pyridoxal phosphate. Deoxypyridoxine phosphate had little effect on the uptake of any vitamers.

## V. NITROGEN METABOLISM

### A. Urea and Proteins

Blood urea nitrogen tended to increase in five subjects during deoxypyridoxine treatment.[530] Blood urea nitrogen concentrations in three patients after an alanine load also tended to increase compared with their response before deoxypyridoxine treatment.

Wayne et al.[538] found no changes in serum protein electrophoresis in four patients who received 300 mg deoxypyridoxine plus about 2 mg vitamin $B_6$ per day for nine weeks. Hodges et al.[199] also failed to observe any changes in serum protein electrophoresis of two prisoners who received a $B_6$-deficient diet plus an oral dose of 100 mg deoxypyridoxine three times daily and 1000 mg omega-methyl pantothenic acid four times daily for 6 weeks.

Urinary protein and sediments were normal in two prisoners who received a $B_6$-deficient diet plus an oral dose of 100 mg deoxypyridoxine three times daily and 1000 mg omega-methyl pantothenic acid four times daily for 6 weeks.[199] In a further seven week study of five men using the same protocol except that the 100 mg dose of deoxypyridoxine was increased to four times daily, serum protein electrophoresis revealed decreased gamma globulin.

Gailani et al.[154] noted occasional changes in blood urea nitrogen and serum protein in patients receiving a $B_6$-deficient diet supplemented in some cases with deoxypyridoxine. Apparently, the changes were not particularly related to deoxypyridoxine administration.

**B. Xanthurenic Acid and Other Tryptophan Metabolites**

Increased excretion of xanthurenic acid following a tryptophan load is probably the most well-documented metabolic effect of deoxypyridoxine in humans. Gellhorn and Jones[162] found no xanthurenic acid in the urine of two cancer patients treated for a few days with a $B_6$-deficient diet plus 2 or 2.5 mg deoxypyridoxine per kilogram per day. One of these patients also received 1 g of tryptophan per day. The authors suggested that perhaps xanthurenic acid was not a metabolite of tryptophan in humans or that the patients were not pyridoxine-deficient. We now know that the first suggestion was incorrect. Considering that the patients did show signs of nonspecific malnutrition, it is possible that a secondary deficiency of some additional nutrients interfered with the normal metabolism of tryptophan. It is also possible that their dosages of deoxypyridoxine and tryptophan were too low to elicit a response, or that their analytical method was not sufficiently sensitive. Mueller and Vilter[346] note that improvements in their method were necessary in order to get good results. Using this improved method, they stated that individuals receiving deoxypyridoxine plus the vitamin B-complex-deficient diet excreted normal amounts of xanthurenic acid except after a tryptophan load when the xanthurenic acid excretion was markedly increased. However, no specific data were reported.

More specific data are provided in two subsequent reports. Glazer et al.[168] examined the effect of deoxypyridoxine on xanthurenic acid excretion in 12 patients with a variety of chronic diseases including rheumatoid arthritis, Parkinsonism, amyotrophic lateral sclerosis, syphilitic amyotrophy, paraplegia due to trauma, chronic lymphatic leukemia, and carcinomatosis. Eight received a diet deficient in the B-complex vitamins and three received the normal hospital diet. All received 200 mg deoxypyridoxine per day. The twelfth subject received the deficient diet but no deoxypyridoxine. Regardless of dietary vitamin content, deoxypyridoxine caused an 8- to 10-fold increase in xanthurenic acid excretion in the 24 hr after a 5 g load of L-tryptophan. There was no change in xanthurenic acid excretion without both the tryptophan and deoxypyridoxine.

Vilter et al.[530] did report more specific data on 34 subjects from a group of 50 who received 50 to 500 mg deoxypyridoxine per day for periods of 3 to 52 weeks. Attempts to correlate dose with response are prevented by their failure to specify the dosages and sampling schedules for the individual patients. However, only one patient showed a decline in xanthurenic acid excretion after deoxypyridoxine treatment, and for the 34 patients the average xanthurenic acid excretion after a tryptophan load was 25-fold higher after deoxypyridoxine treatment than before. Response to a tryptophan load usually returned to normal within 5 days after supplementation with pyridoxine administered orally or as a 1% ointment applied to the skin. There was no correlation between xanthurenic acid excretion and the type or severity of clinical symptoms. Supplementing the B-complex-deficient diet of four patients with 100 g vitamin-free casein per day had no detectable effect on response to a tryptophan load or clinical symptoms.

Based on results from comparing a single patient treated with 75 to 375 mg deoxypyridoxine per day plus an apparently normal diet with four patients treated with isonicotinic acid hydrazide, Price et al.[399] concluded that both treatments caused similar increases in the excretion of kynurenine and 3-hydroxykynurenine after a tryptophan load. However, the increases in the excretion of xanthurenic and kynurenic acids were noticeably greater after treatment with deoxypyridoxine than with isonicotinic acid hydrazide. They noted that normally, almost 50% of the tryptophan load is excreted as the pyridine metabolite of niacin and only about 2% as tryptophan metabolites. However, following treatment with isonicotinic acid hydrazide as much as 40% of the tryptophan appeared as kynurenine and its metabolites suggesting either increased utilization of the kynurenine pathway or interference with the normal flow through it. These data were reiterated by Price.[398]

Wayne et al.[538] also noted increased urinary excretion of xanthurenic acid after a tryptophan load in four patients who received 300 mg deoxypyridoxine plus about 2 mg pyridoxine per day for 9 weeks.

Hodges et al.[199] found the excretion of xanthurenic acid after a tryptophan load to be abnormal in two prisoners who received a $B_6$-deficient diet plus an oral dose of 100 mg deoxypyridoxine three times daily and 1000 mg omega-methyl pantothenic acid four times daily for 6 weeks.

An expanded 6-week study involved five men using the same protocol except that the dose of 100 mg deoxypyridoxine was increased to four times daily. During the control period, xanthurenic acid excretion was not affected by a tryptophan load. However, within 24 hr after initiating deoxypyridoxine treatment, xanthurenic acid excretion following a 10 g tryptophan load increased 20-fold to about 200 mg/day. The excretion of xanthurenic acid after a tryptophan load gradually increased even more as the experiment progressed. This effect disappeared within 24 hr after resuming pyridoxine supplementation. The abnormal response to a tryptophan load also disappeared when the subjects went into negative nitrogen balance. These authors, who used the procedure of Glazer et al.[168] to measure xanthurenic acid, noted that Hodges et al.[199] observed no change in xanthurenic acid excretion during negative nitrogen balance. Faber et al.[137] therefore suggested that perhaps these two methods detect different forms of xanthurenic acid.

Mueller and Iacono[344] reported that xanthurenic acid excretion following a tryptophan load was increased in 17 chronically ill subjects receiving a normal diet plus 300, 400, or 500 mg deoxypyridoxine per day.

Elder and Mengel[133] found that xanthurenic acid excretion was increased before and after a tryptophan load by treating a patient with a $B_6$-deficient diet plus 400 mg deoxypyridoxine per day. However, in contrast to the increased excretion of kynurenic acid reported by Price et al.[399] in this patient the excretion of kynurenic acid was unaltered.

Hansson[181] studied the excretion of 3-hydroxykynurenine and 3-hydroxyanthranilic acid after a tryptophan load in four healthy volunteers treated with 200 to 400 mg deoxypyridoxine plus a normal diet for 3 to 19 days. Excretion of 3-hydroxyanthranilic acid during the 24 hr after tryptophan load was increased about 3- to 5-fold during deoxypyridoxine treatment. Excretion of 3-hydroxyanthranilic acid after a load was also increased in eight out of ten trials. Treatment with pyridoxine restored the values to normal. In controls, the ratio of 3-hydroxyanthranilic acid to 3-hydroxykynurenine in the urine exceeded 1.0 at least once during the 4 hr following a tryptophan load. However, the ratios fell in three subjects after deoxypyridoxine treatment for 13 days or more. In one case in which deoxypyridoxine treatment was terminated 24 hours before the load test, the ratios remained normal. Similar treatment in a second subject failed to indicate that the normal ratio was due to a rapid turn to normal. However, we feel that conflicting data from only two subjects is not adequate to establish that point, particularly in view of our recent demonstration of the rapid excretion of deoxypyridoxine.

In three cases where deoxypyridoxine was given for only four days prior to the tryptophan load test, the ratio of 3-hydroxyanthranilic acid to 3-hydroxykynurenine did not deviate from controls even though the excretion of 3-hydroxykynurenine was markedly increased. The authors therefore concluded that increased excretion of 3-hydroxykynurenine might represent increased flow through the pathway as well as decreased activity of kynureninase. They felt that these observations were consistent with the conclusion of Knox[238a] that increased tryptophan pyrrolase activity is always a necessary condition to the accumulation of tryptophan metabolites in pyridoxine deficiency. They then discussed the possible pitfalls of the normal tryptophan load test and suggested a modified procedure.

Rose and Toseland[419] fed an adult male receiving a normal diet 300 mg deoxypyridoxine per day for 14 days. When first tested after 7 days, the urinary excretion of xanthurenic acid, 3-hydroxykynurenic, 3-hydroxyanthranilic acid, quinolinic acid, and kynurenine was increased 3- to 6-fold after a tryptophan load. The authors commented that the increases in 3-hydroxyanthranilic acid and quinolinic acid were much smaller. While the total micromoles excreted was less for these two metabolites than for the others, the percentage increase compared with control values was not too different for the various metabolites. The ratio between 3-hydroxykynurenine and 3-hydroxyanthranilic acid did rise from 2.3 before treatment to 6.25 after 14 days on deoxypyridoxine. Excretion of kynurenine, 3-hydroxykynurenic, 3-hydroxyanthranilic acid, and quinolinic acid following a tryptophan load returned to normal within 7 days after starting pyridoxine supplementation. However, xanthurenic acid excretion was still abnormal after 7 days of pyridoxine supplementation but had returned to normal by 14 days.

While deoxypyridoxine produces marked increases in xanthurenic excretion in adults, Sartori and Peressini[436] observed no effect in newborn infants. The excretion of xanthurenic acid in the urine was less than 1 mg in the 24 hr following a tryptophan load in 11 infants from 1.5 to 4 months old who had received intramuscular injections of 50 mg (about 10 mg/kg) of deoxypyridoxine daily for five to ten days prior to the tryptophan load. Nine older children (4 to 9 years) who received the same dose (50 mg/day or about 2.5 mg/kg) had xanthurenic acid excretions of 12 to 40 mg in the 24 hours after the tryptophan load. In five adults receiving 100 mg/day (about 1.5 mg/kg) xanthurenic acid, excretion ranged from 50 to 160 mg in the 24 hours following a tryptophan load. There is no mention that $B_6$ intake was restricted. Neither is there any mention of clinical symptoms. The authors concluded that metabolism in infants does not favor the breakdown of significant quantities of tryptophan thus minimizing the appearance of metabolites in the urine independently of the presence or absence of vitamin $B_6$.

Mueller and Vilter[346] reported that excretion of *N*-methylnicotinamide in patients receiving deoxypyridoxine (60 to 150 mg/day) plus a B-complex-deficient diet was in the low normal range. However, they presented no control values for comparison. Data on four subjects who received 200 mg deoxypyridoxine per day intramuscularly in conjunction with a B-complex-deficient diet and three who received deoxypyridoxine plus a normal diet are presented in a subsequent paper.[169] Excretion of *N*-methylnicotinamide was increased by a tryptophan load. However, the response to the load was not altered in a consistent fashion by any of the treatments. These seven patients presumably are the ones referred to in a later paper.[530] Since nicotinamide adenine dinucleotide did correct certain clinical symptoms after nicotinamide alone had failed, the authors suggested that there may have been interference in the conversion of niacin and perhaps other vitamins to their active forms.

Since pyridone excretion increased dramatically following a nicotinamide supplement in a patient receiving deoxypyridoxine plus an apparently normal diet, Price et al.[399] concluded that deoxypyridoxine does not interfere in the conversion of niacin to pyridone. Under normal conditions only about 2% of a tryptophan load appeared as urinary metabolites.

### C. Aminotransferases

Mueller and Iacono[344] reported that serum aspartate aminotransferase activity decreased in thirteen chronically ill patients treated with a normal diet plus 300, 400, or 500 mg deoxypyridoxine per day.

Gailani et al.[154] noted that only one of eleven cancer patients showed a decline in serum aspartate aminotransferase activity while receiving the $B_6$-deficient diet alone. However, all nine patients showed a decline in activity when deoxypyridoxine treatment was combined with the $B_6$-deficient diet. They reported that leukocyte aspartate aminotransferase activity roughly paralleled that of serum but no specific data were given for leukocytes. It seems paradoxical that leukocyte pyridoxal phosphate concentrations could decline to undetectable levels without affecting aminotransferase activity in patients receiving the $B_6$-deficient diet.

### D. Porphyria

Excretion of porphobilinogen and delta-aminolevulinic acid were decreased while a patient with acute intermittent porphyria was treated with a $B_6$-deficient diet plus 400 mg deoxypyridoxine per day.[133]

## VI. HORMONES

### A. Adrenals

In a report involving 50 subjects on widely varying doses of deoxypyridoxine, Vilter et al.,[530] stated that determination of serum sodium, potassium, and 17-ketosteroids, the Kepler-Power water test, and the response of eosinophils, polymorphonuclear neutrophils, and lymphocytes to the intramuscular administration of 40 mg of adrenocorticotropic hormone failed to detect any alteration in adrenal function. Unfortunately, it is not clear whether adrenal function was examined in all 50 subjects.

Hodges et al.[199] did observe an abnormal eosinopenic response to ACTH in two prisoners who received a $B_6$-deficient diet plus an oral dose of 100 mg deoxypyridoxine three times daily and 1000 mg omega-methyl pantothenic acid four times daily for 6 weeks.

### B. Thyroid

Smejkal and Smejkalova[462] studied the interactions between thyroxine, and deoxypyridoxine and aspartate aminotransferase from human leukocytes. Thyroxine caused a 55 to 60% decrease in the aminotransferase activity. This could be almost completely offset by an equimolar amount of pyridoxal phosphate or pyridoxine. Pyridoxamine was slightly less effective. Deoxypyridoxine at the same concentration restored activity to 72% of control values.

## VII. LIPIDS

Mueller and Vilter[346] reported normal blood cholesterol at the height of the deficiency state in three patients receiving deoxypyridoxine (60 to 150 mg/day) plus a B-complex-deficient diet. Vilter et al.[530] also noted that blood lipids, cholesterol, and cholesterol esters were not altered by deoxypyridoxine treatment. Neither was there any change in the response to a high fat, high cholesterol meal. Mueller et al.[344a] again found no consistent change in the serum cholesterol or phospholipids of seven subjects receiving 300 mg deoxypyridoxine per day in conjunction with a normal diet. Three of the seven persons did show a decrease in serum polyunsaturated fatty acids.

However, Mueller and Iacono[344] reported that cholesterol tended to increase in the plasma and erythrocytes of 17 chronically ill patients treated with a normal diet plus 300, 400, or 500 mg deoxypyridoxine per day. A fall in tetraenoic and a rise in trienoic acids was observed. The changes in lipid metabolism were more pronounced in the erythrocytes than in the plasma. The trends were not readily reversed by pyridoxine supplementation. The authors concluded that these results provide the first evidence for the involvement of vitamin $B_6$ in the metabolism of fatty acids in humans.

Gailani et al.[154] reported a 25 to 50% decrease in serum cholesterol in patients receiving a $B_6$-deficient diet with or without a deoxypyridoxine supplement. Supplementation with choline or pyridoxine did not increase the cholesterol level. Return to a normal diet restored normal cholesterol content within 1 to 2 weeks.

## VIII. CARBOHYDRATES

Vilter et al.[530] failed to find any alteration in blood glucose to oral or intravenous glucose tolerance tests as a result of deoxypyridoxine treatment.

Hodges et al.[199] found no abnormalities in urinary sugar or insulin response in two prisoners who received a $B_6$-deficient diet plus an oral dose of 100 mg deoxypyridoxine three times daily and 1000 mg omega-methyl pantothenic acid four times daily. In a further 7-week study of five men using the same protocol except that the 100 mg dose of deoxypyridoxine was increased to four times daily, insulin tolerance tests showed a slightly hypoglycemic response.

## IX. OXALIC ACID

Faber et al.[137] examined oxalic acid excretion in five men who received a diet deficient in $B_6$ and pantothenic acid plus oral doses of 100 mg deoxypyridoxine and 1000 mg omega-methyl pantothenic acid four times per day for 6 weeks. During the deficiency, oxalic acid excretion gradually increased 2- to 3-fold. Excretion was increased about 25 to 50% by a tryptophan load both before and during treatment with deoxypyridoxine. Elder and Mengel[133] reported that oxalate excretion tended to increase in a patient treated with a $B_6$-deficient diet plus 400 mg deoxypyridoxine per day. However, because of marked fluctuations the authors questioned the significance of this observation.

During administration of a $B_6$-deficient diet, oxalate crystals were observed in the urine of 3 out of 17 patients.[154] The authors did not specify whether any of these persons were also receiving deoxypyridoxine.

## X. DOWN'S SYNDROME

Studies by McCoy and others of the effects of deoxypyridoxine in Down's syndrome provided the major evidence for abnormalities of $B_6$ metabolism in this condition. McCoy and Chung[287] examined the effects of deoxypyridoxine (200 mg/day orally for 14 days) on the excretion of several tryptophan metabolites in subjects with Down's syndrome receiving a normal diet plus 0.5 mg pyridoxine per day. There was no difference in the excretion of xanthurenic acid, 3-hydroxykynurenine, kynurenic acid, or kynurenine between the Down's syndrome and control groups before or after a tryptophan load prior to deoxypyridoxine treatment. Excretion of 3-hydroxykynurenine after a tryptophan load was significantly ($p<0.05$) increased in Down's syndrome compared with controls when first tested after seven days of deoxypyridoxine administration. Tests after two weeks showed that the difference persisted both before and after a tryptophan load. Xanthurenic acid excretion also tended to be increased in the group with Down's syndrome but the difference was statistically significant ($p<0.05$) only after a tryptophan load after two weeks of deoxypyridoxine treatment. There were no statistically significant differences between the two groups in the excretion of kynurenic acid or kynurenine.

McCoy et al.[286] also measured oxalic acid excretion in these subjects. There was no difference in baseline excretion. Following a tryptophan load, the subjects with Down's syndrome excreted significantly ($p<0.01$) less oxalic acid than the mentally retarded controls, but after treatment with deoxypyridoxine for 7 to 14 days, the excretion of oxalic acid was significantly higher than in the controls either with or without a tryptophan load. Baseline excretion of xanthurenic acid between the two groups was not significantly different either before or after a tryptophan load. However, after 14 days of treatment with deoxypyridoxine, xanthurenic acid excretion in the Down's syndrome group was significantly higher than in the controls both with and without a tryptophan load.

McCoy and Wehrle[290] examined the effect of deoxypyridoxine excretion (200 mg/day for 10 days) on taurine excretion in four female children (9 to 15 years) with Down's syndrome receiving a normal diet. Taurine excretion before treatment was significantly ($p<0.01$) lower in subjects with Down's syndrome compared with mentally retarded controls matched for age, sex, and weight. In the first 24 hours after starting deoxypyridoxine treatment, taurine excretion in both groups increased ($p<0.01$). By day 2, the excretion dropped below baseline levels ($p<0.01$). By day 10, excretion in both groups was 50 to 100% above baseline ($p<0.05$). Administration of pyridoxine (100 mg i.m.) had no effect. In this case, while persons with Down's syndrome initially had significantly decreased taurine excretion, their response to deoxypyridoxine administration did not differ from that of the mentally retarded controls. Also, the fact that taurine excretion increased following deoxypyridoxine administration differs from the decreased taurine excretion usually observed with a dietary $B_6$ deficiency.

McCoy and England[289] studied the effect of deoxypyridoxine (200 mg/day for 11 days) on 4-pyridoxic acid excretion in eight children (10 to 14 years) with Down's syndrome receiving a normal diet. Pyridoxic acid excretion was not significantly different from that of mentally retarded controls before deoxypyridoxine treatment. Immediately after initiation of deoxypyridoxine treatment, pyridoxic acid excretion increased and then gradually declined. However, in controls, excretion remained well above baseline levels while in the Down's syndrome group it declined below baseline levels. This difference was statistically significant ($p<0.01$). During the 24 hr following an injection of pyridoxine (100 mg i.m.) on day 11, the Down's syndrome group excreted significantly ($p<0.05$) more pyridoxic acid than the controls. The authors concluded that persons with Down's syndrome may have a normal easily dissociable pool and an abnormally small, slowly dissociable pool of $B_6$.

McCoy, Colombini, and Ebadi[288] reviewed the results described above and added one new table of data on excretion of 4-pyridoxic acid in the 24 hr following intramuscular injection of pyridoxal. This was done to determine whether the significantly increased excretion of pyridoxic acid after pyridoxine administration in Down's syndrome[289] was due to increased activity of aldehyde oxidase. Pyridoxic acid excretion in Down's syndrome after the pyridoxal load tended to be lower than in controls but the difference was not statistically significant. This suggested that aldehyde oxidase activity was not elevated in Down's syndrome. The authors cited unpublished results that in liver and brain the main route of metabolism is pyridoxine → pyridoxine phosphate → pyridoxal phosphate, whereas in muscle, the route is pyridoxine → pyridoxal → pyridoxal phosphate. They concluded that the increased pyridoxic acid formation in Down's syndrome could best be explained by increased activity of pyridoxal phosphate phosphatase or pyridoxine oxidase.

Chapter 8

# METABOLISM OF 4′- AND 5′-DEOXYPYRIDOXINE IN VIVO

Because of the many differences between the effects of deoxypyridoxine compared with those of a dietary $B_6$ deficiency, we decided to examine the metabolism of deoxypyridoxine in vivo.[92] Using tritiated deoxypyridoxine, we found that rats excreted 80 to 95% of the injected material within 7.5 hours. Thin layer chromatography of tissue homogenates showed that only the gastrointestinal tract and kidney could contain deoxypyridoxine phosphate:pyridoxal phosphate ratios comparable to those required to produce enzyme inhibition in vitro. Chromatography of the urine revealed a new compound which accounted for about half the dose and was eventually shown to be 4′-deoxypyridoxine-3-sulfate.

Similar studies in humans demonstrated that 50% of a single oral dose (3 to 7.5 mg/kg) appeared in the urine within 6 hr. 4′-Deoxy-5-pyridoxic acid accounted for 50 to 100% of the excreted material. This was the first demonstration of oxidation of the 5′-position in a species other than microorganisms. However, we have not yet conclusively eliminated the intestinal microflora as a source of the 5-acid.

In view of the marked difference between the rat and man in the metabolism of deoxypyridoxine, we examined several other species. The dog, cat, and chicken resemble the rat in producing 4′-deoxypyridoxine-3-sulfate as the major metabolite while the guinea pig, rabbit, pig, and chimpanzee resemble the human in their metabolism of 4′-deoxypyridoxine.[91]

We found that the major urinary metabolite of 5′-deoxypyridoxine in the rat was also the 3-sulfate. Shane and Snell[446] erred in identifying the metabolite as the 4′-sulfate because they were unaware that the 2,4-substituents on the pyridine ring would inhibit arylsulfatase and that addition of sulfate at the 3-position would inhibit manganese dioxide oxidation at the 4′-position. Therefore, they interpreted lack of hydrolysis by arylsulfatase and lack of oxidation by manganese dioxide to indicate substitution at the 4′-position.

Considering that the rat metabolized both 4′- and 5′-deoxypyridoxines in the same way, we were curious to see what would happen in those species which oxidized the 5′-position of 4′-deoxypyridoxine. Since oxidation of the 4′-position is a normal route of pyridoxine metabolism, we expected that the major metabolite of 5′-deoxypyridoxine in these species would be 5′-deoxy-4-pyridoxic acid. While some of this compound was detected in the urine of both the rabbit and human, it was not the sole or major metabolite. In the human, 45% of the excreted material appeared as 5′-deoxypyridoxine-3-sulfate and 30% as the glucuronide. 5′-Deoxy-4-pyridoxic acid accounted for less than 5% of the urinary metabolites. The remainder was a mixture of three unidentified compounds with a free phenol group. Both species also produced another metabolite modified in the 4-position which we have not identified. The fact that in the human, 4′-deoxypyridoxine is not sulfated whereas 5′-deoxypyridoxine is, suggests that they are entering different metabolic pathways. This agrees with the results obtain in microorganisms which also indicated that 4′- and 5′-deoxypyridoxines act by different mechanisms.

McCormick and Snell[285] commented that as of 1961, 5′-deoxypyridoxal had not been tested as an antagonist in animals. Since then Rosen et al.[421] have provided some preliminary data which were discussed in Chapter 3 and we are currently initiating further studies. We have found that 5′-deoxypyridoxal is so reactive that it is rapidly converted to other forms when incorporated in a diet. Therefore, it must be administered by other routes.

Chapter 9

# DISCUSSION

Probably the most obvious fact which emerges from this review is that the effects of 4′-deoxypyridoxine are highly dependent on the experimental conditions, particularly size and timing of the dose, and the amount of vitamin $B_6$ and other nutrients available to both the treated animals and the controls. While such dependence is not unusual, the frequent failure of authors to specify details, such as the $B_6$ content of the diet, suggests that these interactions have often been overlooked in the case of deoxypyridoxine, probably because the fragmented nature of the literature involving this compound has made it very difficult to consolidate the available data. We hope that the preceding chapters have clarified some of these interactions and will permit more effective use of deoxypyridoxine in future studies.

The most well-documented clinical effects of deoxypyridoxine involve the skin, nervous system, leukocytes, and thymus. No generalized disturbance of amino acid or protein metabolism has been demonstrated. If deoxypyridoxine is exerting its major effect through direct interference with amino acid metabolism, it must interfere only with a limited number of enzymes which have not yet been identified. The evidence suggests that glycine metabolism is one area worthy of further study. The fact that glycine is involved in nucleic acid synthesis and that rapidly dividing tissues such as skin are among the most sensitive to deoxypyridoxine leads to the possibility of interference in nucleic acid metabolism. The marked effects on skin and body weight suggest problems with lipid and/or energy metabolism. This possibility is further strengthened by the effect of convulsive doses of deoxypyridoxine on body temperature and brain glucose metabolism.

Since the intestinal flora contribute an unknown amount of vitamin $B_6$, we have considered the possibility that one of the effects of deoxypyridoxine might be to inhibit the intestinal flora. Therefore, we were very interested to discover that Iwanowaska[217] cited Webb as stating that deoxypyridoxine inhibited the microbial production of $B_6$ in animals and man. However, after locating the reference, we found that Webb's statement was, "Deoxypyridoxal has since been found to inhibit certain microorganisms and to produce symptoms of vitamin $B_6$ deficiency in animals, including man." The context of the statement is such that we think Webb was simply listing three separate effects of deoxypyridoxine without meaning to imply any connection between inhibition of microorganisms and production of $B_6$ deficiency symptoms in animals or man. There may well be such a connection, but there is no direct evidence to support it at this time.

While there is no doubt that deoxypyridoxine causes a number of physiological effects, it is clear that these effects do not result from a generalized decrease in pyridoxal phosphate concentrations nor from a general decrease in the activity of pyridoxal phosphate-requiring enzymes. In fact, some of the parameters, such as cysteine desulfhydrase activity, which are most susceptible to a dietary $B_6$ deficiency are unaffected by deoxypyridoxine. Therefore, the original hypothesis that deoxypyridoxine is phosphorylated and subsequently inhibits enzymes requiring pyridoxal phosphate as a cofactor appears to be an over simplification. In addition, the clinical effects of a simple dietary $B_6$ deficiency have not yet been directly attributed to specific enzyme systems. As we attempted to develop explanations for the effects of deoxypyridoxine, several occurrences gradually led us to the conclusion that perhaps our view of the action of vitamin $B_6$ itself was also oversimplified. One of our major concerns has been the fact that deoxypyridoxine produce severe clinical symptoms without depressing pyridoxal phosphate concentrations in most tissues. Yet, pyridoxal phosphate concentrations have been accepted as one of the best indicators of $B_6$ status.

Probably the simplest mechanism for explaining an increase in pyridoxal phosphate concentrations at a time when pyridoxic acid excretion exceeds intake is a shift in the metabolic equilibrium between the other $B_6$ compounds, particularly pyridoxamine phosphate or Shiff-base complexes. It may be that variations in the availability of vitamin $B_6$ cause shifts in these equilibria which in turn causes the response to deoxypyridoxine to vary with the $B_6$ status. Perhaps under certain conditions, the shift caused by deoxypyridoxine is beneficial, thus explaining the reports of improved growth. One of the limitations of most studies of vitamin $B_6$ metabolism is that data is obtained on only one or two of the seven major forms of vitamin $B_6$. Our work with deoxypyridoxine plus the increasing number of clinical conditions which are associated with decreased blood concentrations of pyridoxal phosphate, in spite of apparently normal $B_6$ intakes, have led us to the conclusion that there is a need for more thorough study of the kinetics of the interconversions between the various $B_6$ vitamers. A decrease in pyridoxal phosphate could be caused by reductions in $B_6$ intake, absorption, kinase activity, or oxidase activity, or by increases in phosphatase activity, aldehyde oxidase activity, or formation of Schiff-bases or pyridoxamine. There may also be changes in the number of protein binding sites for pyridoxal and/or pyridoxal phosphate.

Among the key enzymes in the interconversion of the $B_6$ vitamers are the phosphatases. We have been interested in this relationship for some time because of the possibility that the increased leukocyte alkaline phosphatase activity observed in Down's syndrome may have some bearing on the changes we have detected in pyridoxal phosphate concentrations.[295] Andersen et al.[13] showed that pyridoxal phosphate concentrations in erythrocytes increased several-fold when the incubation medium contained sufficient phosphate to inhibit alkaline phosphatase. This suggested that alkaline phosphatase was an important factor in controlling pyridoxal phosphate concentrations. When our phosphatase inhibitor[95] failed to produce similar results in leukocytes,[90a] we tested the inhibitor on erythrocytes and obtained results comparable to Andersen's. Thus, phosphatase apparently does not have the same role in all tissues. Since we found that phosphatase inhibition almost doubled the pyridoxal phosphate concentration of the erythrocytes in 30 min, even without any added $B_6$ in the incubation medium, there would have to be very rapid cycling between pyridoxal and pyridoxal phosphate if the increase in pyridoxal phosphate were due solely to inhibition of hydrolysis of pyridoxal phosphate. We suspected that phosphatase inhibition might also permit accumulation of adenosine triphosphate. Tests confirmed that the concentration of adenosine triphosphate in the erythrocytes doubled after phosphatase inhibition. The concentration of adenosine triphosphate in erythrocytes is so much higher than the concentration of pyridoxal phosphate that only a small fraction of the increase in adenosine triphosphate would have been required to produce the increase in pyridoxal phosphate. Therefore, the increase in pyridoxal phosphate might have resulted from increased synthesis rather than decreased hydrolysis. In either case, the erythrocytes carry a pool of pyridoxal at least equal to the concentration of pyridoxal phosphate. Recent evidence that the balance between these two forms might affect oxygen affinity was discussed in Chapter 3.

It is well documented that pyridoxal and pyridoxal phosphate can bind to various proteins. Webb[539] has suggested that the large portion of $B_6$ bound to nonenzyme protein may obscure changes in the $B_6$ content of various enzymes, thus providing a possible explanation for the changes produced by deoxypyridoxine with little change in tissue $B_6$ concentrations. In addition to serving as a possible reservoir of $B_6$, we feel that perhaps, as was discussed in the case of glucocorticoid receptors and in the case of erythrocytes, $B_6$ bound to nonenzyme proteins may exert a regulatory effect independent of an enzyme cofactor role. Such a role for $B_6$ would open new mechanisms to explain deoxypyridoxine action.

Our experiences with deoxypyridoxine have heightened our awareness of the importance of experimental conditions and led us to suspect that interpretation of tracer studies may be another area of oversimplification. Actually, most cells in the body are rarely exposed to significant amounts of pyridoxine. In most cases, the kinetics of the absorption and metabolism of pyridoxine are quite different from those of pyridoxal and other $B_6$ derivatives. Gaut and Solomon[160] candidly stated that labeled pyridoxine was used as their tracer because it was available. Spector and Greenwald[469] selected pyridoxine phosphate because pyridoxal phosphate readily binds to protein. As was discussed above, we feel that such binding may have some physiological significance. Protein binding would also be expected to alter the kinetics of exchange and uptake. In fact, Andersen et al.[13] noted that inclusion of plasma in their incubation medium increased the release of pyridoxal from erythrocytes compared with cells incubated in physiological saline. Therefore, much of the kinetic data based on pyridoxine may not accurately reflect the in vivo situation in which pyridoxal and pyridoxamine derivatives are predominant. More detailed study of the nature and kinetics of $B_6$-protein interactions and of the interchange between all of the $B_6$ forms should enhance our understanding of this vitamin and provide additional insights into the action of antagonists. Ongoing improvements in liquid chromatography and in the synthesis of labeled $B_6$ compounds should facilitate the conduct of such studies.

In addition to these broad metabolic effects, there are several other aspects of deoxypyridoxine action worthy of further study. We have been intrigued by the fact that some effects such as inhibition of transport systems and increased excretion of xanthurenic acid can be produced and corrected within a few hours while other effects such as the interactions with pregnancy may persist for several weeks after deoxypyridoxine is withdrawn, and in some cases even after normal $B_6$ intake is restored. The great effect of deoxypyridoxine on the immune system, particularly the thymus, suggests that deoxypyridoxine is interfering with a very important process. Further clarification of that action might be useful in understanding the immune system and lead to improvements in the management of a variety of problems possibly including organ transplantation and even cancer.

In using inhibitors or antagonists to study metabolism, the investigator must be constantly aware that these compounds may be influencing results by mechanisms other than the anticipated ones. Because of this potential complication, some researchers eschew the use of antagonists entirely. However, our experience has led us to conclude that if used in carefully designed experiments, the deoxypyridoxines should continue to provide useful information about the metabolism of vitamin $B_6$.

# REFERENCES


1. Survey by the staff of the U.S. Cancer Chemotherapy National Service Center. I. Compilation of Antitumor Data from the Literature. II. Results of Antitumor Screening Tests under the Auspices of the Center, Bethesda, Md., 1958.
2. Summary of results of antitumor tests on 200 selected compounds, *Ann. N.Y. Acad. Sci.*, 76, 965, 1958.
3. **Adron, J., Knox, D., Cowey, C., and Ball, G.**, Studies on the nutrition of marine flatfish. The pyridoxine requirement of turbot *(Scophthalmus maximus)*, *Br. J. Nutr.*, 40, 261, 1978; *Chem. Abstr.*, 90:21144.
4. **Agolini, G. and Caliari, W.**, Substances with antagonistic activity to vitamin $B_6$, *Atti. Soc. Lombarda Sci. Med. Biol.*, 13, 93, 1958; *Chem. Abstr.*, 52:17526a.
5. **Ahrens, M.**, Proton transfer kinetics of 5-deoxypyridoxal, *Biochim. Biophys. Acta*, 320, 86, 1973, *Chem. Abstr.*, 79:112598.
6. **Ahrens, M., Maass, G., Schuster, P., and Winkler, H.**, Kinetic behavior of vitamin $B_6$ compounds. Hydration and proton transfer, *FEBS. Lett.*, 5, 327, 1969, *Chem. Abstr.*, 72:86233.
7. **Ahrens, M., Maass, G., Schuster, P., and Winkler, H.**, Kinetic study of the hydration mechanism of vitamin $B_6$ and related compounds, *J. Am. Chem. Soc.*, 92, 6134, 1970.
8. **Aikawa, J., Reardon, J., and Harms, D.**, Effects of pyridoxine and deoxypyridoxine on magnesium metabolism in the rabbit, *Proc. Soc. Exp. Biol. Med.*, 104, 461, 1960; *Chem. Abstr.*, 54:21481i.
9. **Akov, S. and Guggenheim, K.**, Antimetabolites in the nutrition of *Aedes aegypti* L. larvae. Pyridoxine antagonists, *Comp. Biochem. Physiol.*, 9, 61, 1963; *Chem. Abstr.*, 59:10520c.
10. **Alexander, N. and Greenberg, D. M.**, The biosynthesis of serine, *J. Biol. Chem.*, 214, 821, 1955; *Chem. Abstr.*, 49:11051h.
11. **Amer, M., Abdel-Daim, M., and Abdel-Tawab, G.**, Studies with tryptophan metabolites in vitro. Kynurenine metabolism in liver homogenates of normal and *Schistosoma mansoni*-infested mice, *Biochem. J.*, 104, 656, 1967. *Chem. Abstr.*, 67:52124.
12. **Amer, M., Abdel-Daim, M., and Abdel-Tawab, G.**, Studies with tryptophan metabolite in vitro. II. Effect of tartar emetic on kynurenine metabolism by normal mouse liver, *Biochem. Pharmacol.*, 16, 1227, 1967; *Chem. Abstr.*, 67:80878.
13. **Anderson, B. B., Fulford-Jones C. E., Child, J. A., Beard, M. E. J., and Bateman, C. J. T.**, Conversion of vitamin $B_6$ compounds to active forms in the red blood cell, *J. Clin. Invest.*, 50, 1901, 1971; *Biol. Abstr.*, 53:12049.
14. **Anglesio, E., Hirschberg, E., and Gellhorn, A.**, Effects of carcinostatic and other biologically active compounds in the normal rat. Evaluation as a system for cancer chemotherapy screening, *Cancer Res.*, 18, 113, 1958; *Chem. Abstr.*, 52:20660i.
15. **Argoudelis, C. J. and Kummerow, F. A.**, Synthesis of pyridoxine labelled with tritium at specific positions, *Biochemistry*, 5, 1, 1966; *Chem. Abstr.*, 64:5038.

15a. **Awapara, J.**, Effect of protein depletion on the transaminating activities of some rat organs, *J. Biol. Chem.*, 200, 537, 1953; *Chem. Abstr.*, 47:8201d.

16. **Axelrod, A.**, Vitamins and antibodies, *Nutrition Rev.*, 10, 353, 1952.
17. **Axelrod, A.**, The role of nutritional factors in the antibody responses of the anamnestic process, *Am. J. Clin. Nutr.*, 6, 119, 1958; *Chem. Abstr.*, 53:4464g.
18. **Axelrod, A. E.**, Immune process in vitamin deficiency states, *Am. J. Clin. Nutr.*, 24, 265, 1971; *Chem. Abstr.*, 74:84603.
19. **Axelrod, A., Fisher, B., Fisher, E., Lee, Y., and Walsh, P.**, Effect of a pyridoxine deficiency on skin grafts in the rat, *Science*, 127, 1388, 1958; *Chem. Abstr.*, 52:18803f.
20. **Axelrod, A., Hoppers, S., and Long, D.**, Effects of pyridoxine deficiency upon circulating antibody formation and skin hypersensitivity reactions to diphtheria toxoid in guinea pigs, *J. Nutr.*, 74, 58, 1961. *Chem. Abstr.*, 57:11628b.
21. **Axelrod, A., Lowe, M., and Macins, D.**, Effect of ribonucleic acid extracts upon viability of skin homografts in the rat, *Proc. Soc. Exp. Biol. Med.*, 108, 549, 1961; *Chem. Abstr.*, 56:12269h.
22. **Axelrod, A. and Trakatellis, A.**, Relationship of pyridoxine to immunological phenomena, *Vitam. Horm. (N.Y.)*, 22, 591, 1964; *Biol. Abstr.*, 47:23196.
23. **Axelrod, A. and Trakatellis, A.**, Induction of tolerance to skin homografts by administering splenic cells to pyridoxine-deficient mice, *Proc. Soc. Exp. Biol. Med.*, 116, 206, 1964; *Chem. Abstr.*, 61:6201a.
24. **Axelrod, A., Trakatellis, A., Block, H., and Stineberg, W.**, Effect of pyridoxine deficiency on delayed hypersensitivity in guinea pigs, *J. Nutr.*, 79, 161, 1963; *Chem. Abstr.*, 60:5943.
25. **Ayling, J. and Snell, E.**, Relation of structure to activity of pyridoxal analog as substrates for pyridoxamine pyruvate transaminase, *Biochemistry*, 7, 1626, 1968; *Chem. Abstr.*, 69:258.

26. **Bader, J. P. and Morgan, H. R.,** A comparison of cytopathology caused by myxoviruses. II. Induced resistance of cells to the cytopathic action of myxoviruses, *J. Immunol.,* 87, 90, 1961; *Chem. Abstr.,* 56:13414f.
27. **Bain, J. A. and Williams, H. L.,** Concentrations of $B_6$ vitamers in tissues and tissue fluids, in *Inhibition in the Nervous System and Gamma-Aminobutyric Acid,* Roberts, E., Baxter, C. F., van Harrevald, A., Wiersma, C. A. G., Ross, W. R., and Killam, K. F., Eds., Permagon Press, Elmsford, N.Y., 1959, 275; *Chem. Abstr.,* 58:12927a.
28. **Balyakina, M., Rubtsov, I., Zhdanovich, E., and Preobrazhenskii, N.,** Synthesis of 2,4-dimethyl-3-hydroxy-5-hydroxymethylpyridine, *Zh. Prikl. Khim. (Leningrad),* 39, 916, 1966; *Chem. Abstr.,* 65:2216e.
29. **Bardhan, J. C.,** Chemistry of 1,3-dicarbonyl compounds. I. The mechanism of the cyanoacetamide and cyanoacetic ester condensations, *J. Chem. Soc.,* 2223, 1929; *Chem. Abstr.,* 24:617.
30. **Bayoumi, R., Kirwan, J., and Smith, W.,** Effects of dietary vitamin $B_6$ deficiency and 4-deoxypyridoxine on alpha-aminobutyric acid metabolism in rat brain, *J. Neurochem.,* 19, 569, 1972; *Chem. Abstr.,* 76:152398.
31. **Bazhulina, N., Lomakin, A., Morozov, Y., Savin, F., Cherkashina, L., Karpeiskii, M., and Florentev, V.,** Absorption properties of pyridoxal, pyridoxal-5-phosphate and 5-deoxypyridoxal, *Mol. Biol.,* 4, 899, 1970; *Chem. Abstr.,* 74:28153.
32. **Beare, J., Beaton, J., and McHenry, E.,** Studies on vitamin $B_6$. III. Carcass composition of the vitamin $B_6$ deficient rat, *J. Biol. Chem.,* 202, 589, 1953; *Chem. Abstr.,* 47:9452f.
33. **Beare, J., Beaton, J., Smith, F., and McHenry, E.,** Conjoint effects of pyridoxine deficiency and anterior pituitary growth hormone administration in the rat, *Endocrinology,* 52, 396, 1953; *Chem. Abstr.,* 49:2211g.
34. **Beare, J., Beaton, J., Smith, F., White, J., and McHenry, E.,** Effects of adrenalectomy and of adrenal cortical extract in pyridoxine deprived and supplemented rats, *Am. J. Physiol.,* 175, 21, 1953; *Chem. Abstr.,* 48:2872i.
35. **Beaton, J., Beare, J., Beaton, G., and McHenry, E.,** Studies on vitamin $B_6$. IV. The effect of vitamin $B_6$ on protein synthesis and maintenance in the rat, *J. Biol. Chem.,* 204, 715, 1953; *Chem. Abstr.,* 48:832d.
36. **Beaton, J., Beare, J., White, J., and McHenry, E.,** Studies on vitamin $B_6$. I. Biochemical changes in vitamin $B_6$ deficiency in rats, *J. Biol. Chem.,* 200, 715, 1953; *Chem. Abstr.,* 47:8856e.
37. **Beaton, J. and Goodwin, M.,** Renal glutaminase activities in vitamin $B_6$ deficient rats, *J. Biol. Chem.,* 212, 195, 1955; *Chem. Abstr.,* 49:4824h.
38. **Beaton, J. R., Goodwin, M. E., Ozawa, G., and McHenry, E. W.,** The effects of vitamin $B_6$ and deoxypyridoxine in urea formation in rats, *Arch. Biochem. Biophys.,* 51, 94, 1954; *Chem. Abstr.,* 48:11590a.
39. **Beaton, G. and McHenry, E.,** Vitamin $B_6$ levels in rat tissues, *Brit. J. Nutr.,* 7, 357, 1953; *Chem. Abstr.,* 48:3498d.
40. **Beaton, J., Smith, F., and McHenry, E.,** Studies on vitamin $B_6$. II. The metabolism of amino acids in the vitamin $B_6$ deficient rat, *J. Biol. Chem.,* 201, 587, 1953; *Chem. Abstr.,* 47:8856f.
41. **Beiler, J. M. and Martin J.,** Inhibition of the action of tyrosine decarboxylase by phosphorylated desoxypyridoxine, *J. Biol. Chem.,* 169, 345, 1947; *Chem. Abstr.,* 41:6293a.
42. **Benesch, R., Benesch, R. E., Edalji, R., and Suzuki, T.,** 5-Deoxypyridoxal as a potential anti-sickling agent, *Proc. Natl. Acad. Sci. U.S.A.,* 74, 1721, 1977; *Chem. Abstr.,* 87:62414.
43. **Benesch, R. E., Benesch, R., Renthal, R. D., and Maeda, N.,** Affinity labeling of the polyphosphate binding site of hemoglobin, *Biochemistry,* 11, 3576, 1972; *Chem. Abstr.,* 77:148805.
44. **Benesch, R. E., Yung, S., Suzuki, T., Bauer, C., and Benesch, R.,** Pyridoxal compounds as specific reagents for the alpha and beta *N*-termini of hemoglobin, *Proc. Natl. Acad. Sci. U.S.A.,* 70, 2595, 1973; *Chem. Abstr.,* 79:133799v.
45. **Bennett, R., Burger, A., and Umbreit, W.,** Vitamin $B_6$ phosphonic acids, *J. Med. Pharm. Chem.,* 1, 213, 1959; *Chem. Abstr.,* 54:5642i.
46. **Bento, R. S. and Felipe Da Silva, J. A.,** Experimental pyridoxine deficiency and skin homografts, *Arq. Port. Bioquim.,* 5, 317, 1962; *Chem. Abstr.,* 58:7181g.
47. **Bergamini, A., Bonati, B., and Freddi, S.,** Amino acid blood pattern in rabbits treated with deoxypyridoxine, *Acta Vitaminol.,* 11, 63, 1957; *Chem. Abstr.,* 51:16914g.
48. **Bhagavan, H. and Stewart, C.,** Tissue pyridoxal 5-phosphate content and avoidance behavior in 4-deoxypyridoxine treated rats, *Int. J. Vitam. Nutri. Res.,* 44, 357, 1974; *Chem. Abstr.,* 82:56366.
49. **Bianchi Donnasibilla, L. and Cortesi, M.,** Immunization power of experimental pyridoxine deficiency caused by deoxypyridoxine or deficient diets, *Acta Vitaminol.,* 8, 269, 1954; *Chem. Abstr.,* 49:10492.
50. **Bianchi Donnasibilla, L. and Cortesi, M.,** Immunizing power in deoxypyridoxine induced experimental pyridoxine deficiency, *Acta Vitaminol.,* 9, 219, 1955; *Chem. Abstr.,* 50:8022a.
51. **Bianchi Donnasibilla, L. and Maggiora, L.,** Deoxypyridoxine induced pyridoxine deficiency in the rat. The peripheral blood and bone marrow, *Acta Vitaminol.,* 9, 211, 1955; *Chem. Abstr.,* 50:8021i.

52. Bianchi Donnasibilla, L. and Mainardi, L., Measurement of the pyridoxine-deoxypyridoxine activity ratio by means of xanthurenic acid index, *Acta Vitaminol.*, 20, 59, 1966; *Chem. Abstr.*, 65:14086d.
53. Bieber, S. and Hitchings, G., The developing frog embryo and growth inhibitors, *Cancer Res. Suppl.*, 3, 80, 1955; *Chem. Abstr.*, 49:16226g.
54. Black, C. and San Pietro, A., Vitamin $B_6$ activity in photosynthetic reactions, *Arch. Biochem. Biophys.*, 128, 482, 1968; *Chem. Abstr.*, 70:951.
55. Blakley, R., Interconversion of serine and glycine: participation of pyridoxal phosphate, *Biochem. J.*, 61, 315, 1955; *Chem. Abstr.*, 50:2695h.
56. Blaustein, M. and Schneiderman, H., A brief survey of the effects of potential antimetabolites and enzymes on the development of giant silkmoths, *J. Insect Physiol.*, 5, 143, 1960; *Chem. Abstr.*, 55:8673i.
57. Blethen, S., Boeker, E., and Snell, E., Arginine decarboxylase from *Escherichia coli*. I. Purification and specificity for substrates and coenzyme, *J. Biol. Chem.*, 243, 1671, 1968; *Chem. Abstr.*, 68:93147.
58. Blum, M., Cunningham, W., and Thanassi, J., Copper (II)-catalyzed peptide bond formation in the reaction of 5-deoxypyridoxal and alpha-phenyl-alpha-aminomalonic acid, *Bioorg. Chem.*, 5, 415, 1976; *Chem. Abstr.*, 86:116260.
59. Blum, M. and Thanassi, J., Metal ion induced reaction specificity in vitamin $B_6$ model systems. The effects of zinc and copper ions on the 5-deoxypyridoxal catalyzed reactions of alpha-phenyl-alpha-aminomalonic acid, *Bioorg. Chem.*, 6, 31, 1977; *Chem. Abstr.*, 86:185099.
60. Bodian, D., Poliomyelitis in an uninoculated rhesus monkey and in orally inoculated monkeys receiving desoxypyridoxine, *Am. J. Hyg.*, 48, 87, 1948; *Biol. Abstr.*, 23:4539.
61. Bohonos, N., Hutchings, B. L., and Peterson, W. H., Pyridoxine nutrition of lactic acid bacteria, *J. Bacteriol.*, 44, 479, 1942; *Chem. Abstr.*, 37:662.
62. Boll, W., Inhibition of growth of excised tomato roots by deoxypyridoxine and its reversal by pyridoxine, *Science*, 120, 991, 1954; *Chem. Abstr.*, 49:6382b.
63. Boone, I., Turney, D., and Langham, W., Isoniazid in vitamin $B_6$ deficient rats, *J. Lab. Clin. Med.*, 46, 549, 1955; *Chem. Abstr.*, 50:2773h.
64. Boullin, D., Effect of deoxypyridoxine on signs of pyridoxine deficiency and intestinal 5-hydroxytryptamine, *Brit. J. Nutr.*, 18, 319, 1964; *Chem. Abstr.*, 62:6854f.
65. Bradner, W. and Clarke, D., Inhibition of anaerobic bacteria as a screen for antitumor agents, *Cancer Res.*, 18, 299, 1958; *Chem. Abstr.*, 52:20663c.
66. Braunstein, A., Vilenkina, G., and Brusova, L., Participation of pyridoxal phosphate in active transport of amino acids through cell membranes, *Vopr. Med. Khim.*, 9, 475, 1963; *Chem. Abstr.*, 60:14927.
67. Brockman, R., Thomson, J., Schabel, F., and Skipper, H., The inhibition of growth of sarcoma 180 by combinations of vitamin $B_6$ antagonists and acid hydrazides, *Cancer Res.*, 16, 788, 1956; *Chem. Abstr.*, 50:17123g.
68. Brown, M. and Pike, R., Changes in blood volume and thiocyanate space in vitamin $B_6$ deficient rats, *Fed. Proc. Fed. Am. Soc. Exp. Biol.*, 17, (Abstr.), 472, 1958.
69. Brown, M. and Pike, R., Blood pressure and thiocyanate space in the vitamin $B_6$ deficient rat during pregnancy, *J. Nutr.*, 70, 453, 1960; *Biol. Abstr.*, 35:40571.
70. Brown, M. and Pike, R., Blood volume and serum protein in the deoxypyridoxine fed rat during pregnancy, *J. Nutr.*, 71, 191, 1960; *Chem. Abstr.*, 56:14691i.
71. Bruce, W. F. and Coover, H. W., Pyridine derivatives. I. 3-cyano-4-ethoxymethyl-6-methyl-2-pyridone and some related transformation products, *J. Am. Chem. Soc.*, 66, 2092, 1944; *Chem. Abstr.*, 39:1167.
72. Brucker, W., Effects of growth inhibitors and X-rays on the growth of pea roots, *Wiss. Z. Ernst Moritz Arndt-Univ. Greifswald. Math. Naturwiss. Reihe*, 15, 41, 1966; *Chem. Abstr.*, 70:657.
73. Buell, M. and Hansen, R., Reaction of pyridoxal-5-phosphate with aminothiols, *J. Am. Chem. Soc.*, 82, 6042, 1960; *Chem. Abstr.*, 55:11489h.
74. Buesing, K. H., Significance of vitamin $B_6$ for the synthesis of immune bodies in the rat, *Int. Z. Vitaminforsch.*, 22, 313, 1950; *Chem. Abstr.*, 45:2552c.
75. Bukin, Y. V., Mechanism of the toxic effect upon the central nervous system of antagonists of pyridoxine; isonicotinoyl hydrazide, toxopyrimidine and 3-deoxypyridoxal, in *Pyridoxal Catalysis: Enzymes and Model Systems, Proc. 2nd Int. Symp.*, Snell, E. E., Ed., Interscience, New York, 1966, 721; *Chem. Abstr.*, 70:2032.
76. Cake, M. H., Disorbo, D. M., and Litwack, G., Effect of pyridoxal phosphate on the DNA binding site of activated hepatic glucocorticoid receptor, *J. Biol. Chem.*, 253, 4886, 1978; *Chem. Abstr.*, 89:102394m.
77. Caldwell, E. and McHenry, E., Deoxypyridoxine and liver transamination in the rat, *Arch. Biochem. Biophys.*, 45, 466, 1953; *Chem. Abstr.*, 47:11391a.

78. **Carsiotis, M. and Suskind, S.**, The role of pyridoxal phosphate in the aldolytic activity of tryptophan synthetase from *Neurospora crassa*, *J. Biol. Chem.*, 239, 4227, 1964; *Chem. Abstr.*, 61:16361.
79. **Carter, C. W. and Phizackerley, P. J. R.**, The influence of pyridoxine on fat metabolism in the rat, *Biochem. J.*, 49, 227, 1951; *Chem. Abstr.*, 46:590c.
80. **Cerecedo, L., Lombardo, M., Reddy, D., and Travers, J.**, Nucleic acids and their components in tumor bearing mice during growth and regression of tumors, *Proc. Soc. Exp. Biol. Med.*, 80, 648, 1952; *Chem. Abstr.*, 46:11396c.
81. **Chan, M. M. and Kare, M. R.**, Effect of vitamin $B_6$ deficiency on preference for several taste solutions in the rat, *J. Nutr.*, 109, 339, 1979; *Chem. Abstr.*, 90:166919.
81a. **Chatterjee, G. G. and Halder, D. K.**, Accumulation of certain amino acids during the inhibition of growth of *Vibrio cholera* by different anti-vitamins, *Naturwissenschaften*, 47, 86, 1960; *Chem. Abstr.*, 54:17533f.
82. **Chern, C. and Beutler, E.**, Purification and characterization of human erythrocyte pyridoxine kinase, *Clin. Chim. Acta*, 61, 353, 1975; *Chem. Abstr.*, 83:110366.
82a. **Chern, C. and Beutler, E.**, Biochemical and electrophoretic studies of erythrocyte pyridoxine kinase in white and black Americans, *Am. J. Hum. Genet.*, 28, 9, 1976; *Biol. Abstr.*, 61:66624.
83. **Chiancone, F., Ginoulhiac, E., Tenconi, L., and Bianchi Donnasibilla, L.**, Xanthurenic index in pyridoxine deficiency produced by diet or deoxypyridoxine. Influence of riboflavine, *Boll. Soc. Ital., Biol. Sper.*, 29, 523, 1953; *Chem. Abstr.*, 49:7674.
84. **Chiao, J. and Peterson, W.**, Some factors affecting the inhibitory action of thiamine on the growth of *Saccharomyces carlsbergensis*, *Arch. Biochem. and Biophys.*, 64, 115, 1956; *Chem. Abstr.*, 51:553c.
85. **Christensen, H. and Riggs, T.**, Structural evidence for chelation and Shiff's base formation in amino acid transfer into cells, *J. Biol. Chem.*, 220, 265, 1956; *Chem. Abstr.*, 50:10835e.
86. **Christensen, H., Riggs, T., Aspen, A., and Mothon, S.**, Concentrative uptake of amino acids by free-cell neoplasms, *Ann. N.Y. Acad. Sci.*, 63, 983, 1956; *Chem. Abstr.*, 50:13249h.
87. **Christensen, H., Riggs, T., and Coyne, B.**, Effects of pyridoxal and indoleacetate on cell uptake of amino acids and potassium, *J. Biol. Chem.*, 209, 413, 1954; *Chem. Abstr.*, 48:10812i.
88. **Cidlowski, J. and Thanassi, J.**, Extraction of nuclear glucocorticoid-receptor complexes with pyridoxal phosphate, *Biochem. Biophys. Res. Commun.*, 82, 1140, 1978; *Chem. Abstr.*, 89:86783e.
89. **Clark, I. and Stoerk, H.**, The uptake of P32 by nucleic acids of lymphoid tissue undergoing atrophy, *J. Biol. Chem.*, 222, 285, 1956; *Chem. Abstr.*, 50:17080h.
90. **Clarke, D.**, Mouse sarcoma 180, *Cancer Res. Suppl.*, 3, 14, 1955; *Chem. Abstr.*, 49:16226c.
90a. **Coburn, S.**, unpublished observations.
91. **Coburn, S. and Mahuren, D.**, Major urinary metabolites of 4′- and 5′-deoxypyridoxines in various species, *IRCS Med. Sci.*, 7, 556, 1979.
92. **Coburn, S., Mahuren, D., and Sallay, S.**, In vivo metabolism of 4′-deoxypyridoxine in rat and man, *J. Biol. Chem.*, 251, 1646, 1976; *Chem. Abstr.*, 84:162244.
93. **Coburn, S. P., Mahuren, J. D., Schaltenbrand, W. G., Wostmann, B. S., and Madsen, D.**, The effects of vitamin $B_6$ deficiency and 4′-deoxypyridoxine on pyrodoxal phosphate concentrations, pyrodoxine kinase, and other aspects of metabolism in the rat, *J. Nutr*, in press.
94. **Coburn, S., Mahuren, J., Schaltenbrand, W., and Sallay, S.**, Identification of 5-deoxypyridoxine-3-sulfate as the major urinary metabolite of 5-deoxypyridoxine in rats with comments on the inhibition of arylsulfatase activity and manganese dioxide oxidation by neighboring groups, *Biochem. Biophys. Res. Commun.*, 80, 942, 1978, *Chem. Abstr.*, 88:167328.
95. **Coburn, S. and Schaltenbrand, W.**, Simplified preparation of a phosphatase inhibitor and further studies of its action, *Biochem. J.*, 171, 485, 1978; *Chem. Abstr.*, 89:142468.
95a. **Coburn, S. P., Schaltenbrand, W. E., Leykauf, R. E., Bendixen, G. E., and Townsend, D.**, Decreased erythrocyte fragility in black Americans, *IRCS Med. Sci.*, 7, 240, 1979.
96. **Colleter, J., Giraud, J., and Dallet, P.**, Copper complexes of pyridoxine and similar molecules, *Trav. Soc. Pharm. Montpellier*, 33, 289, 1973; *Chem. Abstr.*, 80:8790.
97. **Colombini, C. and Celon, E.**, Synthesis, purification and storage of pyridoxine-4,5-$^{14}C$ at high specific activity, *Gazz. Chim. Ital.*, 99, 526, 1969; *Chem. Abstr.*, 71:61154.
98. **Conti, G. and Milio, G.**, Action of vitamin $B_6$ and its antivitamin on the myocardium of the fowl embryo cultured in vitro, *J. Embryol. Exp. Morphol.*, 14, 53, 1965; *Chem. Abstr.*, 63:18872.
99. **Conti, G. and Milio, G.**, Effect of vitamin and antivitamin $B_6$ on the intestine of chick embryo grown in vitro, *Arch. Biol.*, 77, 79, 1966; *Chem. Abstr.*, 67:72226.
100. **Contractor, S. F. and Shane, B.**, 4-Pyridoxic acid 5′-phosphate: metabolite of pyridoxal in the rat, *Biochem. Biophys. Res. Commun.*, 39, 1175, 1970; *Chem. Abstr.*, 73:63865.
101. **Coover, H. and Bowman, N.**, Substituted pyridine compounds, U.S. Patent 2,481,573, 1949, *Chem. Abstr.*, 44:668e.
102. **Cordoba, F., Gonzales, C., and Rivera, P.**, Antibodies against pyridoxal-5′-phosphate and pyridoxamine-5′-phosphate, *Biochim. Biophys. Acta*, 127, 151, 1966; *Chem. Abstr.*, 65:17520c.

103. **Corey, E. and Schmidt, G.**, Useful procedures for the oxidation of alcohols involving pyridinium dichromate in aprotic media, *Tetrahedron Lett.*, 399, 1979.

104. **Cote, L., Oleson, J., and Williams, J.**, Nicotinamide inhibitors, *Proc. Soc. Exp. Biol. Med.*, 80, 434, 1952; *Chem. Abstr.*, 46:9714d.

105. **Cravens, W.**, Vitamin deficiencies and antagonists, *Ann. N. Y. Acad. Sci.*, 55, 188, 1952; *Chem. Abstr.*, 47:2845i.

106. **Cravens, W. and Snell, E.**, Deoxypyridoxine and development of the chick embryo, *Fed. Proc. Fed. Am. Soc. Exp. Biol.*, 8, 380, 1949; *Biol. Abstr.*, 23:23233.

107. **Cravens, W. and Snell, E.**, Effects of deoxypyridoxine and vitamin $B_6$ on development of the chick embryo, *Proc. Soc. Exp. Biol. Med.*, 71, 73, 1949; *Chem. Abstr.*, 43:6293f.

108. **Cunningham, W. C. and Thanassi, J. W.**, Synthesis of 5-deoxypyridoxal derivatives, *Experientia*, 35, 451, 1979; *Chem. Abstr.*, 91:5081.

109. **Cushing, R. and Morgan, H.**, Effects of some metabolic analogs on growth of mumps and influenza viruses in tissue cultures, *Proc. Soc. Exp. Biol. Med.*, 79, 497, 1952; *Chem. Abstr.*, 46:6187b.

110. **Davis, S. D.**, Immunodeficiency and runting syndrome in rats from congenital pyridoxine deficiency, *Nature (London)*, 251, 548, 1974; *Chem. Abstr.*, 82:84860.

111. **Davis, S., Nelson, T., and Shepard, T.**, Teratogenicity of vitamin $B_6$ deficiency: omphalocele skeletal and neural defects and splenic hypoplasia, *Science*, 169, 1329, 1970; *Chem. Abstr.*, 74:1647.

112. **DeJongh, D., Perricone, S., Gay, M., and Korytnyk, W.**, Mass spectrometry of vitamin $B_6$: different forms of the vitamins, its metabolites, antimetabolites and analogs, *Org. Mass Spectrom.*, 1, 151, 1968; *Chem. Abstr.*, 69:76389.

113. **DeLorme, C., Lupien, P., and Despointes, R.**, Influence of vitamin $B_6$ on the renin-angiotensin system in rats, *J. Nutr.*, 105, 1192, 1975; *Chem. Abstr.*, 83:162505.

113a. **Dempsey, W. B. and Snell, E. E.**, Pyridoxamine-pyruvate transaminase. II. Characteristics of the enzyme, *Biochemistry*, 2, 1414, 1963.

114. **Derenzo, E. and Cerecedo, L.**, Effects of deoxy- and methoxypyridoxine in the mouse, *Proc. Soc. Exp. Biol. Med.*, 73, 356, 1950; *Chem. Abstr.*, 44:6497i.

115. **Dietrich, L. and Borries, E.**, Comparison of the affinity of pyridoxal phosphate and deoxypyridoxine phosphate for apo-cysteine desulfhydrase, *Arch. Biochem. Biophys.*, 64, 512, 1956; *Chem. Abstr.*, 51:2912d.

116. **Dietrich, L. and Shapiro, D.**, Antagonist studies: comparison of deoxypyridoxine and pyridoxine deficiency therapy on various $B_6$ enzymes, *Proc. Soc. Exp. Biol. Med.*, 84, 555, 1953; *Chem. Abstr.*, 48:3493a.

117. **Dietrich, L. and Shapiro, D.**, Testosterone augmentation of deoxypyridoxine antagonism of various vitamin $B_6$ containing enzyme systems in tumor and host tissue, *Cancer Res.*, 15, 133, 1955; *Chem. Abstr.*, 49:16225a.

118. **Di Paolo, J. and Moore, G.**, The use of ascites tumors in solid agar for screening of cancer chemotherapeutic agents, *Ann. N.Y. Acad. Sci.*, 76, 870, 1958; *Chem. Abstr.*, 53:10573b.

119. **Dixon, M. and Webb, E.**, *Enzymes*, Academic Press, New York, 1964, 319; *Chem. Abstr.*, 60:16152b.

120. **Dobbelstein, H., Baumgaertner, R., Schubert, G., and Thoenes, G.**, Immunosuppression under vitamin $B_6$ deficiency. Experimental studies with skin transplants in inbred mice, *Res. Exp. Med.*, 169, 189, 1977; *Chem. Abstr.*, 86:153796.

121. **Dobbelstein, H., Pielsticker, K., Schrieber, M. A., Schubert, G., Thoenes, G. H.**, Immunsuppression by vitamin $B_6$ antagonists — prolonged survival of inbred rats with renal transplants, *Res. Exp. Med.*, 175, 169, 1979; *Biol. Abstr.*, 69:18753.

122. **Doctor, V.**, Hematologic changes in rats following deoxypyridoxine, aminopterin, oxythlamine and 6,7-dimethyl-9 (2-acetoxyethyl) isoalloxazine, *Blood*, 14, 1244, 1959; *Biol. Abstr.*, 35:18789.

123. **Doktorova, N., Karpeiskii, N., Padyukova, N., and Florentev, V.**, Vitamin $B_6$ analogs. IX. Synthesis of 5-deoxy and 5-*O*-methyl analogs, *Khim. Geterotsikl. Soedin.*, 7, 365, 1971; *Chem. Abstr.*, 76:3655.

124. **Dowhan, W. and Snell, E.**, D-serine dehydratase from *Escherichia coli*. III. Resolution of pyridoxal-5-phosphate and coenzyme specificity, *J. Biol. Chem.*, 245, 4629, 1970; *Chem. Abstr.*, 73:105752.

125. **Druyan, R. and Haeger-Aronsen, B.**, Effect of conditioned pyridoxine deficiency on excretion of porphyrin precursors in normal and experimentally porphyric rabbits, *J. Lab. Clin. Med.*, 66, 122, 1965; *Chem. Abstr.*, 63:8791h.

126. **Dubnick, B., Leeson, G., and Scott, C.**, Effect of forms of vitamin $B_6$ on acute toxicity of hydrazines, *Toxicol. and Appl. Pharmacol.*, 2, 403, 1960; *Chem. Abstr.*, 54:21508f.

127. **Eagle, H. and Foley, G.**, Cytotoxicity in human cell cultures as a primary screen for the detection of antitumor agents, *Cancer Res.*, 18, 1017, 1958.

128. **Easton, E., Simpson, J., Martin, J., and Campbell, D.**, Tyrosinemia induced by a pyridoxine antagonist, deoxypyridoxine, *Clin. Chem.*, 18, 161, 1972; *Chem. Abstr.*, 76:135913.

129. **Ebadi, M. and Klangkalya, B.**, On the mechanism of pyridoxal phosphate-related convulsions as implicated in enhanced transport of GABA, *Neuropharmacology*, 18, 301, 1979; *Chem. Abstr.*, 91:155141.

130. **Ebadi, M., McCoy, E., and Kugel, R.**, Interrelations between pyridoxal phosphate and pyridoxal kinase in rabbit brain, *J. Neurochem.*, 17, 941, 1970; *Chem. Abstr.*, 73:74524.
130a. **Ebadi, M.**, personal communication.
131. **Efimovsky, O., Fourneau, J., Jacquier, R., and Stoven, O.**, Structure of pyridoxilate, the condensation product of glyoxylic acid and pyridoxine, *Chim. Ther.*, 4, 89, 1969; *Chem. Abstr.*, 71:70470.
132. **Efremova, L., Florentev, V., and Goryachenkova, E.**, Preparation of the apoenzymes of serine sulfhydrase and cystathionine beta-synthase and their interactions with pyridoxal-5-phosphate and its analogs, *Mol. Biol.*, 8, 154, 1974; *Chem. Abstr.*, 80:129793.
133. **Elder, T. and Mengel, C.**, Effect of pyridoxine deficiency on porphyrin precursor excretion in acute intermittent porphyria, *Am. J. Med.*, 41, 369, 1966; *Chem. Abstr.*, 65:19112b.
134. **Emerson, G.**, The antivitamin $B_6$ activity of deoxypyridoxine, *Fed. Proc. Fed. Am. Soc. Biol.*, 6, 406, 1947.
135. **Emerson, G.**, Some metabolite antimetabolite interrelations, *Int. Cong. Biochem.*, 224, 1952; *Chem. Abstr.*, 50:2777c.
136. **Evangelopoulos, A. and Sizer, I.**, Glutamic-aspartic transaminase. Reaction of holoenzyme with substrates and of apoenzyme with vitamin $B_6$ derivatives, *J. Biol. Chem.*, 240, 2983, 1965; *Chem. Abstr.*, 63:3241g.
137. **Faber, S., Feitler, W., Bleiler, R., Ohlson, M., and Hodges, R.**, The effects of an induced pyridoxine and pantothenic acid deficiency on excretions of oxalic acid and xanthurenic acids in the urine, *Am. J. Clin. Nutr.*, 12, 406, 1963; *Chem. Abstr.*, 59:10518c.
138. **Feldmann, K., Zeisel, H. J., and Helmreich, E. J. M.**, Complementation of subunits from glycogen phosphorylases of frog and rabbit skeletal muscle and rabbit liver, *Eur. J. Biochem.*, 65, 285, 1976; *Chem. Abstr.*, 85:16350.
139. **Firestone, R. A., Harris, E. E., and Reuter, W.**, Synthesis of pyridoxine by Diels-Alder reactions with 4-methyl-5-alkoxyoxazoles, *Tetrahedron*, 23, 943, 1967; *Chem. Abstr.*, 66:65373.
140. **Firouzabadi, H. and Ghaderi, E.**, Barium manganate. An efficient oxidizing reagent for oxidation of primary and secondary alcohols to carbonyl compounds, *Tetrahedron Lett.*, 839, 1978; *Chem. Abstr.*, 89:108773.
141. **Fisher, B., Axelrod, A., Fisher, E., Lee, S., and Calvanese, N.**, The favorable effect of pyridoxine deficiency on skin homograft survival, *Surgery*, 44, 149, 1958.
142. **Fisher, B.**, Skin homografts in pyridoxine deficient humans, *Transplant Bull.*, 6, 432, 1959.
143. **Fisher, B. and Schewe, E.**, Further observations on skin homografts in pyridoxine deficient animals, *Ann. Surg.*, 155, 457, 1962.
144. **Foley, G., McCarthy, R., Binns, V., Snell, E., Guirard, B., Kidder, G., Dewey, V., and Thayer, P.**, Comparative study of the use of microorganisms in the screening of potential antitumor agents, *Ann. N.Y. Acad. Sci.*, 76, 413, 1958; *Chem. Abstr.*, 53:9357f.
145. **Fonda, M.**, Interaction of pyridoxal analogs with glutamate apodecarboxylase and aspartate apoaminotransferase, *J. Biol. Chem.*, 246, 2230, 1971; *Chem. Abstr.*, 74:135523.
146. **Forse, R. and Webster, D.**, The effect of pyridoxine deficiency on gastric secretion and blood, in Sur. Forum Proc. 36th Clin. Congr. Am. Coll. Surgeons, American College of Surgeons, Chicago, Ill., 1950, 95; *Chem. Abstr.*, 45:5263e.
147. **Fridhandler, L. and Quastel, J.**, Absorption of amino acids from isolated surviving intestine, *Arch. Biochem. Biophys.*, 56, 424, 1955; *Chem. Abstr.*, 49:12635d.
148. **Fujimoto, S.**, Effect of pyridoxal phosphate on toxicity and antitumor activity of mitomycin C and 4-deoxypyridoxine hydrochloride in rats. Preliminary observations, *Cancer Chemother. Rep.*, 50, 313, 1966; *Chem. Abstr.*, 65:11174f.
149. **Fujita, A., Nose, Y., and Kuratani, K.**, Second type of bacterial thiaminase, *J. Vitaminol.*, 1, 1, 1954; *Chem. Abstr.*, 49:8339a.
150. **Furbish, F., Fonda, M., and Metzler, D.**, Reaction of apoaspartate aminotransferase with analogs of pyridoxal phosphate, *Biochemistry*, 8, 5169, 1969; *Chem. Abstr.*, 72:51184.
151. **Gadekar, S., Frederick, J., and De Renzo, E.**, Synthesis and biological activity of pyridoxine analogs, *J. Med. Pharm. Chem.*, 5, 531, 1962; *Chem. Abstr.*, 57:8544g.
152. **Gailani, S. and Holland, J.**, Biochemical and clinical studies of $B_6$ deficiency in man, *Proc. Am. Assoc. Cancer Res.*, 5, 20, 1964.
153. **Gailani, S. and Holland, J.**, Biochemical and clinical studies of vitamin $B_6$ deficiency in man, *Fed. Proc. Fed. Am. Soc. Exp. Biol.*, 24, 314, 1965.
154. **Gailani, S., Holland, J., Nussbaum, A., and Olson, K.**, Clinical and biochemical studies of pyridoxine deficiency in patients with neoplastic disease, *Cancer (New York)*, 21, 975, 1968; *Biol. Abstr.*, 49:100449.
155. **Gailani, S., Roque, A., and Holland, J.**, The effect of pyridoxine deficiency on carcinoid syndrome. Study of a case of metastatic bronchial adenoma treated with induced vitamin $B_6$ deficiency, *Ann. Intern. Med.*, 65, 1044, 1966; *Biol. Abstr.*, 48:44264.

156. **Gambassi, G.**, Synthesis of cholesterol and phospholipids in humans and animals from compounds labelled with 14C, *Minerva Nucl.*, 3, 367, 1959; *Chem. Abstr.*, 55:17788a.

157. **Gandullia, E. and Cordone, G.**, Relations between immunological processes and pyridoxine deficiency, *Minerva Pediatr.*, 15, 528, 1963; *Chem. Abstr.*, 64:7194f.

158. **Garattini, S. and Vegeto, A.**, Vitamin $B_6$ deficiency and inhibition of tumor growth, *Giorn. Ital. Chemioter.*, 2, 41, 1955; *Chem. Abstr.*, 50:15786e.

159. **Gardner, T., Smith, F., Wenis, E., and Lee, J.**, The synthesis of compounds for the chemotherapy of tuberculosis. I. Heterocyclic thiosemicarbazide derivatives, *J. Org. Chem.*, 16, 1121, 1951; *Chem. Abstr.*, 46:3536d.

160. **Gaut, Z. and Solomon, H.**, Phosphorylation of pyridoxine by human blood platelets. Effects of structure analogs and metabolic inhibitors, *Biochem. Pharmacol.*, 21, 2395, 1972; *Chem. Abstr.*, 77:135790.

161. **Gellhorn, A. and Hirschberg, E., Eds.**, Diverse systems for cancer chemotherapy screening. I. Summary of results and general correlations, *Cancer Res. Suppl.*, 3, 1, 1955; *Chem. Abstr.*, 49:16225i.

162. **Gellhorn, A. and Jones, L.**, Pyridoxine deficient diet and deoxypyridoxine in the therapy of lymphosarcoma and acute leukemia in man, *Blood*, 4, 60, 1949; *Chem. Abstr.*, 47:4507.

163. **Gershoff, S. and Faragalla, F.**, Endogenous oxalate synthesis and glycine, serine and deoxypyridoxine interrelations in vitamin $B_6$ deficient rats, *J. Biol. Chem.*, 234, 2391, 1959; *Chem. Abstr.*, 53:22319c.

164. **Gershoff, S., Gill, T., Simonian, S., and Steinberg, A.**, Effects of amino acid deficiencies on antibody formation in the rat, *J. Nutr.*, 95, 184, 1968; *Chem. Abstr.*, 69:33910.

165. **Gillette, R.**, The action of deoxypyridoxine in chick fibroblasts in vitro, *Exp. Cell Res.*, 9, 589, 1955; *Chem. Abstr.*, 50:9541b.

166. **Gilmer, P. and Kirsch, J.**, Pyridoxamine-pyruvate transaminase. II. Temperature-jump and stopped-flow kinetic investigation of the rates and mechanism of the reaction of 5-deoxypyridoxal with the enzyme, *Biochemistry*, 16, 5246, 1977; *Chem. Abstr.*, 87:196334.

167. **Gilmer, P., McIntire, W., and Kirsch, J.**, Pyridoxamine pyruvate transaminase. I. Determination of the active site stoichiometry and the pH dependence of the dissociation constant for 5′-deoxypyridoxal, *Biochemistry*, 16, 5241, 1977; *Chem. Abstr.*, 87:196406h.

168. **Glazer, H., Fichter, B., Mueller, J., and Vilter, R.**, The effect of pyridoxine deficiency induced by deoxypyridoxine in human subjects upon the excretion of $N^1$-methylnicotinamide after tryptophan administration, *Arch. Biochem. Biophys.*, 32, 424, 1951; *Chem. Abstr.*, *46:1633h*.

169. **Glazer, H., Mueller, J., Thompson, C., Hawkins, V., and Vilter, R.**, A study of urinary excretion of xanthurenic acid and other tryptophan metabolites in human beings with pyridoxine deficiency induced by deoxypyridoxine, *Arch. Biochem.*, 33, 243, 1951; *Chem. Abstr.*, 46:1633i.

170. **Goldsmith, E.**, Comparative effects of folic acid and pyridoxine antagonist and partial inanition on the response of target organs to sex hormones, *Cong. Int. Biochim. 2nd Cong.*, 470, 1952; *Chem. Abstr.*, 48:7163b.

171. **Goryachenkova, E., Polyakova, L., Efremova, L., and Florentev, V.**, Interaction of pyridoxal phosphate analogs with apoenzymes of gamma-cystathionase and serine sulfhydrase, *Biophys. Res. Commun.*, 55, 1021, 1973; *Chem. Abstr.*, 80:45126.

172. **Gothilf, S. and Waites, R.**, Inhibition of growth and increased mortality of Mexican bean beetle larvae fed with thiamine and pyridoxine antagonists and reversal of effect with vitamin supplementation, *Entomol. Exp. Appl.*, 11, 261, 1968; *Chem. Abstr.*, 70:19168.

173. **Gregoire, C.**, Effects of pyridoxine deficiency on the normal and neoplastic lymphoid tissue of the rat, *Arch. Int. Pharmacodynamie*, 78, 313, 1949; *Chem. Abstr.*, 43:4362f.

174. **Grimaldi, T., Lapesa, M., Curci, E., and Semeraro, N.**, Effects of deoxypyridoxine on the formation and development of some experimental neoplasias. I. Ascites tumor in mice, *Pathologica*, 63, 89, 1971; *Chem. Abstr.*, 77:56536.

175. **Gross, W., Ring, K., and Heinz, E.**, Positive feedback regulation of amino acid transport in *Streptomyces hydrogenans*, *Arch. Biochem. Biophys.*, 137, 253, 1970; *Chem. Abstr.*, 72:129532.

176. **Guha, S., Ghosh, B., and Ghosh, J.**, Brain metabolism. II. Endogenous ammonia formation in rat brain, *Ann. Biochem. Exp. Med.*, 19, 255, 1959; *Chem. Abstr.*, 54:13448c.

177. **Gupta, S., Prasanna, H., Viswanathan, L., and Venkitasubramanian, T. A.**, Synthesis of aflatoxins by the non-growing mycelia of *Aspergillus parasiticus* and the effect of inhibitors, *J. Gen. Microbiol.*, 91, 417, 1975; *Chem. Abstr.*, 84:69761.

178. **Gustafson, T. and Horstadius, S.**, Vegetalization and animalization in the sea urchin egg induced by antimetabolites, preliminary note, *Exp. Cell Res. Suppl.*, 3, 170, 1955; *Chem. Abstr.*, 50:3655c.

179. **Haekkinen, H. and Kulonen, E.**, Ethanol intoxication and gamma-aminobutyric acid, *J. Neurochem.*, 27, 631, 1976; *Chem. Abstr.*, 86:38373.

180. **Halverson, A. and Hendrick, C.**, Negative relation of the selenium protective factor and the antivitamin $B_6$ principle of linseed oil meal, *Proc. S. D. Acad. Sci.*, 33, 95, 1954; *Chem. Abstr.*, 50:2054a.

181. **Hansson, O.**, Effect of 4-deoxypyridoxine on the initial excretion rates of 3-hydroxy-kynurenine and 3-hydroxyanthranilic acid after tryptophan loading, *Acta. Soc. Med. Ups.*, 73, 19, 1968; *Chem. Abstr.*, 69:42500.
181a. **Harfenist, M., Bavely, A., and Lazier, W.**, The oxidation of allyl and benzyl alcohols to the aldehydes, *J. Org. Chem.*, 19, 1608, 1954; *Chem. Abstr.*, 49:15789b.
182. **Harris, C. M., Johnson, R. J., and Metzler, D. E.**, Band shape analysis and resolution of electronic spectra of pyridoxal phosphate and other 3-hydroxypyridine-4-aldehydes, *Biochim. Biophys. Acta*, 421, 181, 1976; *Chem. Abstr.*, 84:120911.
183. **Harris, D.**, Interaction of thiamine and pyridoxine in Neurospora. II. Competition between pyridoxine and the pyrimidine precursor of thiamine, *Arch. Biochem. Biophys.*, 60, 35, 1956; *Chem. Abstr.*, 50:6543c.
184. **Harris, S.**, Vitamin $B_6$. II. Reactions and derivative, *J. Am. Chem. Soc.*, 62, 3203, 1940; *Chem. Abstr.*, 35:114.
185. **Harris, S.**, Chemistry of vitamin $B_6$. IV. Reactions in solutions at elevated temperatures, *J. Am. Chem. Soc.*, 63, 3363, 1941; *Chem. Abstr.*, 36:1041.
186. **Harris, S. A. and Folkers, K.**, Synthesis of vitamin $B_6$. II. *J. Am. Chem. Soc.*, 61, 3307, 1939.
187. **Harris, S., Webb, T. D., and Folkers, K.**, Chemistry of vitamin $B_6$. I. Tautomerism, *J. Am. Chem. Soc.*, 62, 3198, 1940; *Chem. Abstr.*, 35:114.
188. **Harris, S. A., Harris, E. E., Peterson, E. R., and Rogers, E. F.**, Synthesis of vitamin $B_6$. III. 3-Deoxypyridoxine, *J. Med. Chem.*, 10, 261, 1967; *Chem. Abstr.*, 70:3787.
189. **Hartsook, E., Hershberger, T., and French, C.**, The effect of deoxypyridoxine or isoniazid upon mineral retention and liver enzyme activities of pyridoxine deficient male rats, *J. Nutr.*, 65, 547, 1958; *Chem. Abstr.*, 52:4550g.
190. **Hartsook, E. and Hershberger, T.**, A study of the effect of deoxypyridoxine on isonicotinic hydrazide upon tissue potassium and sodium content of pyridoxine deficient male rats, *J. Nutr.*, 72, 297, 1960; *Biol. Abstr.*, 36:19731.
191. **Hartsook, E. and Hershberger, T.**, Tissue cation content of rats. Effect of supplemental dietary amino acids in pyridoxine and potassium deficiencies, *J. Nutr.*, 95, 31, 1968; *Chem. Abstr.*, 69:9155.
192. **Hawkins, W. and Evans, M.**, White blood cells and lymphoid tissue in vitamin $B_6$ insufficiency, *Am. J. Physiol.*, 170, 160, 1952; *Chem. Abstr.*, 46:11368h.
193. **Heidelberger, C. and Keller, R.**, The effects of twenty-nine compounds on nucleic acid and protein biosynthesis in slices of Flexner-Jobling carcinoma and rat spleen, *Cancer Res. Suppl.*, 3, 106, 1955; *Chem. Abstr.*, 49:16226i.
194. **Helgebostad, A., Svenkerud, R., and Ender, F.**, Sterility in mink induced experimentally by deficiency of vitamin $B_6$, *Acta Vet. Scand.*, 4, 228, 1963; *Chem. Abstr.*, 60:7207a.
195. **Heyl, D., Harris, S., and Folkers, K.**, Chemistry of vitamin $B_6$. IX. Derivatives of 5-deoxypyridoxine, *J. Am. Chem. Soc.*, 75, 653, 1953; *Chem. Abstr.*, 48:1355g.
196. **Heyl, D., Luz, E., Harris, S., and Folkers, K.**, Chemistry of vitamin $B_6$. XI. Homologs of 4-deoxypyridoxine, *J. Am. Chem. Soc.*, 75, 4080, 1953; *Chem. Abstr.*, 49:8272b.
197. **Hiroaka, E.**, Influence of antivitamins on microorganisms. III. Inhibitory effect of isonicotinoylhydrazide and 4-deoxypyridoxine on the growth of lactic acid bacteria and their reversal by various vitamins, *Yakugaku Zasshi*, 77, 1324, 1957; *Chem. Abstr.*, 52:3911e.
198. **Hitchings, G. and Elion, G.**, The lactobacillus screening test, *Cancer Res. Suppl.*, 3, 66, 1955; *Chem. Abstr.*, 49:16226f.
199. **Hodges, R., Bean, W., Ohlson, M., and Bleiler, R.**, Factors affecting human antibody response. V. Combined deficiencies of pantothenic acid and pyridoxine, *Am. J. Clin. Nutr.*, 11, 187, 1962; *Chem. Abstr.*, 58:14497g.
200. **Holman, R.**, Function and metabolism of essential fatty acids, in *Biochemical Problems of Liquids*, Popjak, G. and Le Breton, E., Eds., Butterworths, London, 1956, 463.
201. **Holman, R.**, Factors influencing the requirements for polyunsaturated fatty acids, *Am. J. Clin. Nutr.*, 8, 403, 1960; *Chem. Abstr.*, 54:25093g.
202. **Holtz, P. and Palm, D.**, Pharmacological aspects of vitamin $B_6$, *Pharmacol. Rev.*, 16, 113, 1964.
203. **Horton, R. and Meldrum, B.**, Seizures induced by allylglycine, 3-mercaptopropionic acid and 4-deoxypyridoxine in mice and photosensitive baboons, and different modes of inhibition of cerebral glutamic acid decarboxylase, *Brit. J. Pharmacol.*, 49, 52, 1973; *Chem. Abstr.*, 80:55780.
204. **Horton, R. W. and Meldrum, B. S.**, Preconvulsive changes in brain glucose metabolism following drugs inhibiting glutamate decarboxylase, *J. Neurochem.*, 26, 805, 1976; *Chem. Abstr.*, 85:171643.
205. **Hruban, Z. and Wissler, R.**, Effect of beta-3-thienyl alanine and deoxypyridoxine on the growth of the Murphy-Sturm lymphosarcoma, *Cancer Res.*, 20, 1530, 1960; *Chem. Abstr.*, 55:4790d.
206. **Hruban, Z., Wissler, R., and Slesers, A.**, Effects of beta-3-thienylalanine and deoxypyridoxine on tumor growth. II. Effects on the tumor bearing host, *Lab. Invest.*, 11, 382, 1962; *Chem. Abstr.*, 57:11815.

207. **Hsu, J., Davis, R., and Chow, B.**, Electrolyte imbalance in $B_6$ deficient rats, *J. Biol. Chem.*, 230, 889, 1958; *Chem. Abstr.*, 52:12120a.
208. **Hsu, J. and Kawin, B.**, Effect of pyridoxine deficiency on blood volume measured by radioiodinated serum albumin and on organ weights in rats, *Proc. Soc. Exp. Biol. Med.*, 109, 222, 1962; *Chem. Abstr.*, 57:1346e.
209. **Huidobro, F., Contreras, E., and Croxatto, R.**, Morphine. IV. Effects of some metabolites on the abstinence syndrome to morphine in white mice, *Arch. Intern. Pharmacodyn.*, 146, 455, 1963; *Chem. Abstr.*, 60:8512b.
210. **Hurwitz, J.**, The enzymic phosphorylation of vitamin $B_6$ derivatives and their effects on tyrosine decarboxylase, *Biochem. Biophys. Acta*, 9, 496, 1952; *Chem. Abstr.*, 47:2229i.
211. **Hurwitz, J.**, Enzymic phosphorylation of vitamin $B_6$ analogs and their effect on tyrosine decarboxylase, *J. Biol. Chem.*, 217, 513, 1955; *Chem. Abstr.*, 50:4267i.
212. **Hurwitz, J.**, Metabolism of vitamin $B_6$ and its derivatives, *Nutrition Symp. Series No. 13*, National Vitamin Foundation, New York, 1956, 49.
213. **Ichiba, A., Michi, K., and Emoto, S.**, Pyridine series. III. Synthesis of 2-methyl-4,5-bis (hydroxymethyl) pyridine, *Sci. Papers Inst. Phys. Chem. Res.*, 39, 126, 1941; *Chem. Abstr.*, 41:6246g.
214. **Ikawa, M. and Snell, E.** Oxidative deamination of amino acids by pyridoxal and metal salts, *J. Am. Chem. Soc.*, 76, 4900, 1954; *Chem. Abstr.*, 49:1847f.
215. **Ikehata, H.**, Thiaminase of *Bacillus aneurinolyticus.* V. Isolation of the destructive products of thiamine by intracellular thiaminase, *Vitamins*, 10, 233, 1956; *Chem. Abstr.*, 50:14860a.
216. **Itiba, A. and Emoto, S.**, Pyridine series. I. Synthesis of 2-methyl-3-hydroxy-4,5-pyridine-dicarboxylic acid, *Sci. Papers Inst. Phys. Chem. Res.*, 38, 347, 1941; *Chem. Abstr.*, 35:6960.
217. **Iwanowska, J.**, Effect of 4-deoxypyridoxine on the metabolism of amino acids in rats with hypo and normovitaminosis $B_6$, *Acta. Physiol. Pol.*, 23, 325, 1972; *Chem. Abstr.*, 77:135226.
218. **Iwata, C.**, 5-Deoxypyridoxal, *Biochem. Prep.*, 12, 117, 1968; *Chem. Abstr.*, 69:59050.
219. **Iwata, C. and Metzler, D.**, Synthesis of vitamin $B_6$ derivatives. III. 3-(4-Formyl-3-hydroxy-2-methyl-5-pyridyl) propionic acid, an analog of pyridoxal phosphate, *J. Heterocyclic Chem.*, 4, 319, 1967; *Chem. Abstr.*, 67:99961.
220. **Jacobs, F., Coen, J., and Hillman, R.**, Influence of pyridoxine, pyridoxal phosphate, deoxypyridoxine and 2,4 dinitrophenol on methionine absorption, *J. Biol. Chem.*, 235, 1372, 1960; *Chem. Abstr.*, 54:21503b.
221. **Jacobs, F. and Hillman, R.**, Intestinal absorption of the methionine isomers as affected by deoxypyridoxine, *J. Biol. Chem.*, 232, 445, 1958; *Chem. Abstr.*, 52:16547g.
222. **Jacobs, F. and Tarnasky, W.**, Primary and secondary transport systems for amino acids in the intact intestine, *J. Am. Med. Assoc.*, 183, 765, 1963; *Chem. Abstr.*, 58:14525.
223. **Jaffe, I., Merryman, P., and Ehrenfield, E.**, Antiviral effect of D-penicillamine, *Postgrad. Med. J. Suppl.*, 50, 50, 1974; *Chem. Abstr.*, 82:51397.
224. **Jannes, J. and Lahikainen, T.**, Changes in the microbial amino acid metabolism induced by 4-deoxypyridoxine, *Ann. Med. Exp. Biol. Fenn.*, 37, 320, 1959; *Chem. Abstr.*, 54:4767b.
225. **Jannes, J. and Saris, N.**, Effect of 4-deoxypyridoxine on tryptophan metabolism in *Escherichia coli*, *Ann. Med. Exp. Fenn.*, 38, 442, 1960; *Chem. Abstr.*, 55:12530g.
226. **Jensen, W., Ashenbrucker, H., Cartwright, G., and Wintrobe, M.**, The uptake in vitro of radioactive iron by avian erythrocytes, *J. Lab. Clin. Med.*, 42, 833, 1953; *Chem. Abstr.*, 48:3524a.
227. **Johnson, R. and Metzler, D.**, Analyzing spectra of vitamin $B_6$ derivatives, in *Meth. Enzymol.*, 18A, 433, 1970; *Chem. Abstr.*, 74:117863.
228. **Johnson, W. and McColl, J.**, Potentiation of tumor growth inhibition from deoxypyridoxine testosterone or 6-mercaptopurine by 6-aminoicotinamide, U.S. Patent 3,148,115, 1964; *Chem. Abstr.*, 61:14479c.
229. **Johnston, F., White, S., Halsey, M., and Sevigny, Sister St. J.**, Effect of deoxypyridoxine on the amount of pyridoxal phosphate in the liver and leukocytes of rats and on leukocyte number, size and type, *J. Nutr.*, 88, 51, 1966; *Chem. Abstr.*, 64:7105b.
230. **Jones, R. G. and Kornfeld, E. C.**, Lithium aluminum hydride reduction of pyridine carboxylic esters: synthesis of vitamin $B_6$, *J. Am. Chem. Soc.*, 73, 107, 1951; *Chem. Abstr.*, 46:983b.
231. **Karasek, M.**, Effects of water soluble antimetabolites on epithelial cell growth in vitro, *Psoriasis Proc. Int. Symp.*, Farber, E. M., Ed., Stanford University Press, Stanford, CA, 1971, 271; *Chem. Abstr.*, 79:61356.
232. **Karnofsky, D.**, Assay of chemotherapeutic agents on the developing chick embryo, *Cancer Res. Suppl.*, 3, 83, 1955; *Chem. Abstr.*, 49:16226g.
233. **Karnofsky, D., Stock, C., Ridgway, L., and Patterson, P.**, The toxicity of vitamin $B_6$, 4-deoxypyridoxine and 4-methoxymethyl pyridoxine alone and in combination to the chick embryo, *J. Biol. Chem.*, 182, 471, 1950; *Chem. Abstr.*, 44:4586g.

234. **Kawakita, N., Kataoka, M., Morisawa, Y., and Yusa, T.**, Inhibitory action of alpha$^4$-norpyridoxine-5′-phosphate on aspartate amino transferase, tyrosine aminotransferase and tryptophanase, *Vitamins*, 51, 43, 1977; *Chem. Abstr.*, 86:135529.

235. **Kawakita, N. and Oyama, M.**, Studies on vitamin $B_6$ antagonists. III. Inhibitory action of alpha$^4$-norpyridoxine on the growth of *Saccharomyces carlsbergensis* 4228 in vitamin $B_1$ and vitamin $B_6$ free media, *Vitamins*, 51, 345, 1977, *Chem. Abstr.*, 87:130290.

236. **Kawakita, N., Oyama, M., and Yusa, T.**, Antivitamin $B_6$ activity of alpha$^4$-norpyridoxine on microorganisms and rat, *Vitamins*, 51, 15, 1977; *Chem. Abstr.*, 86:134131.

237. **Klein, J.**, Depression of heme formation and production of free porphyrin in duck erythrocytes, *Am. J. Physiol.*, 203, 971, 1962; *Chem. Abstr.*, 58:3709a.

238. **Klinghardt, G. W.**, Untersuchungen zur Neurotoxizitat des Isonicotinsaurehydrazids und ihrer Verhutung, *Verh. Dtsch. Ges. Inn. Med.*, 64, 352, 1958.

238a. **Knox, W. E.**, The relation of liver kynureninase activity to tryptophan metabolism in pyridoxine deficiency, *Biochem. J.*, 53, 379, 1953; *Chem. Abstr.*, 47:4974f.

239. **Koerner, W. and Nowak, H.**, Quantitative determination of the biological anti-vitamin $B_6$ effect of pyridoxal antagonists, *Int. Z. Vitaminforsch.*, 37, 89, 1967; *Chem. Abstr.*, 66:92820.

240. **Koide, T. and Furuta, K.**, Effect of vitamin $B_6$ (especially pyridoxal phosphate) on carbohydrate metabolism, *Nagoya Igaku*, 93, 283, 1970; *Chem. Abstr.*, 76:81150.

241. **Koivusalo, M., Penttila, I., Raina, A., and Tenhunen, R.**, Effect of hydrazides on liver alanine and aspartate transaminases and lactate dehydrogenase in developing chick embryo, *Acta Physiol. Scand.*, 57, 454, 1963; *Chem. Abstr.*, 59:5658c.

242. **Kolodynska, Z. and Wieniawski, W.**, Synthesis of 4-deoxypyridoxine hydrochloride, *Acta Pol. Pharm.*, 26, 271, 1969; *Chem. Abstr.*, 72:43358.

243. **Korytnyk, W.**, Gas chromatography of vitamin $B_6$, in *Meth. Enzymol.*, 18A, 500, 1970; *Chem. Abstr.*, 74:121128.

244. **Korytnyk, W. and Ahrens, H.**, Nuclear magnetic resonance spectroscopy of vitamin $B_6$, *Meth. Enzymol.*, 18A, 475, 1970; *Chem. Abstr.*, 74:118103.

245. **Korytnyk, W. and Ahrens, H.**, 5-Homopyridoxals, 5-thriopyridoxal and related compounds. Synthesis, tautomerism and biological properties, *J. Med. Chem.*, 14, 947, 1971; *Chem. Abstr.*, 76:21072.

246. **Korytnyk, W., Angelino, N., Lachmann, B., and Potti, P.**, Substrate specificity of pyridoxine dehydrogense from yeast and the synthesis and biological activities of 5-vinyl and 5-ethynyl analogs of pyridoxol, *J. Med. Chem.*, 15, 1262, 1972; *Chem. Abstr.*, 78:53225.

247. **Korytnyk, W., Fricke, G., and Paul, B.**, Pyridoxine chemistry. Gas chromatography of compounds in the vitamin $B_6$ group, *Anal. Biochem.*, 17, 66, 1966; *Chem. Abstr.*, 65:18971.

248. **Korytnyk, W., Ghosh, A. C., Potti, P. G. G., and Srivastava, S. C.**, Synthesis and biological properties of 4-amino and 4-bromo-4-norpyridoxal. New approaches for modification of the 4 position of vitamin $B_6$, *J. Med. Chem.*, 19, 999, 1976; *Chem. Abstr.*, 85:56535.

249. **Korytnyk, W., Grindey, G., and Lachmann, B.**, 4-Vinyl analog of pyridoxal, a potent antagonist of vitamin $B_6$, *J. Med. Chem.*, 16, 865, 1973; *Chem. Abstr.*, 79:132815.

250. **Korytnyk, W., Hakala, M., Potti, P. G. G., Angelino, N., and Chang, S.**, Inhibitory activity of 4-vinyl analogs of pyridoxal enzyme and cell culture studies, *Biochemistry*, 15, 5458, *Chem. Abstr.*, 86:52064.

251. **Korytnyk, W. and Ikawa, M.**, Synthesis of vitamin $B_6$ analogs, *Meth. Enzymol.*, 18A, 524, 1970; *Chem. Abstr.*, 74:99731.

252. **Korytnyk, W. and Paul, B.**, Acyl migration and selective esterification in pyridoxol, *J. Org. Chem.*, 32, 3791, 1967; *Chem. Abstr.*, 68:28937.

253. **Korytnyk, W., Paul, B., Bloch, A., and Nichol, C.**, Synthesis and antagonist properties of pyridoxal analogs modified in the 5-position, *J. Med. Chem.*, 10, 345, 1967; *Chem. Abstr.*, 66:114256.

254. **Korytnyk, W. and Potti, P.**, Antagonists of vitamin $B_6$. Simultaneous and stepwise modifications of the 2 and 4 positions, *J. Med. Chem.*, 20, 1, 1977; *Chem. Abstr.*, 86:25838.

255. **Korytnyk, W. and Potti, P.**, 4-Halovinyl and 4-ethynyl-4-deformylpyridoxal derivatives and related analogs as potentially irreversible antagonists of vitamin $B_6$, *J. Med. Chem.*, 20, 567, 1977; *Chem. Abstr.*, 87:33421.

256. **Korytnyk, W. and Singh, R.**, Proton magnetic resonance spectra of compounds in the vitamin $B_6$ group, *J. Am. Chem. Soc.*, 85, 2813, 1963; *Chem. Abstr.*, 59:9763a.

256a. **Korytnyk, W.**, personal communication.

257. **Kosower, N. and Rock, R.**, Seizures in experimental porphyria, *Nature (London)*, 217, 565, 1968; *Chem. Abstr.*, 68:86051.

258. **Kreisky, S.**, Preparation of pyridoxine derivatives especially sulfur analogs, *Monatshefte fuer Chemie*, 89, 685, 1958; *Chem. Abstr.*, 53:11370h.

259. **Kring, J., Ebisuzake, K., William, J., and Elvehjem, C.**, The influence of vitamin $B_6$ on the formation of liver pyridine nucleotides, *J. Biol. Chem.*, 195, 591, 1952; *Chem. Abstr.*, 47:10073c.

260. **Krishnaswamy, K. and Rao, S.**, Failure to produce atherosclerosis in *Macaca radiata* on a high methionine, high fat pyridoxine deficient diet, *Atherosclerosis*, 27, 253, 1977; *Chem. Abstr.*, 87:83454.
261. **Kumar, S. and Mahadevan, S.**, 3-Indoleacetaldoxime hydro-lyase: a pyridoxal-5′-phosphate-activated enzyme, *Arch. Biochem. Biophys.*, 103, 516, 1963; *Chem. Abstr.*, 60:5826f.
262. **Kuroda, T.**, Synthetic studies of vitamin $B_6$ derivatives. I. Synthesis of sulfuric esters of vitamin $B_6$ group, *Vitamins*, 28, 21, 1963; *Chem. Abstr.*, 62:515f.
263. **Kuroda, T.**, Synthetic studies of vitamin $B_6$ derivatives. V. Synthesis of 5-deoxy derivatives of the vitamin $B_6$ group, *Vitamins*, 29, 116, 1964; *Chem. Abstr.*, 62:515g.
264. **Lampert, P. and Schochet, S.**, Electron microscopic observations on experimental spongy degeneration of the cerebellar white matter, *J. Neuropath. Exp. Neurol.*, 27, 210, 1968.
265. **Lanigan, M. and Anderson, K.**, Action of antimetabolites on the growth of *Desulfovibrio desulufricans*, Scientific Studies, St. Bonaventure University, Olean, N.Y., 21, 80, 1963; *Chem. Abstr.*, 60:16247h.
266. **LaPesa, M., Grimaldi, T., Curci, E., and Semeraro, N.**, Influence of some vitamin $B_6$ antimetabolites on the induction and development of solid IRE reticulosarcoma in rats. I. Deoxypyridoxine, *Pathologica*, 62, 61, 1970, *Chem. Abstr.*, 75:74591.
267. **Lauro, V., Cucchia, G., Delli Ponti, E., and Marinucci, S.** Effects of antivitamin $B_6$ on the reproductive function of rats, *Arch. Ostet. Ginecol.*, 73, 953, 1968; *Chem. Abstr.*, 71:79464.
268. **Lauro, V., Cucchia, G., Delli Ponti, E., and Marinucci, S.**, Experimental study of the effects of antivitamin $B_6$ on reproduction in rats, *Ann. Fac. Med. Chir. Uni. Stud. Perugia*, 61, 257, 1969; *Chem. Abstr.*, 74:30610.
269. **Lauro, V. and Dominici, C.**, Effects of antivitamins on the embryos of rodents in vitro, *Scritti in Onore Del. Prof. E. Debiasi*, p. 641, 1970.
270. **Lauro, V., Dominici, C., Santilli-Giornelli, F. E., and Marinucci, S.**, Effects of antivitamins (galactoflavine, deoxypyridoxine, 6-aminonicotinamide and 4-aminomethylpteroylglutamic acid) on nidation: comparative experimental indexes in three kinds of rodents, *Arch. Ostet. Ginecol.*, 75, 1, 1970; *Chem. Abstr.*, 74:51244.
271. **Lehninger, A. L.**, *Biochemistry*, 2nd. ed., Worth Publ., New York, 1975, 197; *Chem. Abstr.*, 83:189571.
272. **LePage, G. and Greenlees, J.**, Incorporation of glycine-2-$^{14}$C into ascites tumor cell purines as a biological test system, *Cancer Res. Suppl.*, 3, 102, 1955; *Chem. Abstr.*, 49:16226i.
273. **Levine, R., Smith, E., and Clark, B.**, Effect of pyridoxal and other compounds on the mechanical activity and the sodium and potassium contents of isolated rabbit atria, *J. Pharmacol. Exp. Ther.*, 128, 159, 1960; *Chem. Abstr.*, 54:7890g.
274. **Levinson, H., Barelkovsky, J., and Bar Ilan, A.**, Nutritional effects of vitamin omission and antivitamin administration on development and longevity of the hidebeetle *(Dermestes maculatus)*, *J. Stored Prod. Res.*, 3, 345, 1967; *Chem. Abstr.*, 68:66900.
275. **Levinson, Z. and Bergmann, E.**, Vitamin deficiencies in the housefly produced by antivitamins, *J. Insect Physiol.*, 3, 293, 1959; *Chem. Abstr.*, 54:13465b.
276. **Li, T. K., Lumeng, L., and Veitch, R. L.**, Regulation of pyridoxal-5-phosphate metabolism in liver, *Biochem. Biophys. Res. Commun.*, 61, 677, 1974; *Chem. Abstr.*, 82:84864.
277. **Ling, C., Hegsted, M., and Stare, F.**, The effect of pyridoxine deficiency on the tryptophan-niacin transformation in rats, *J. Biol. Chem.*, 174, 803, 1948; *Chem. Abstr.*, 42:8280e.
278. **Loefer, J.**, Effect of pyridoxine and deoxypyridoxine on rat fibrosarcoma grafts, *Cancer Res.*, 11, 481, 1951; *Chem. Abstr.*, 45:9702e.
279. **Long, R. and Morrison, A.**, The synthesis of 4-deoxypyridoxine phosphates, *J. Chem. Soc.*, 3854, 1954; *Chem. Abstr.*, 49:12456g.
280. **Loveless, L., Spoerl, E., and Weisman, T.**, A survey of effects of chemicals on division and growth of yeast and *Escherichia coli*, *J. Bacteriol.*, 68, 637, 1954; *Chem. Abstr.*, 49:4079i.
281. **Lowe, M. and Axelrod, A. E.**, Studies on specificity of action of ribonucleic acid extracts upon viability of skin homografts in the rat, *Transplantation*, 2, 82, 1964.
282. **Lunn, A. and Morton, R.**, Ultraviolet absorption spectra of pyridoxine and related compounds, *Analyst*, 77, 718, 1952, *Chem. Abstr.*, 47:1010e.
283. **Maass, G., Ahrens, M., Schuster, P., and Winkler, H.**, Kinetic study of the hydration mechanism of vitamin $B_6$ and related compounds, *J. Am. Chem. Soc.*, 92, 6134, 1970; *Chem. Abstr.*, 74:365.
284. **McCasland, G., Gottwald, L., and Furat, A.**, 4,5-dihalo and 3-amino analogs of pyridoxine. New route to 4-deoxypyridoxine, *J. Org. Chem.*, 26, 3541, 1961; *Chem. Abstr.*, 56:11556f.
285. **McCormick, D. and Snell, E.**, Pyridoxal phosphokinases. II. Effects of inhibitors, *J. Biol. Chem.*, 236, 2085, 1961; *Chem. Abstr.*, 55:24875h.
286. **McCoy, E., Anast, C., and Naylor, J.**, The excretion of oxalic acid following deoxypyridoxine and tryptophan administration in mongoloid and nonmongoloid subjects, *J. Pediatr.*, 65, 208, 1964.
287. **McCoy, E. and Chung, S.**, Excretion of tryptophan metabolites following deoxypyridoxine administration in mongoloid and nonmongoloid patients, *J. Pediatr.*, 64, 227, 1964.

288. **McCoy, E., Colombini C., and Ebadi, M.**, Metabolism of vitamin $B_6$ in Down's syndrome, *Ann. N.Y. Acad. Sci.*, 166, 116, 1969; *Chem. Abstr.*, 72:1555.
289. **McCoy, E. and England, J.**, Excretion of 4-pyridoxic acid during deoxypyridoxine and pyridoxine administration to mongoloid and nonmongoloid subjects, *J. Nutr.*, 96, 525, 1968; *Chem. Abstr.*, 70:46115.
290. **McCoy, E. and Wehrle, H.**, The excretion of taurine during deoxypyridoxine administration in Down's syndrome patients and controls, *Proc. Soc. Exp. Biol. Med.*, 123, 170, 1966; *Chem. Abstr.*, 65:20731b.
291. **McDonald, G., McIver, J., Llamas, A., and Sky-Peck, H.**, Tumor cytotoxicity—quantitative changes in deoxyribonucleic acid (DNA) synthesis as an indicator, *Arch. Surg. (Chicago)*, 92, 541, 1966; *Chem. Abstr.*, 64:20485.
292. **Maggi, V., Magaro, M., Sensi, S., Greco, A., and Gambassi, G.**, Influence of treatment with 4-deoxypyridoxine on the synthesis of cholesterol and phospholipids from acetate-1-$^{14}C$ by rat liver, aorta and adrenals, *Boll. Soc. Ital. Biol. Sper.*, 35, 864, 1959; *Chem. Abstr.*, 55:16814c.
293. **Maggi, V., Sensi, S., Greco, A., and Magaro, M.**, Cholesterol phospholipids and glycogen of the myocardium of the left ventricle of rats treated with 4-deoxypyridoxine, *Boll. Soc. Ital. Biol. Sper.*, 35, 564, 1959; *Chem. Abstr.*, 55:16814a.
294. **Mahler, H. R. and Cordes, E. H.**, *Biological Chemistry*, Harper & Row, New York, 1966; 250; *Chem. Abtr.*, 65:7527g.
295. **Mahuren, J. D. and Coburn, S. P.**, Pyridoxal phosphate in lymphocytes, polymorphonuclear leukocytes and platelets in Down's syndrome, *Am. J. Clin. Nutr.*, 27, 521, 1974.
295a. **Mahuren, J. D. and Coburn, S.**, unpublished results.
295b. **Mancera, O., Rosenkranz, G., and Sondheimer, F.**, Steroids. XLVI. Synthesis of II-beta-hydroxytestosterone all II-oxytestosterone, *J. Chem. Soc.*, 2189, 1953; *Chem. Abstr.*, 48:8803a.
296. **Martell, A.**, Schiff bases of pyridoxal analogs: molecular species in solution, *Int. Union Biochem. Symp. Series*, 30, 13, 1962; *Chem. Abstr.*, 62:6011f.
297. **Martin, G., Avakian, S., and Moss, J.**, Studies of pyridoxine displacement, *J. Biol. Chem.*, 174, 495, 1948; *Chem. Abstr.*, 42:6430f.
298. **Martin, G., and Beiler, J.**, Inhibition of the action of DOPA decarboxylase, *Arch. Biochem.*, 15, 201, 1947; *Chem. Abstr.*, 42:1975e.
299. **Marvel, C. S. and Dreger, E. E.**, Ethyl acetopyruvate, in *Organic Synthesis*, Vol. 1, 2nd ed., Gilman, H. and Blatt, A. H., Eds., John Wiley & Sons, New York, 1961, 238.
300. **Matano, S.**, Biochemical studies on vitamin $B_6$ deficient mouse brain with special reference to experimental epilepsy. I. Cholinesterase activity and the content of sodium, potassium and water, *Okayama Igaku Zasshi*, 73, 145, 1961; *Chem. Abstr.*, 57:13108a.
301. **Matano, S.**, Biochemical studies on vitamin $B_6$ deficient mouse brain with special reference to experimental epilepsy. II. Free amino acid, *Okayama Igaku Zasshi*, 73, 153, 1961; *Chem. Abstr.*, 57:13108b.
302. **Matsumoto, S.**, Electrolytic manufacture of 4-deoxypyridoxine, Japanese Patent 41,582, 1970; *Chem. Abstr.*, 75:5719.
302a. **Maynert, E. W.**, personal communication
303. **Maynert, E. W. and Kaji, H. K.**, Relation of brain gamma-aminobutyric acid to convulsions, *J. Pharmacol. Exp. Ther.*, 137, 114, 1962; *Chem. Abstr.*, 57:11813i.
303a. **Mehansho, H., Hamm, M. W., and Henderson, L. M.**, Transport and metabolism of pyridoxal and pyridoxal phosphate in the small intestine of the rat, *J. Nutr.*, 109, 1542, 1979.
304. **Meister, A.**, Enzymatic transfer of alpha-amino groups, *Science*, 120, 43, 1954; *Chem. Abstr.*, 48:11605i.
305. **Meister, A., Morris, H., and Tice, S.**, Effect of vitamin $B_6$ deficiency on hepatic transaminase and cysteine desulfhydrase systems, *Proc. Soc. Exp. Biol. Med.*, 82, 301, 1953; *Chem. Abstr.*, 47:6004i.
306. **Meister, A., Sober, H., and Peterson, E.**, Studies on the coenzyme activation of glutamic-aspartic transaminase, *J. Biol. Chem.*, 206, 89, 1954; *Chem. Abstr.*, 48:3433i.
307. **Meister, A., Sober, H., and Tice, S.**, Enzymatic decarboxylation of aspartic acid to alanine, *J. Biol. Chem.*, 189, 577, 1951; *Chem. Abstr.*, 45:6232i.
308. **Mekhanik, M., Florentev, V., and Torchinski, Y.**, Interaction of glutamic acid apodecarboxylase with pyridoxal phosphate analogs, *Mol. Biol. (U.S.S.R.)*, 6, 586, 1972; *Chem. Abstr.*, 78:1429.
309. **Mekhanik, M., Torchinski, Y., Florentev, V., and Karpeiskii, M.**, Interaction of the apoenzyme of L-glutamate decarboxylase with pyridoxal phosphate analogs, *FEBS Lett.*, 13, 177, 1971; *Chem. Abstr.*, 74:135554.
310. **Melching, H., Abe, M., and Streffer, C.**, Studies on biological radiation protection. 59. The meaning of radiation caused changes in the content of metabolites to the survival rate of mice, *Strahlentherapie*, 125, 352, 1964; *Chem. Abstr.*, 62:9438c.

311. **Meldrum, B. and Horton, R.**, Convulsive effects of 4-deoxypyridoxine and of bicuculline in photosensitive baboons and in rhesus monkeys, *Brain Res.*, 35, 419, 1971; *Chem. Abstr.*, 76:121849.

312. **Melius, P. and Marshall, D. L.**, Synthesis and enzymological activity of some pyridoxine analogs, *J. Med. Chem.*, 10, 1157, 1967; *Chem. Abstr.*, 68:499.

313. **Mercer, N., Bowen, P., and Johnston, F.**, Effect of age, vitamin $B_6$ deficiency, isoniazid and deoxypyridoxine on the urinary taurine of the rat, *J. Nutr.*, 90, 13, 1966; *Chem. Abstr.*, 65:20578.

314. **Merck, E.**, Preparation of compounds having the effect of adermine, German Patent, 707,266, 1941; *Chem. Abstr.*, 37:502.

315. **Metzler, D., Harris, C., Johnson, R., Siano, D., and Thomson, J.**, Spectra of 3-hydroxypyridines: band-shape analysis and evaluation of tautomeric equilibria, *Biochemistry*, 12, 5377, 1973; *Chem. Abstr.*, 80:45137.

316. **Metzler, D., Ikawa, M., and Snell, E.**, A general mechanism for vitamin $B_6$ catalyzed reactions, *J. Am. Chem. Soc.*, 76, 648, 1954; *Chem. Abstr.*, 49:1722b.

317. **Metzler, D. and Snell, E.**, Deamination of serine. I. Catalytic deamination of serine and cystine by pyridoxal and metal salts, *J. Biol. Chem.*, 198, 353, 1952; *Chem. Abstr.*, 47:652f.

318. **Metzler, D. and Snell, E.**, Spectra and ionization constants of the vitamin $B_6$ group and related 3-hydroxypyridine derivatives, *J. Am. Chem. Soc.*, 77, 2431, 1955; *Chem. Abstr.*, 49:11415d.

319. **Mickelson, M. and Flippin, R.**, The use of an amino acid analog in the therapy of mouse sarcoma 180, *Arch. Biochem. Biophys.*, 64, 246, 1956; *Chem. Abstr.*, 51:613i.

320. **Mickelson, M. and Flippin, R.**, Effects of certain amino acid derivatives on microorganisms and sarcoma 180, *J. Natl. Cancer Inst.*, 20, 495, 1958; *Chem. Abstr.*, 52:11250b.

321. **Middleton, H. M.**, Uptake of pyridoxine hydrochloride by the rat jejunal mucosa in vitro, *J. Nutr.*, 107, 126, 1977; *Biol. Abstr.*, 64:19516.

322. **Middleton, H. M.**, Jejunal phosphorylation and dephosphorylation of absorbed pyridoxine HCl in vitro, *Am. J. Physiol.*, 235, E272, 1978; *Chem. Abstr.*, 89:212612.

323. **Middleton, H. M.**, In vivo absorption and phosphorylation of pyridoxine HCl in rat jejunum, *Gastroenterology*, 76, 43, 1979; *Chem. Abstr.*, 90:100869.

324. **Mihich, E. and Nichol, C.**, Differences in the selective antitumor effects of 4-deoxypyridoxine and dietary pyridoxine deficiency, *Cancer Res.*, 25, 153, 1965; *Chem. Abstr.*, 62:12334.

325. **Mihich, E. and Nichol, C.**, Antitumor effects of methylglyoxal bis (*N*-methyl-thiosemicarbazone) and their potentiation in pyridoxine-deficient animals, *Cancer Res.*, 25, 794, 1965; *Chem. Abstr.*, 63:6203b.

326. **Mihich, E., Simpson, C., Rosen, F., Gailani, S., and Nichol, C.**, Comparative toxicology of 4-deoxypyridoxine in animals previously or concurrently fed pyridoxine deficient diets, *Proc. Am. Assoc. Cancer Res.*, 5, 45, 1964.

327. **Milholland, R. and Rosen, F.**, Role of tryptophan pyrrolase adaptation in the excretion of xanthurenic acid by rats deficient in vitamin $B_6$, *Am. J. Clin. Nutr.*, 24, 740, 1971; *Chem. Abstr.*, 75:85914.

328. **Milholland, R., Rosen, F., and Nichol, C.**, Inhibitory effects of 4-deoxypyridoxine on the endogenous activity and induction of tyrosine-alpha-ketoglutarate transaminase by cortisol, *Ann. N.Y. Acad. Sci.*, 166, 126, 1969; *Chem. Abstr.*, 71:122076.

329. **Miller, T.**, Cleft palate formation: a role for pyridoxine in the closure of the secondary palate in mice, *Teratology*, 6, 351, 1972; *Chem. Abstr.*, 78:41945.

330. **Moeller, E. F.**, Biological determination of vitamin $B_6$ with use of lactic acid bacteria, *Angew Chem.*, 52, 466, 1939.

331. **Moeller, E. F.**, Growth requirements of lactic acid bacteria, *Z. Physiol. Chem.*, 260, 246, 1939; *Chem. Abstr.*, 33:9353.

332. **Moeller, E. F.**, Food and growth substances of lactic acid bacteria, *Angew Chem.*, 53, 204, 1940; *Chem. Abstr.*, 34:6666.

333. **Moeller, E. F., Zima, O., Jung, F., and Moll, T.**, Biological comparison of synthetic and natural adermin., *Naturwissenschaften*, 27, 228, 1939; *Chem. Abstr.*, 33:6917.

334. **Molina, J. and Alexander, M.**, The effect of antimetabolites on nodulation and growth of leguminous plants, *Can. J. Microbiol.*, 13, 819, 1967; *Chem. Abstr.*, 67:63148.

335. **Mora, J. and Snell, E.**, The uptake of amino acids by cells and protoplasts of *Streptococcus faecalis*, *Biochemistry*, 2, 136, 1963; *Chem. Abstr.*, 58:6012c.

336. **Morgan, H.**, Factors related to the growth of psittacosis virus (strain 6BC). IV. Certain amino acids, vitamins and other substances, *J. Exp. Med.*, 99, 451, 1954; *Chem. Abstr.*, 48:12878c.

337. **Mori, T., Mitabiya, A., and Nakanishi, S.**, Vitamin $B_6$ and its derivatives, Japanese Patent 27,813, 1965, *Chem. Abstr.*, 64:11182e.

338. **Morino, Y., Wada, H., Morisue, T., Sakamoto, Y., and Ichihara, K.**, Enzymic studies on pyridoxine metabolism. II. A pyridoxine-oxidizing enzyme system, *J. Biochem. (Tokyo)*, 48, 18, 1960; *Chem. Abstr.*, 55:2747e.

339. Morisawa, Y., Kataoka, M., Sakamoto, T., and Saito, F., Studies on anticoccidial agents. VI. Modification of the 2-position of 4-deoxypyridoxol and alpha 4-norpyridoxol, *Argric. Biol. Chem.*, 40, 101, 1976; *Chem. Abstr.*, 84:179979.

340. Morisawa, Y., Kataoka, M., Watanabe, T., Kitano, N., and Matsuzawa, T., Anticoccidial agents. I. Synthesis and anticoccidial activity of 4-deoxypyridoxol and its esters, *J. Med. Chem.*, 17, 1083, 1974; *Chem. Abstr.*, 82:25774.

341. Morisawa, Y., Kataoka, M., Watanabe, T., Kitano, N., and Matsuzawa, T., Anticoccidial agents. IV. Modification at the 5-position of 4-deoxypyridoxol and alpha[4]-norpyridoxol, *Agric. Biol. Chem.*, 39, 1275, 1975; *Chem. Abstr.*, 83:163949.

342. Morisawa, Y., Kataoka, M., Watanabe, T., and Kitano, N., Anticoccidiosis reagents, *Japan Kokai*, 75, 12, 215; *Chem. Abstr.*, 83:131469.

343. Morisue, I., Morino, T., Sakamoto, Y., and Ichihara, K. L., Enzymic studies on pyridoxine metabolism. III. Pyridoxine phosphate oxidase, *J. Biochem. (Tokyo)*, 48, 28, 1960; *Chem. Abstr.*, 55:2747f.

344. Mueller, J. and Iacono, J., Effect of deoxypyridoxine induced vitamin $B_6$ deficiency on polyunsaturated fatty acid metabolism in human beings, *Am. J. Clin. Nutr.*, 12, 358, 1963; *Chem. Abstr.*, 59:9129g.

344a. Mueller, J. F., Iacono, J., Will, J. J., and Repasky, W., The effect of deoxypyridoxine induced vitamin $B_6$ deficiency on polyunsaturated fatty acid metabolism in human beings, *J. Lab. Clin. Med.*, 54, 929, 1959.

345. Mueller, J. F. and Vilter, R., Pyridoxine deficiency in human beings induced with deoxypyridoxine, *J. Lab. Clin. Med.*, 34 (Abstr.), 1730, 1949.

346. Mueller, J. F. and Vilter, R., Pyridoxine deficiency in human beings induced with deoxypyridoxine, *J. Clin. Invest.*, 29, 193, 1950; *Chem. Abstr.*, 44:7949i.

347. Mueller, J. F., Weir, D., and Heinle, R., Relation of pyridoxine deficiency to adrenal function in production of leucocytes in mice, *Proc. Soc. Exp. Biol. Med.*, 77, 312, 1951; *Chem. Abstr.*, 45:9646g.

348. Muhlradt, P. and Snell, E., Vitamin $B_6$ analogs. An improved synthesis of 5-deoxypyridoxal, *J. Med. Chem.*, 10, 129, 1967; *Chem. Abstr.*, 66:115560.

349. Mulligan, J. and Snell, E., Transport and metabolism of vitamin $B_6$ in *Salmonella typhimurium LT2*, *J. Biol. Chem.*, 251, 1052, 1976; *Biol. Abstr.*, 61:68801.

350. Mulligan, J. H. and Snell, E. E., Transport and metabolism of vitamin $B_6$ in lactic acid bacteria, *J. Biol. Chem.*, 252, 835, 1977; *Chem. Abstr.*, 86:102904.

351. Munck, B., Amino acid transport by the small intestine of the rat. An evaluation of the possibility of vitamin $B_6$ as an intestinal amino acid carrier, *Biochem. Biophys. Acta*, 94, 136, 1965; *Chem. Abstr.*, 62:8206c.

352. Murakami, Y. and Makino, K., Convulsive action of castrix, *Biochem. Pharmacol.*, 22, 2247, 1973; *Chem. Abstr.*, 80:22863.

353. Murakami, Y. and Matsuda, M., The effect of castrix on the distribution of $B_6$ vitamers in mouse brain, *Biochem. Pharmacol.*, 25, 639, 1976; *Chem. Abstr.*, 85:29199.

354. Murakami, Y., Murakami, K., and Makino, K., Convulsive action of castrix (2-chloro-4-dimethylamino-6-methylpyrimidine), *Biochem. Pharmacol.*, 21, 277, 1972; *Chem. Abstr.*, 76:108837.

355. Mushett, C., Stebbins, R., and Barton, M., Studies on the pathologic effects produced by two analogs of pyridoxine, *Trans. N.Y. Acad. Sci.*, 9, 291, 1947; *Chem. Abstr.*, 41:6992h.

356. Mushett, C., Stebbins, R., and Barton, M., Histologic changes produced by 2,4-dimethyl-3-hydroxy-5-hydroxymethyl pyridoxine, an analogue of pyridoxine, *Am. J. Med. Sci.*, 213, 509, 1947.

357. Musini, A. and Tenconi, T. L., Parallel study of anatomical-clinical and biochemical lesions in rats deficient in vitamin A or pyridoxine, *Acta Vitaminol.*, 7, 200, 1953; *Chem. Abstr.*, 48:13851i.

358. Nagano, K. and Metzler, D. E., Machine computation of equilibrium constants and plotting of spectra of individual ionic species in the pyridoxal alanine system, *J. Am. Chem. Soc.*, 89, 2891, 1967; *Chem. Abstr.*, 67:26281.

359. Naito, T. and Ueno, K., Synthesis of vitamin $B_6$ derivatives — catalytic reduction of hydroxymethyl group substituted in pyridine ring, *Yakugaku Zasshi*, 79, 1277, 1959; *Chem. Abstr.*, 54:4566d.

360. Nakamura, H. and Inoue H., Determination of water-soluble vitamins by high-performance liquid chromatography, *Koshien Daigaku Kiyo*, 5, 1, 1977, *Chem. Abstr.*, 86:67734.

361. Nakao, A. and Greenberg, D., Cofactor requirements for the incorporation of $H_2$ $^{14}CO$ and serine-3-$^{14}C$ into methionine, *J. Am. Chem. Soc.*, 77, 6715, 1955; *Chem. Abstr.*, 50:5812g.

362. Neary, J. T., Meneely, R. L., Grever, M. R., and Diven, W. F., Interaction between biogenic amines and pyridoxal, pyridoxal phosphate, and pyridoxal kinase, *Arch. Biochem. Biophys.*, 151, 42, 1972; *Chem. Abstr.*, 77:58222.

363. Nelson, M., and Evans, H., Effect of deoxypyridoxine on reproduction in the rat, *Proc. Soc. Exp. Biol. Med.*, 68, 274, 1948; *Chem. Abstr.*, 42:7846c.

364. Nelson, M. and Evans, H., Effect of pyridoxine deficiency on reproduction in the rat, *J. Nutr.*, 43, 281, 1951; *Biol. Abstr.*, 25:30343.

365. **Nishiwaki, J.**, Uptake of vitamin $B_6$ by ascites sarcoma cells. II. Uptake of pyridoxal and pyridoxal phosphate, *Vitamin*, 42, 229, 1970; *Chem. Abstr.*, 74:51569.

366. **Nishiwaki, J.**, Uptake of vitamin $B_6$ by ascites sarcoma cells. III. Relation between uptake of vitamin $B_6$ and intracellular metabolism in the cells, *Vitamin*, 42, 238, 1970; *Chem. Abstr.*, 74:51568.

367. **Novogrodsky, A. and Meister, A.**, Protective effect of 4′-deoxypyridoxine-5′-phosphate on aspartate beta-decarboxylase: evidence for an additional function of vitamin $B_6$, *Biochim. Biophys. Acta*, 85, 170, 1964; *Chem. Abstr.*, 60:16148c.

368. **Nozaki, J.**, Convulsion by deoxypyridoxine. I. Experimental studies on the convulsion in mice, *Vitamin*, 14, 310, 1958; *Chem. Abstr.*, 54:21474e.

368a. **Nozaki, J.**, Convulsion by deoxypyridoxine. II. The effect of deoxypyridoxine on glutamic decarboxylase in the brain of mice, *Vitamin*, 14, 1958; *Chem. Abstr.*, 54:21474f.*

369. **Olivard, J. and Snell, E.**, Growth and enzymic activities of vitamin B analogs. I. D-Alanine synthesis, *J. Biol. Chem.*, 213, 203, 1955; *Chem. Abstr.*, 49:8387f.

370. **Olsen, N. and Martindale, W.**, Studies on chronic vitamin $B_6$ deficiency in the rat. I. Changes in the intact animal, *J. Nutr.*, 53, 317, 1954; *Chem. Abstr.*, 48:12942c.

371. **Olsen, N. and Martindale, W.**, Studies on chronic vitamin $B_6$ deficiency in the rat. II. Changes in tissue metabolism, *J. Nutr.*, 53, 329, 1954; *Chem. Abstr.*, 48:12942c.

372. **Ono, S., Hirota, Y., Ito, Y., Watanabe, T., and Kunii, A.**, Mastocytoma and cancerostatic agent, *Saishin Igaku*, 18, 669, 1963; *Chem. Abstr.*, 59:9222c.

373. **Osetrova, S.**, Role of antivitamins after homoplastic skin transplants, *K. Mekh. Deistviya. Vitam. Zhivotn. Rast. Organizmy.*, 27, 1971; *Chem. Abstr.*, 79:73799.

374. **Osetrova, S. and Eingorn, A.**, Effect of tubazid and deoxypyridoxine on the preservation of skin homotransplants, *Farmakol. Toksikol.*, 35, 318, 319, 1972; *Chem. Abstr.*, 77:56648.

375. **Ott, W.**, Antipyridoxine activity of 2,4-dimethyl-3-hydroxy-5-hydroxymethylpyridine in the chick, *Proc. Soc. Exp. Biol. Med.*, 61, 125, 1946; *Chem. Abstr.*, 40:2873.

376. **Ott, K.**, Antipyridoxine activity of methoxypyridoxine in the chick, *Proc. Soc. Exp. Biol. Med.*, 66, 215, 1947.

377. **Oya, N.**, Pyridoxine uptake in *Escherichia coli*, *Vitamin*, 41, 222, 1970; *Chem. Abstr.*, 73:1101.

378. **Ozawa, T., Sano, M., and Tomita, I.**, Biological and coenzymic activity of vitamin $B_6$ derivatives. I. Activities of $B_6$ analogs modified at 3 or 5 position of pyridoxol in growth of *Saccharomyces carlsbergensis*, *Vitamin*, 46, 59, 1972; *Chem. Abstr.*, 77:135591.

379. **Ozawa, T., Sano, M., and Tomita, I.**, Biological and coenzymic activity of $B_6$ derivatives. II. Chemical and biological properties of 3-aminopyridoxal, *Vitamin*, 46, 97, 1972; *Chem. Abstr.*, 77:149024.

380. **Padyukova, N. S., Florentev, V. L., and Karpeiskii, M. Y.**, Synthesis and physicochemical properties of pyridoxal keto analogs and their derivatives, *Izv. Akad. Nauk. SSSR. Ser. Khim.*, 1101, 1973; *Chem. Abstr.*, 79:91922.

381. **Pal, P.**, Effect of vitamin $B_6$ and its analog on the swelling of rat liver mitochondria, *Nature (London)*, 200, 173, 1963; *Chem. Abstr.*, 60:2203.

382. **Pappas, G.**, Effects of vitamin antimetabolites on *Lebistes reticulatus*, *Zoologica*, 51, 77, 1966; *Chem. Abstr.*, 66:62970.

383. **Parkes, A.**, Dietary factors in the homograft reaction, *Nature (London)*, 184, 699, 1959; *Biol. Abstr.*, 35:12652.

384. **Patzer, E. and Hilker, D.**, New reagent for vitamin $B_6$ derivative formation in gas chromatography, *J. Chromatogr.*, 135, 489, 1977; *Biol. Abstr.*, 65:62301.

385. **Paul, K., Kumlien, A., and Brody, S.**, Eosinophil granulocytes. II. The effect of cytostatics on the uterine response to estrogen, *Experientia*, 22, 799, 1966; *Chem. Abstr.*, 66:35166.

386. **Peluffo, R., Brenner, R., and Mercuri, O.**, Action of linoleic and arachidonic acids upon the eicosatrienoic acid level in rat heart and liver, *J. Nutr.*, 81, 110, 1963; *Chem. Abstr.*, 60:2111b.

387. **Peterson, E. and Sober, H.**, Preparation of crystalline phosphorylated derivatives of vitamin $B_6$, *J. Am. Chem. Soc.*, 76, 169, 1954; *J. Biol. Chem. Soc.*, 206, 89, 1954; *Chem. Abstr.*, 49:9645f.

388. **Pfeiffer, C.**, Some observations regarding hydrazides and convulsions, in *Inhibition in the Nervous System and Gamma-Aminobutyric Acid*, Roberts, E., Baxter, C. F., van Harraveld, A., Wiersma, C. A. G., Ross, W. R., and Killam, K F., Eds., Pergamon Press, Elmsford, N.Y., 1959, 324.

389. **Pflug, W. and Lingens, F.**, Vitamin $B_6$ biosynthesis in *Bacillus subtilis*, *Hoppe Seyler's Z. Physiol. Chem.*, 359, 559, 1978; *Chem. Abstr.*, 89:56226.

390. **Picton, C.**, Studies of the Metabolism of 3-Hydroxy-2,4,5-Trihydroxymethylpyridine, Ph.D. thesis, University of Leeds, Leeds, England, 1975.

391. **Pike, R., and Brown, M.**, Changes in hemoglobin, hematocrit and plasma protein in vitamin $B_6$ deficient rats during pregnancy, *J. Nutr.*, 68, 551, 1959; *Chem. Abstr.*, 53:22307c.


The citation in *Chemical Abstracts* lacks page numbers and we have been unable to confirm this reference from any other source.

392. **Pike, R. and Kirksey, A.**, Some effects of isonicotinic acid hydrazide-induced vitamin $B_6$ deficiency in pregnant rats, *J. Nutr.*, 68, 561, 1959; *Chem. Abstr.*, 53:22307d.
393. **Platzer, E. G. and Kassis, J. A.**, Pyridoxine kinase in *Plasmodium lophurae* and duckling erythrocytes, *J. Protozool.*, 25, 556, 1978; *Biol. Abstr.*, 68:10881.
394. **Pocker, A.**, Synthesis of 2-nor-2-formyl pyridoxal-5-phosphate, a bifunctional reagent specific for the cofactor site in proteins, *J. Org. Chem.*, 38, 4295, 1973; *Chem. Abstr.*, 80:14815.
395. **Pol, G., Hartog, C., Meyknecht, E., Keller, G., Wetemans, L., and Fernik, J.**, Vitamins A, E, $B_6$ and lipid metabolism: interrelation in the rat, *Intern. Z. Vitaminforsch.*, 35, 364, 1965; *Chem. Abstr.*, 64:14661.
396. **Porter, C., Clark, I., and Silber, R.**, The effect of pyridoxine analogs on tryptophan metabolism in the rat, *J. Biol. Chem.*, 167, 573, 1947; *Chem. Abstr.*, 41:3215h.
397. **Pressey, R.**, Inhibition of invertases by pyridoxal and its analogs, *Biochem. Biophys. Acta*, 159, 414, 1968; *Chem. Abstr.*, 69:16373.
398. **Price, J.**, Tryptophan metabolism in human subjects, *Tanabe Amino Acid Res. Found. Symp.*, 1, 227, 1964; *Chem. Abstr.*, 63:6112.
399. **Price, J., Brown, R., and Larson, F.**, Quantitative studies on human urinary metabolites of tryptophan as affected by isoniazid and deoxypyridoxine, *J. Clin. Invest.*, 36, 1600, 1957; *Chem. Abstr.*, 52:7527c.
400. **Prostenik, M. and Butula, I.**, Selective reduction of some halogen substituted pyridines of the vitamin $B_6$ series, *Chem. Ind. (Berlin)*, 110, 2106, 1977; *Chem. Abstr.*, 87:102135.
401. **Rabinowitz, J. and Snell, E.**, Vitamin $B_6$ antagonists and growth of microorganisms. I. 4-Deoxypyridoxine, *Arch. Biochem. Biophys.*, 43, 399, 1953; *Chem. Abstr.*, 47:9414h.
402. **Rabinowitz, J. and Snell, E.**, Vitamin $B_6$ antagonists and growth of microorganisms. II. 5-Deoxypyridoxal and related compounds, *Arch. Biochem. Biophys.*, 43, 408, 1953; *Chem. Abstr.*, 47:9415a.
403. **Rasmussen, A.**, Plasma amino acid responses in the rat to dietary versus antagonist-induced vitamin $B_6$ deficiency, *Univ. Del. Agr. Exp. Sta. Bull.*, 408, 1974; *Chem. Abstr.*, 81:134936.
404. **Reddy, S. K., Reynolds, M. S., and Price, J. M.**, The determination of 4-pyridoxic acid in human urine, *J. Biol. Chem.*, 233, 691, 1958; *Chem. Abstr.*, 53:2338d.
405. **Reeves, R., Meleney, H., and Frye, W.**, Cultivation of *Entamoeba histolytica* with penicillin-inhibited *Bacteroides symbiosus* cells. I. Pyridoxine requirement, *Am. J. Hyg.*, 69, 25, 1959; *Chem. Abstr.*, 53:15194a.
406. **Reiber, H.**, Photochemical reactions of vitamin $B_6$ compounds, isolation and properties of products, *Biochem. Biophys. Acta*, 279, 310, 1972; *Chem. Abstr.*, 78:1336.
406a. **Reiber, H.**, personal communication.
407. **Reiber, H.**, Vitamin $B_6$-catalized beta-elimination of serine and orthophosphoserine. Qualitative and quantitative aspects of catalytic influences at the rate limiting step, a comparison with the rate of enzymic beta-elimination, *Biochim. Biophys. Acta*, 444, 734, 1976; *Chem. Abstr.*, 86:13178.
408. **Reider, M. J. and Elderfield, R. C.**, Studies in the pyridine series. V. Reactions involving the ortho effect in certain beta, gamma substituted pyridines, *J. Org. Chem.*, 7, 286, 1942; *Chem. Abstr.*, 36:5173.
409. **Reio, L.**, Supplementary data for the paper-chromatographic separation and identification of phenol derivatives and related compounds of biochemical interest, using a reference system, *J. Chromatogr.*, 4, 458, 1960; *Chem. Abstr.*, 55:18552f.
410. **Remaggi, P.**, Taste and vitamin $B_6$, *Boll. Mal. Orecch. Gola Naso*, 77, 361, 1959.
411. **Richter, R.**, Antimetabolites and fetal development, *Helv. Physiol. Pharmacol. Acta*, 18, C46, 1960; *Chem. Abstr.*, 54:15682i.
412. **Riggs, T. and Walker, L.**, Diminished uptake of alpha-aminoisobutyric acid by tissues of vitamin $B_6$ deficient rats, *J. Biol. Chem.*, 233, 132, 1958; *Chem. Abstr.*, 53:521c.
413. **Robbins, W. J. and Ma, R.**, Specificity of pyridoxine for *Ceratostomella ulmi*, *Bull. Torrey Bot. Club*, 69, 342, 1942; *Chem. Abstr.*, 36:4854.
414. **Roberts, E., Younger, F., and Frankel, S.**, Influence of dietary pyridoxine on glutamic decarboxylase activity of brain, *J. Biol. Chem.*, 191, 277, 1951; *Chem. Abstr.*, 46:172b.
415. **Roberts, L. and Mong, F.**, Developmental physiology of cestodes. XIII. Vitamin $B_6$ requirement of *Hymenolepis diminuta* during in vitro cultivation *J. Parasitol.*, 59, 101, 1973; *Chem. Abstr.*, 78:146561.
416. **Robson, L. C. and Schwarz, M. R.**, Vitamin $B_6$ deficiency and the lymphoid system. I. Effects on cellular immunity and in vitro incorporation of $^3$H-uridine by small lymphocytes, *Cell Immunol.*, 16, 135, 1975; *Biol. Abstr.*, 59:66918.
417. **Robson, L. C. and Schwarz, M. R.**, Vitamin $B_6$ deficiency and the lymphoid system. II. Effects of vitamin $B_6$ deficiency in utero on the immunological competence of the offspring, *Cell Immunol.*, 16, 145, 1975; *Biol. Abstr.*, 59:66918.
418. **Rodwell, V. W., Volcani, B. E., Ikawa, M., and Snell, E. E.**, Bacterial oxidation of Vitamin $B_6$. Isopyridoxal and 5-pyridoxic acid, *J. Biol. Chem.* 233, 1548, 1958; *Chem. Abstr.*, 53:8299g.

419. **Rose, D. and Toseland, P.**, Urinary excretion of quinolinic acid and other tryptophan metabolites after deoxypyridoxine or oral contraceptive administration, *Metab. Clin. Exp.*, 22, 165, 1973; *Chem. Abstr.*, 78:38207.

420. **Rosen, F.**, Effect of isonicotinic acid hydrazide on niacin and pyridoxine metabolism in rats, *Proc. Soc. Exp. Biol. Med.*, 88, 243, 1955; *Chem. Abstr.*, 49:7076h.

421. **Rosen, F., Mihich, E., and Nichol, C.**, Selective metabolic and chemotherapeutic effects of vitamin $B_6$ antimetabolites, *Vitam. Horm. (N.Y.)*, 22, 609, 1964; *Chem. Abstr.*, 62:13402d.

422. **Rosen, F. and Milholland, R.**, Effects of 4-deoxypyridoxine on peripheral lymphocytes and the growth of two lymphosarcomas, *Proc. Am. Assoc. Cancer Res.*, 4, 58, 1963.

423. **Rosen, F., Milholland, R. K., and Nichol, C. A.**, Studies on the inhibition of L-glutamic acid decarboxylase in vivo, in *Inhibition in the Nervous System and Gamma-Aminobutyric Acid*, Roberts, E., Baxter, C. F., van Harreveld, A., Wiersma, C. A. G., Ross, W. R., and Killam, K. F., Eds., Permagon Press, Elmsford, N.Y., 1959, 338; *Chem. Abstr.*, 61:16661a.

424. **Rosen, F., Milholland, R. and Nichol, C.**, Biochemical studies on the mode of action of 4-deoxypyridoxine, in *3rd Symp. Pyridoxal Enzymes*, Yamada, K., Ed., Maruzen, Tokyo, 1967, 177; *Chem. Abstr.*, 70:86126.

425. **Ross, M. and Pike, R.**, The relationship of vitamin $B_6$ to protein metabolism during pregnancy in the rat, *J. Nutr.*, 58, 251, 1956; *Chem. Abstr.*, 50:7975b.

426. **Ross, M. and Pike, R.**, The relationship of vitamin $B_6$ to serum protein and nonprotein nitrogen in the rat during pregnancy, *J. Nutr.*, 60, 211, 1956; *Chem. Abstr.*, 51:2961d.

427. **Ross, M. and Pike, R.**, Vitamin $B_6$ and reproduction, *J. Am. Diet. Assoc.*, 33, 42, 1957; *Chem. Abstr.*, 52:2195i.

428. **Runyan, T. and Gershoff, S.**, Glycine metabolism in vitamin $B_6$ deficient and deoxypyridoxine treated rats, *J. Nutr.*, 98, 113, 1969; *Chem. Abstr.*, 71:19949.

428a. **Sakami, W.**, The biochemical relationship between glycine and serine in *Amino Acid Metabolism*, McElroy, W. B., Glass, H. B., Eds., Johns Hopkins Press, Baltimore, 1955, 658; *Chem. Abstr.*, 49:9703b.

429. **Sakuragi, T., Argoudelis, C., and Kummerow, F.**, Some effects of antagonists and analogs of vitamin $B_6$ in yeast, *Arch. Biochem. Biophys.*, 89, 160, 1960; *Chem. Abstr.*, 54:25021a.

430. **Sakuragi, T. and Kummerow, F.**, The biological utilization of various fat soluble esters of pyridoxine and 4-deoxypyridoxine by rats, *J. Nutr.*, 58, 557, 1956; *Chem. Abstr.*, 50:15771h.

431. **Sakuragi, T. and Kummerow, F.**, Synthesis of long chain fatty acid derivatives of vitamin $B_6$ group, *J. Am. Chem. Soc.*, 78, 839, 1956; *Chem. Abstr.*, 50:13006f.

432. **Sakuragi, T. and Kummerow, F.**, Behavior of various antivitamin $B_6$ compounds in *Saccharomyces carlsbergensis*, *Arch. Biochem. Biophys.*, 82, 89, 1959; *Chem. Abstr.*, 53:15205f.

433. **Sakurai, Y., Shiroishi, M., and Masuhara, T.**, The vitamin $B_6$ group. I. The physiological effect of pyridoxine and allied compounds on albino rats, *Rept. Food Res. Inst.*, 5, 63, 1951; *Chem. Abstr.*, 49:16104g.

434. **Sanders, L., Cetorelli, J., and Winefordner, J.**, Phosphorescence characteristics of several antimetabolites, *Talanta*, 16, 407, 1969; *Chem. Abstr.*, 70:103598.

435. **Sano, M., Tomita, I., Ozawa, T., Ikekawa, T., Tanaki, M., and Fukuoka, F.**, Antitumor effect of polysaccharides and its relationship to biogenetic amines in tumor tissues under vitamin $B_6$ deficient feeding, *Chem. Pharm. Bull.*, 21, 2090, 1973; *Chem. Abstr.*, 80:22764.

436. **Sartori, E. and Peressini, A.**, Antivitamin $B_6$ and urinary xanthurenic acid from tryptophan administration in children, *Acta Paediatr. Lat.*, 6, 929, 1953; *Chem. Abstr.*, 48:13005b.

437. **Sassenrath, E.**, Total packed cell volume as a measure of growth retardation in mouse ascites tumors, *Ann. N.Y. Acad. Sci.*, 76, 601, 1958; *Biol. Abstr.*, 33:33958.

438. **Sassenrath, E., Costa, F., and Greenberg, D.**, Growth inhibition data from a primary screen against ascites tumors, *Cancer Res.*, 18, 341, 1958; *Chem. Abstr.*, 53:18275h.

439. **Sauberlich, H. E.**, Active compounds and antagonists and biochemical systems and biochemical detection of deficiency, in *The Vitamins*, Sebrell, W. H. and Harns, R., Eds, Academic Press, New York, 1968, 40.

440. **Savin, F. A., Bazhulina, N. P., and Yu, V.**, Morozov, Pi-electronic structure of various ionic forms of 3-hydroxypyridine and some of its derivatives, *Mol. Biol.*, 7, 674, 1973; *Chem. Abstr.*, 80:47282.

441. **Sawaya, C., Horton, R., and Meldrum, B.**, Transmitter synthesis and convulsant drugs. Effects of pyridoxal phosphate antagonists and allylglycine, *Biochem. Pharmacol.*, 27, 475, 1978. *Chem. Abstr.*, 89:16872.

442. **Schreiner, A., Slinger, W., Hawkins, V., and Vilter, R.**, Seborrheic dermatitis, a local metabolic defect involving pyridoxine, *J. Lab. Clin. Med.*, 40, 121, 1952; *Chem. Abstr.*, 47:207e.

442a. **Schrodt, G., Dizon, F., Peskoe, L., Howell, R., Robie, C., and Stevenson, T.**, The effect of pyridoxine antagonists in the carcinoid syndrome, *Am. Practitioner Dig. Treat.*, 11, 750, 1960.

443. Schultz, J., Rothman, N., and Aronson, M., The growth and morphogenesis of *Drosophila melanogaster* as criteria for screening tests, *Cancer Res. Suppl.* 3, 86, 1955; *Chem. Abstr.*, 49:16226h.
444. Scott, T. A. and Hockney, R. C., Synthesis of vitamin $B_6$ by a mutant of *Escherichia coli* K12 and the action of 4-deoxypyridoxine, *J. Gen. Microbiol.*, 110, 285, 1979; *Chem. Abstr.*, 91:71475.
445. Shane, B. and Snell, E., The nature of pyridoxine oxidase from rabbit liver preparations, *Arch. Biochem. Biophys.*, 153, 333, 1972; *Chem. Abstr.*, 78:54665.
446. Shane, B. and Snell, E., Metabolism of 5′-deoxypyridoxine in rats, 5′-deoxypyridoxine, 4′-sulfate as a major urinary metabolite, *Biochem. Biophys. Res. Commun.*, 66, 1294, 1975; *Chem. Abstr.*, 84:26474.
447. Shane, B. and Snell, E., Transport and metabolism of vitamin $B_6$ in the yeast *Saccharomyces carlsbergensis* 4228, *J. Biol. Chem.*, 251, 1042, 1976; *Biol. Abstr.*, 61:69135.
448. Shapiro, D., Dietrich, L., and Shils, M., Quantitative biochemical differences between tumor and host as a basis for cancer chemotherapy. II. Riboflavin, *Cancer Res.*, 16, 575, 1956; *Chem. Abstr.*, 50:17186e.
449. Shapiro, D., Fugmann, R., and Dietrich, L., Partial regression of a solid tumor with multi-compound chemotherapy, *Cancer Res.*, 17, 1067, 1957; *Chem. Abstr.*, 52:5623e.
450. Shapiro, D. and Gellhorn, A., Combinations of chemical compounds in experimental cancer therapy, *Cancer Res.*, 11, 35, 1951; *Chem. Abstr.*, 45:8655d.
451. Shapiro, D. and Shils, M., Some vitamin hormone relationships in cancer chemotherapy, *Proc. Am. Assoc. Cancer Res.*, 1, 49, 1953.
452. Shapiro, D., Shils, M., and Dietrich, L., Quantitative biochemical differences between tumor and host as a basis for cancer chemotherapy. I. Vitamin $B_6$, *Cancer Res.*, 13, 703, 1953; *Chem. Abstr.*, 48:8957.
453. Shwartzmann, G. and Hift, H., Transamination reaction in normal and $B_6$ deficient hamsters, *J. Nutr.*, 44, 575, 1951; *Chem. Abstr.*, 45:10324a.
454. Shyamala, M. and Bhat, J., Antimetabolites in the nutrition of the silkworm *Bombyx mori L.*, I. Niacin and pyridoxine antagonists, *J. Insect. Physiol.*, 2, 137, 1958; *Chem. Abstr.*, 52:17539h.
455. Sieburth, J. and McLaren, B., Growth studies with the cockroach *Periplaneta americana* fed vitamin deficient diets substituted with corresponding antivitamins, *Ann. Entomol. Soc. Am.*, 46, 43, 1953; *Chem. Abstr.*, 47:11579b.
456. Silva, M., Morphologic changes in pyridoxine deficiency: experimental study, *G.E.N.* 14, 307, 1960; *Chem. Abstr.*, 55:3744h.
457. Singh, R. and Korytnyk, W., Pyridoxine chemistry. VII. Some modifications in the 4-position of pyridoxal, *J. Med. Chem.*, 8, 116, 1965; *Chem. Abstr.*, 62:5251.
458. Sinha, A. and Chatterjee, G., Effect of certain metabolic inhibitors on the growth of parent and a pyrithiamine mutant *Staphylococcus aureus*, *Ind. J. Biochem.*, 5, 73, 1968; *Chem. Abstr.*, 69:74700.
459. Skipper, H. and Thomson, J., Test of compounds against certain mouse tumors and leukemias, *Cancer Res. Suppl.*, 2, 147, 1955.
460. Skipper, H. and Thomson, J., A preliminary study of the influence of amino acid deficiencies on experimental cancer chemotherapy, in *CIBA Found. Symp. on Amino Acids and Peptides with Antimetabolic Activity*, Little, Brown, Boston, 1958, 38; *Chem. Abstr.*, 53:14170b.
461. Skipper, H., Thomson, J., and Schabel, F., Jr., Experimental evaluation of potential anticancer agents. XI. Quantitative response of the sarcoma 180 system to deoxypyridoxine, *Cancer Chemotherapy Rep.*, 29, 63, 1963; *Chem. Abstr.*, 59:14459e.
462. Smejkal, V. and Smejkalova, E., Effect of thyroxine and related compounds on aspartate-oxoglutarate transaminase activity as studied in human leukocytes in vitro, *Enzymologia*, 33, 320, 1967; *Chem. Abstr.*, 68:36344.
463. Smith, C., Lummis, W., and Grady, J., An improved tissue culture assay. II. Cytotoxicity studies with antibiotics, chemicals and solvents, *Cancer Res.*, 19, 847, 1959; *Chem. Abstr.*, 53:22437d.
464. Snell, E., The vitamin $B_6$ group. I. Formation of additional members from pyridoxine and evidence concerning their structures, *J. Am. Chem. Soc.*, 66, 2082, 1944; *Chem. Abstr.*, 39:1166.
465. Snell, E., Chemical structure in relation to biological activities of vitamin $B_6$, *Vitam. Horm.*, 16, 77, 1958; *Chem. Abstr.*, 53:544a.
466. Snell, E., and Ayling, J., Relation of transaminase activity to structure of pyridoxal and its analogs, in *3rd Symp. Pyridoxal Enzymes*, Yamata, K., Ed., Maruzen, Tokyo, 1967, 5; *Chem. Abstr.*, 70:84551.
467. Snodgrass, S. and Iversen, L., Effects of amino-oxyacetic acid on tritium labeled GABA uptake by rat brain slices, *J. Neurochem.*, 20, 431, 1973; *Chem. Abstr.*, 79:797.
468. Spector, R., Vitamin $B_6$ transport in the central nervous system in vitro studies, *J. Neurochem.*, 30, 889, 1978; *Chem. Abstr.*, 89:143975.
469. Spector, R. and Greenwald, L., Transport and metabolism of vitamin $B_6$ in rabbit brain and choroid plexus, *J. Biol. Chem.*, 253, 2373, 1978; *Chem. Abstr.*, 88:167378.

470. **Spector, R. and Shikuma, S.**, The stability of vitamin $B_6$ accumulation and pyridoxal kinase activity in rabbit brain and choroid plexus, *J. Neurochem.*, 31, 1403, 1978; *Chem. Abstr.*, 90:165441.

471. **Spiera, H. and Vallarino, R.**, Serum kynurenine in rheumatoid arthritis, *J. Clin. Invest.*, 48, 856, 1969; *Chem. Abstr.*, 71:28572.

472. **Stinebring, W., Trakatellis, A., and Axelrod, A.**, Studies on systemic reactivity to purified protein derivative and endotoxin. II. Systemic reactivity to endotoxin in pyridoxine deficient guinea pigs and its relation to systemic reactivity to PPD, *J. Immunol.*, 91, 46, 1963; *Chem. Abstr.*, 59:15668f.

473. **Stoerk, H.**, Effects of calcium deficiency and pyridoxine deficiency on thymic atrophy, *Proc. Soc. Exp. Biol. Med.*, 62, 90, 1946; *Chem. Abstr.*, 40:4778.

474. **Stoerk, H.**, The regression of lymphosarcoma implants in pyridoxine deficient mice, *J. Biol. Chem.*, 171, 437, 1947; *Chem. Abstr.*, 42:2004a.

475. **Stoerk, H.**, Deoxypyridoxine—morphological and functional changes in acute pyridoxine deficiency, *Fed. Proc. Fed. Am. Soc. Exp. Biol.*, 7, 281, 1948.

476. **Stoerk, H.**, Growth retardation of lymphosarcoma implants in pyridoxine deficient rats by testosterone and cortisone, *Proc. Soc. Exp. Biol. Med.*, 74, 798, 1950; *Chem. Abstr.*, 44:10072c.

477. **Stoerk, H.**, Desoxypyridoxine observations in acute pyridoxine deficiency, *Ann. N.Y. Acad. Sci.*, 52, 1302, 1950; *Chem. Abstr.*, 44:10835b.

478. **Stoerk, H., Eisen, H., and John, H.**, Impairment of antibody response in pyridoxine deficient rats, *J. Exp. Med.*, 85, 365, 1947; *Chem. Abstr.*, 41:4550a.

479. **Stoerk, H. and Zucker, T.**, Nutritional effects on the development and atrophy of the thymus, *Proc. Soc. Exp. Biol. Med.*, 56, 151, 1944; *Chem. Abstr.*, 38:5265.

480. **Stokes, J. L., Larsen, A., Woodward, C. R., and Foster, J. W.**, A neurospora assay for pyridoxine, *J. Biol. Chem.*, 150, 17, 1943; *Chem. Abstr.*, 37:6690.

481. **Stoll, R., Bodit, F., and Maraud, R.**, Teratogenic activity of semicarbazide — role of vitamin $B_6$ and biogenic amines, *C. R. Soc. Biol.*, 164, 1011, 1970; *Chem. Abstr.*, 74:96083.

482. **Stone, W.**, Effects of alterations in the metabolism of alpha-aminobutyrate on convulsant potencies, *Epilepsia*, 18, 507, 1977; *Biol. Abstr.*, 65:41402.

483. **Struck, R., Shealey, Y., and Montgomery, J.**, Vitamin $B_6$ analogs. IV. 4-Deoxyisopyridoxal and the phosphonic acid analog of 4-deoxypyridoxine phosphate, *J. Med. Chem.*, 14, 568, 1971; *Chem. Abstr.*, 75:76948.

484. **Suda, T.**, Participation of pyridoxal phosphate in phosphatidylserine decarboxylase, *Vitamin*, 49, 53, 1975; *Chem. Abstr.*, 82:134772.

485. **Suda, T. and Matsuda, M.**, Nonenzymic decarboxylation of alpha-keto acids by pyridoxal phosphate and manganous ion. I. Characteristics and properties of the reaction, *Jikeikai Med. J.*, 22, 137, 1975; *Chem. Abstr.*, 84:26998.

486. **Sugiura, K.**, Relative sensitivity of the solid and ascites forms of sarcoma 180 and Ehrlich carcinoma to inhibitory compounds, *Ann. N.Y. Acad. Sci.*, 76, 575, 1958; *Chem. Abstr.*, 53:10573a.

487. **Sugiura, K. and Creech, H.**, Merits of ascites tumors for chemotherapeutic screening, *Ann. N.Y. Acad. Sci.*, 63, 962, 1956; *Chem. Abstr.*, 50:13284d.

488. **Summers, W. A.**, The effect of 2,4-dimethyl-3-hydroxy-5-hydroxymethyl pyridine upon experimental toxoplasmosis in mice, *Exp. Parasitol.*, 6, 194, 1957; *Biol. Abstr.*, 32:28046.

489. **Sundaram, E., Ranganathan, G., and Sarma, P.**, The interrelationship among vitamins and amino acids. Influence of deoxypyridoxine on the biosynthesis of nicotinic and ascorbic acids in germinating pulses, *Curr. Sci.*, 20, 122, 1951; *Chem. Abstr.*, 46:4064d.

490. **Swell, L., Law, M. D., Schools, P. E., and Treadwell, C. R.**, Tissue lipid fatty acid composition in pyridoxine deficient rats, *J. Nutr.*, 74, 148, 1961; *Biol. Abstr.*, 36:59490.

491. **Szmelcman, S. and Guggenheim, K.**, The effect of deoxypyridoxine on portal vein levels of amino acids after a protein meal, *Ann. Biol. Anim. Biochim. Biophys.*, 3, 299, 1963; *Chem. Abstr.*, 60:9667.

492. **Taborsky, R.**, Preparation of 5-hydroxy-4,6-dimethyl-3-pyridinemethanol (4-deoxypyridoxine) by the use of hydrazine, *J. Org. Chem.*, 26, 596, 1961; *Chem. Abstr.*, 55:16561b.

493. **Tadera, K., Yasumoto, K., Tsuji, H., and Mitsuda, H.**, Automated analysis of vitamin $B_6$ complex by ion-exchange column chromatography. I. Diazotized 5-chloroaniline-2,4-disulfonylchloride as a color producing reagent, *Vitamin*, 48, 69, 1974; *Chem. Abstr.*, 80:142651.

494. **Talageri, V.**, Some biochemical observations in induced pyridoxine deficiency in rats, *J. Postgrad. Med. (Bombay)*, 7, 175, 1961; *Biol. Abstr.*, 38:13152.

495. **Tani, H.**, Reaction of n-alkoxypyridinium derivatives. V., *Yakugaku Zasshi*, 81, 182, 1961; *Chem. Abstr.*, 55:14450e.

496. **Tarnowski, G. and Stock, C.**, Effects of combinations of azaserine and of 6-diazo-5-oxo-L-norleucine with purine analogs and other antimetabolites on the growth of two mouse mammary carcinomas, *Cancer Res.*, 17, 1033, 1957; *Chem. Abstr.*, 52:5657d.

497. **Tasaki, J. and Yokota, H.** Effect of deoxypyridoxine on the intestinal amino acid absorption in chickens in situ, *Nippon Chikusan Gakkai-Ho*, 41, 104, 1970; *Chem. Abstr.*, 76:81708.

498. **Tate, S. and Meister, A.**, Effects of various vitamin $B_6$ 5′-phosphate derivatives on the structure and activities of L-aspartate beta-decarboxylase, *Biochemistry*, 8, 1056, 1969; *Chem. Abstr.*, 70:111863.
499. **Tedeschi, G.**, Action of deoxypyridoxine phosphate ester on benzopyrene induced sarcoma in the rat, *3 Giornata Sci., Consiglio Nazl. Richerche Convegno Vitamine Milan*, Consiglio Nazionale delle Ricerche, Rome, 1953, 642; *Chem. Abstr.*, 49:4104g.
500. **Thanassi, J.**, Aminomalonic acid. Spontaneous decarboxylation and reaction with 5-deoxypyridoxal, *Biochemistry*, 9, 525, 1970; *Chem. Abstr.*, 72:90848.
501. **Thanassi, J.**, Alpha-methylamino-malonate dependent reaction of 5-deoxypyridoxal, *Biochemistry*, 11, 2909, 1972; *Chem. Abstr.*, 77:71712.
502. **Thanassi, J.**, Vitamin $B_6$ catalysis alpha-phenylaminomalonate-catalyzed reactions of 5-deoxypyridoxal, *Biochemistry*, 12, 5109, 1973; *Chem. Abstr.*, 80:44803.
503. **Thanassi, J.**, Vitamin $B_6$-catalized reactions of alpha-amino and alpha-amino-alpha methyldiethylmalonate, *Bioorg. Chem.*, 4, 132, 1975; *Chem. Abstr.*, 83:114853.
504. **Tomita, O., Brooks, H., and Metzler, D.**, Synthesis of vitamin $B_6$ derivatives. II. 3-Hydroxy-4-hydroxymethyl-2-methyl-5-pyridine acetic acid and related substances, *J. Heterocyclic Chem.*, 3, 178, 1966; *Chem. Abstr.*, 65:5436g.
505. **Trakatellis, A. and Axelrod, A.**, Effect of pyridoxine deficiency on the induction of immune tolerance in mice, *Proc. Soc. Exp. Biol. Med.*, 132, 46, 1969; *Biol. Abstr.*, 51:56639.
506. **Trakatellis, A., Axelrod, A., and Montjar, M.**, Studies on nucleic acid metabolism in pyridoxine deficient rats, *Fed. Proc. Fed. Am. Soc. Exp.* Biol., 23, 137, 1964.
507. **Trakatellis, A., Stinebring, W., and Axelrod, A.**, Studies on systemic reactivity to purified protein derivative and endotoxin. I. Systemic reactivity to PPD in pyridoxine deficient guinea pigs, *J. Immunol.*, 91, 39, 1963; *Chem. Abstr.*, 59:15668d.
508. **Tsubosaka, M.**, Substrate specificity of pyridoxal kinase of mouse brain, *Vitamin*, 35, 7, 1967; *Chem. Abstr.*, 66:52489.
509. **Tsubosaka, M. and Makino, K.**, Study on the pyridoxal kinase of mouse brain, *J. Vitaminol.*, 15, 131, 1969; *Biol. Abstr.*, 51:43081.
510. **Tsuji, M., Nishiwaki, J., and Yamada, K.**, Transport of vitamin $B_6$ in human erythrocytes and in ascites-sarcoma cells, *7th Int. Cong. Clin. Chem.*, Roth, M., Ed., S. Karger, Basel, 1969, 362; *Chem. Abstr.*, 75:73801.
511. **Tsuji, M., Nishiwaki, J., and Yamada, K.**, Uptake of vitamin $B_6$ by ascites-sarcoma cells. I. Uptake of pyridoxine, *Vitamin*, 42, 222, 1970; *Chem. Abstr.*, 74:51567.
512. **Tsuji, S.**, Changes in serum and thymus transaminase activity in AKR mice with lymphoid leukemia, *Gann*, 57, 99, 1966; *Chem. Abstr.*, 65:4396.
513. **Tsuji, T.**, Mechanism of intestinal absorption of vitamin $B_6$. Uptake of tritium labeled pyridoxol by inverted segments of rat intestine, *Vitamin*, 47, 453, 1973; *Chem. Abstr.*, 80:13061.
514. **Tsuji, T. and Yamada, R.**, Mechanism of intestinal absorption of vitamin $B_6$. Studies with the everted sacs of rat intestine, *Vitamin*, 48, 471, 1974; *Chem. Abstr.*, 82:41078.
515. **Tsuji, T. and Yamada, R.**, Mechanism of intestinal absorption of vitamin $B_6$. III. Studies of intestinal absorption of pyridoxal with everted sacs of rat intestine, *Vitamin*, 50, 97, 1976; *Chem. Abstr.*, 84:144553.
516. **Tsuji, T. and Yamada, R.**, Mechanism of intestinal absorption of vitamin $B_6$. IV. Studies of intestinal absorption of pyridoxamine with everted sacs of rat intestine, *Vitamin*, 50, 103, 1976; *Chem. Abstr.*, 84:144554.
517. **Tsuji, T., Yamada, R., and Nose, Y.**, Intestinal absorption of vitamin $B_6$. I. Pyridoxol uptake by rat intestinal tissue, *J. Nutr. Sci. Vitaminol.*, 19, 401, 1973; *Chem. Abstr.*, 81:47891.
518. **Tsunoo, S., Horisaka, K., Takahashi, S., and Okazaki, M.**, Convulsions induced by vitamin $B_6$ antagonists and brain serotonin, *Igaku to Seibutsugaku*, 85, 221, 1972; *Chem. Abstr.*, 81:103.
519. **Tuszynska, S.**, 4-Deoxypyridoxine antagonism against pyridoxine as investigated with *Saccharomyces carlsbergensis*, *Acta. Pol. Pharm.*, 26, 583, 1969; *Chem. Abstr.*, 72:108204.
520. **Tyzzer, E. E.**, *J. Med. Res.*, 37, 307, 1917.
521. **Umbreit, W.**, Vitamin $B_6$ antagonists, *Am. J. Clin. Nutr.*, 3, 291, 1955; *Chem. Abstr.*, 49:12629h.
522. **Umbreit, W., and Waddell, J.**, Mode of action of deoxypyridoxine, *Proc. Soc. Exp. Biol. Med.*, 70, 293, 1949; *Chem. Abstr.*, 43:4710i.
523. **Unna, K.**, Antidermatitic effect of vitamin $B_6$ analogues, *Proc. Soc. Exp. Biol. Med.*, 43, 122, 1940; *Chem. Abstr.*, 34:2039.
524. **Van Bijsterveld, O. P.**, In vitro phagocytosis in pyridoxine deficiency, *J. Med. Microbiol.*, 4, 165, 1971; *Biol. Abstr.*, 53:50044.
525. **Van Wagtendonk, H. and Wibaut, J.**, Derivatives of pyridine and quinoline 52. Synthesis of 2,4-dimethyl-3-hydroxy-5-hydroxymethyl pyridine, *Recl. Trav. Chim. Pays Bas.*, 61, 728, 1942; *Chem. Abstr.*, 38:3284.

526. **Vaughan, D. and Winders, R.**, Studies with Tritium Labeled Pyridoxine, U.S. Air Force Systems Command Arctic Aeromed, Lab. Tech. Note, Fort Wainwright, Alaska, AAL-TR-64-29, 1965, 1. *Chem. Abstr.*, 63:13756.

527. **Vaughan, D. and Winders, R.**, Excretion of tritium labeled pyridoxine by pyridoxine deficient rats, *Metab. Clin. Exp.*, 15, 676, 1966; *Chem. Abstr.*, 65:12626.

528. **Victor, M. and Adams, R. D.**, The neuropathology of experimental vitamin $B_6$ deficiency in monkeys, *Am. J. Clin. Nutr.*, 4, 346, 1956; *Chem. Abstr.*, 50:15777b.

529. **Vilter, R.**, The metabolism of vitamin $B_6$ in human beings, *Am. J. Clin. Nutr.*, 4, 378, 1956; *Chem. Abstr.*, 50:15777e.

530. **Vilter, R., Mueller, J., Glazer, H., Jarrold, T., Abraham, J., Thompson, C., and Hawkins, V.**, The effect of vitamin $B_6$ deficiency induced by deoxypyridoxine in human beings, *J. Lab. Clin. Med.*, 42, 335, 1953; *Chem. Abstr.*, 48:3493c.

531. **Vilter, R. and Schreiner, W.**, The skin lesions of human pyridoxine deficiency and their relationship to seborrheic dermatitis of the sicca type, *Natl. Vitamin Found. Nutr. Symp.*, 5, 104, 1952; *Chem. Abstr.*, 49:7664h.

532. **Vogel, W., Snyder, R., and Schulman, M.**, Effects of aromatic and nonaromatic model compounds and drugs on enzymic activities, *J. Pharmacol. Exp. Ther.*, 146, 66, 1964; *Chem. Abstr.*, 62:1914a.

533. **Wada, H. and Morino, Y.**, Comparative studies on glutamic oxalacetic transaminases from the mitochondrial and soluble fractions of mammalian tissues, *Vitam. Horm.*, 22, 411, 1964; *Chem. Abstr.*, 62:13402b.

534. **Wada, H. and Snell, E.**, The enzymatic oxidation of pyridoxine and pyridoxamine phosphates, *J. Biol. Chem.*, 236, 2089, 1961; *Chem. Abstr.*, 55:27481b.

535. **Wagner, A. F. and Folkers, K.**, *Vitamins and Coenzymes*, John Wiley & Sons, New York, 1964, 160, 188; *Chem. Abstr.*, 60:9530g.

536. **Wagner-Jauregg, T., Fischer, J.**, Kinetin, nucleic acids and vitamin $B_6$ as antagonists for inhibitors of *Neurospora* and *Lactobacillus casei*, *Arzneim. Forsch.*, 24, 278, 1974; *Chem. Abstr.*, 81:10095.

537. **Waldi, D.**, Spray reagents for thin layer chromatography, *Thin Layer Chromatography*, Stahl, E., Ed., Academic Press, New York, 1965, 489.

538. **Wayne, L., Will, J., Friedman, B., Becker, L., and Vilter, R.**, Vitamin $B_6$ in internal medicine, *Arch. Int. Med.*, 101, 143, 1958; *Chem. Abstr.*, 52:10306i.

539. **Webb, J. L.**, *Enzyme and Metabolic Inhibitors*, Vol. 2, Academic Press, New York, 1966, 562, *Chem. Abstr.*, 65:2569a.

540. **Wei, K. T., Bell, R. R., and Haskell, B. E.**, Regarding 4-pyridoxic acid 5-phosphate in animal tissue, *Biochem. Biophys. Res. Commun.*, 48, 1671, 1972; *Chem. Abstr.*, 77:150034.

541. **Weir, D.**, Effect of deoxypyridoxine on acute lymphatic leukemia of adults, *J. Clin. Invest.*, 31, 670, 1952.

542. **Weir, D. and Heinle, R.**, Similarity of hematologic effect of pyridoxine deficiency, cortisone and a myeloid metaplasia factor of human urine, *Proc. Soc. Exp. Biol. Med.*, 75, 655, 1950; *Chem. Abstr.*, 45:2581c.

543. **Weir, D. and Heinle, R.**, Some studies on the Role of Certain of the B Vitamins in Leukopoiesis, *Nutritional Symposium Series No. 5, National Vitamin Foundation*, 1952, 70; *Chem. Abstr.*, 49:7664c.

544. **Weir, D., Heinle, R., and Welch, A.**, Role of pyridoxine in the production of leucocytes in normal and leukemic mice , *Proc. Soc. Exp. Biol. Med.*, 72, 457, 1949; *Chem. Abstr.*, 44:3587i.

545. **Weir, D. and Morningstar, W.**, The effect of pyridoxine deficiency induced by deoxypyridoxine on acute lymphatic leukemia of adults, *Blood*, 9, 173, 1954; *Biol. Abstr.*, 28:15813.

546. **Wendt, G. and Bernhart, F.** The structure of a sulfur-containing compound with vitamin $B_6$ activity, *Arch. Biochem. Biophys.*, 88, 270, 1960; *Chem. Abstr.*, 54:22621e.

547. **Wheeler, G., Schabel, F., and Skipper, H.**, Potentiated inhibition of *Escherichia coli* by certain combinations of agents, *Proc. Soc. Exp. Biol. Med.*, 92, 396, 1956; *Chem. Abstr.*, 50:15703.

548. **White, R. S. and Dempsey, W. B.**, Purification and properties of vitamin $B_6$ kinase from *Escherichia coli* B, *Biochemistry*, 9, 4057, 1970; *Chem. Abstr.*, 74:557.

549. **Wibaut, J., Uhlenbroek, J., Kooyman, E., and Kettenes, D.**, Preparation of 3-hydroxy-5-hydroxymethyl-2,4-dimethyl pyridine, *Recl. Trav. Chim. Pays Bas.*, 79, 481, 1960; *Chem. Abstr.*, 55:1609e.

550. **Williams, H. and Wiegand, R.**, Xanthurentic acid excretion and $B_6$ deficiency in hydrazide treated dogs, *J. Pharmacol. Exp. Ther.*, 113, 54, 1955.

551. **Williams, R. J., Eakin, R. E., Beerstecher, E., and Shive, W.**, *The Biochemistry of the B Vitamins*, Reinhold, New York, 1950.

552. **Willis-Carr, J. and St. Pierre, R.**, Effects of vitamin $B_6$ deficiency on thymic epithelial cells and T lymphocyte differentiation, *J. Immunol.*, 120, 1153, 1978; *Chem. Abstr.*, 88:188655.

553. **Woolley, D.**, Recent advances in the study of biological competition between structurally related compounds, *Physiol. Rev.*, 27, 308, 1947; *Chem. Abstr.*, 41:4579a.

554. **Wooley, J. and Murphy, M.**, Metabolic studies on $T_2$ *Escherichia coli* bacteriophage. I. A study of deoxypyridoxine inhibition and its reversal, *J. Biol. Chem.*, 178, 869, 1949; *Chem. Abstr.*, 43:6697f.

555. **Wuest, H. M., Bigot, J. A., DeBoer, T. J., Van der Wal, B., and Wibaut, J. P.**, Pyridine derivatives CV an improved synthesis of vitamin $B_6$, *Rec. Trav. Chim.*, 78, 226, 1959; *Chem. Abstr.*, 53:21929f.

556. **Yamada, K., Hayami, S., and Sawaki, S.**, Influences of deoxypyridoxine, hydroxymethylpyrimidine and isonicotinoyl hydrazide upon the content of coenzyme A in the liver of the rat, *Vitamin*, 11, 267, 1956; *Chem. Abstr.*, 51:15728f.

557. **Yamada, K. and Tsuji, J.**, Transport of vitamin $B_6$ into human erythrocytes. IV. Effects of incubation temperature, glycolysis inhibitor and vitamin $B_6$ analogs on the transport of vitamin $B_6$ into human erythrocytes, *Vitamin*, 37, 472, 1968; *Chem. Abstr.*, 69:17223.

558. **Yamada, K. and Tsuji, M.**, Transport of vitamin $B_6$ in human erythrocytes. II. Effects of incubation temperature, glycolytic inhibitor, and vitamin $B_6$ analogs on transport of vitamin $B_6$ in human erythrocytes, *J. Vitaminol.*, 16, 237, 1970; *Chem. Abstr.*, 74:1896.

559. **Yamashita, J.**, Action site of antagonists of vitamin $B_6$ in the central nervous system of frogs and cockroaches, *J. Nutr. Sci. Vitaminol.*, 21, 163, 1975; *Chem. Abstr.*, 84:54747.

560. **Yamashita, J.**, Convulsive seizure induced by intracerebral injection of semicarbazide in the mouse, *J. Nutr. Sci. Vitaminol.*, 22, 1, 1976; *Chem. Abstr.*, 85:41797.

560a. **Yang, I.-Y. B.**, personal communication.

561. **Yonezawa, T. and Iwanami, H.**, In vitro study of vitamin B deficiency of nervous tissue produced by antivitamins, *Proc. 5th Int. Cong. Neuropath.*, Luthy, F. and Bischoff, A., Eds., Excerpta Medica Fountation, New York, 1965, 868.

562. **Yonezawa, T., Mori, T., and Nakatani, Y.**, Effects of pyridoxine deficiency in nervous tissue maintained in vitro, *Ann. N.Y. Acad. Sci.*, 166, 146, 1969; *Chem. Abstr.*, 71:121047.

563. **Yoshida, T., Yunoki, N., Nakazima, Y., Kaito, T., and Anmo, T.**, Simultaneous determination of vitamin $B_6$ group in blood by high-performance liquid chromatography (HPLC), *Analytical Journal of the Pharmaceutical Society of Japan*, 98, 1319, 1978; *Chem. Abstr.*, 90:35781.

564. **Yoshikawa, T., Ishikawa, F., and Naito, T.**, Synthesis of 3-pyridinols. III. Synthesis of pyridoxine skeletons from 4-methyloxazole, *Chem. Pharm. Bull.*, 13, 878, 1965; *Chem. Abstr.*, 63:11489g.

565. **Yuspa, H. and Blaustein, A.**, Inhibitory effect of pyridoxine deficiency on growth of a transplanted tumor in rats, *Proc. Soc. Exp. Biol. Med.*, 123, 398, 1966; *Biol. Abstr.*, 48:65316.

566. **Zavenyagina, T. and Bukin, Y.**, Effects of $B_6$ vitamins and pyridoxine antagonists on some physiological and biochemical characteristics of developing pea and wheat seedlings, *Pyridoxal Catalysis: Enzymes and Model Systems, Proc. 2nd Int. Symp.*, Snell, E. E., Ed., Interscience, New York, 1966, 663; *Chem. Abstr.*, 70:9393.

567. **Zervas, N., Hori, H., and Rosoff, C.**, Experimental inhibition of serotonin by antibiotic:prevention of cerebral vasospasm, *J. Neurosurg.*, 41, 59, 1974; *Chem. Abstr.*, 82:80463.

568. **Zhdanovich, N. and Udalov, Y.**, Prophylactic and therapeutic effectiveness of some vitamins in chlorophos poisoning, *Gig. Tr. Prof. Zabol.*, 14, 37, 1970; *Chem. Abstr.*, 73:23602.

569. **Zima, O. and Jung, F.**, *Chem. Zentralbl.*, 112, 2353, 1941.

# AUTHOR INDEX

## A



## B



## C

## D



## E



## F



## G

## H



## I



## J

K



L



M

## N



## O

## P



## Q



## R



## S

## T



## U



## V

# SUBJECT INDEX

## A



## B



## C



## D

## E

## F



## G



## H



## I

## Q



## R



## S



## T

## U



## V



## W



## X